AF594974

Pigments and Dyes in Archaeological and Historical Objects—Scientific Analyses and Conservation Challenges

Pigments and Dyes in Archaeological and Historical Objects—Scientific Analyses and Conservation Challenges

Editor

Diego Tamburini

MDPI • Basel • Beijing • Wuhan • Barcelona • Belgrade • Manchester • Tokyo • Cluj • Tianjin

Editor
Diego Tamburini
Department of Scientific Research
The British Museum
UK

Editorial Office
MDPI
St. Alban-Anlage 66
4052 Basel, Switzerland

This is a reprint of articles from the Special Issue published online in the open access journal *Heritage* (ISSN 2571-9408) (available at: https://www.mdpi.com/journal/heritage/special_issues/pigments_and_dyes).

For citation purposes, cite each article independently as indicated on the article page online and as indicated below:

LastName, A.A.; LastName, B.B.; LastName, C.C. Article Title. *Journal Name* **Year**, *Volume Number*, Page Range.

ISBN 978-3-0365-3134-2 (Hbk)
ISBN 978-3-0365-3135-9 (PDF)

Contents

About the Editor

Diego Tamburini—Scientist: Polymers and Modern Organic Materials - is an analytical chemist by training and obtained his PhD in Chemistry and Materials Science from the University of Pisa in 2015 with a thesis on the evaluation of the degradation state of archaeological wood by analytical pyrolysis couple to gas chromatography and mass spectrometry. He joined the Department of Scientific Research of the British Museum in 2016 with an Andrew W. Mellon Postdoctoral Fellowship focusing on the application of liquid chromatography tandem mass spectrometry to the identification of natural dyes in historical and archaeological textiles. His main project focused on the palette of Asian dyes used in the Dunhuang (China) textiles of the Sir Aurel Stein collection at the British Museum. In 2020, he moved to the Department of Conservation and Scientific Research of the Freer Gallery of Art and Arthur M. Sackler Gallery (National Museum of Asian Art, Smithsonian Institution) as a Smithsonian Postdoctoral Fellow with a project focused on the dye analysis of Central Asian 19th-century ikat textiles. After a 6-month Postdoctoral Fellowship at the Centre for Scientific Studies in the Arts (NU-ACCESS - Northwestern University/Art Institute of Chicago), he joined the British Museum again in 2021 in the role of Scientist: Polymers and Modern Organic Materials. His main responsibility is to answer questions about the origin, technology and stability of natural and synthetic polymers as well as other organic materials in the collection mostly using mass spectrometric techniques.

heritage

MDPI

Editorial

Colour Analysis: An Introduction to the Power of Studying Pigments and Dyes in Archaeological and Historical Objects

Diego Tamburini

Department of Scientific Research, The British Museum, Great Russell Street, London WC1B 3DG, UK; DTamburini@britishmuseum.org

Nature offers a myriad of colours and the desire to replicate them is intrinsic to human nature. People from past civilisations have, in fact, searched for natural sources of colour since prehistoric times and have looked for ways to apply them to surfaces and make them last as long as possible. These ancient people found out that some minerals could be ground to obtain a fine powder and that some plants, insects and molluscs yielded coloured solutions upon immersion in hot water. The former group of materials is referred to as mineral or inorganic pigments, whereas the latter group of materials is referred to as natural or organic dyes.

Pigments are completely or nearly insoluble in water and must be applied to surfaces by mixing them with a binding medium. Dyes can be applied to fibres in several ways in a water solution. The use of mordanting agents is often required. Additionally, organic dyes can be processed into insoluble pigments by precipitating them onto an inorganic substrate, thus creating the so-called lake pigments.

Since their discoveries and first applications, pigments and dyes have contributed to some of the most spectacular forms of art and craftsmanship, from dyed textiles to painted masterpieces. Painters, weavers, dyers, artists and craftsmen have developed their skills over millennia to respond to their own creativity, stimuli from the natural world, availability of local resources, as well as to everyday life needs, introduction of new materials, market requests and even fashion. Terms such as Tyrian purple, lapis lazuli and cochineal, to mention only a few, resonate in the collective imagination, evoking images of priceless royal dresses, tireless caravanserai and fearless explorers.

This constant evolution and exchange of knowledge between different peoples have led to a plethora of natural materials used to create and apply colour. These can drastically vary based on geographical regions and time periods, so that the study of these materials requires a high level of expertise and knowledge from the researchers tasked with their identification. The scenario becomes even more complex in the second half of the 19th century, when scientific advances revealed that colour was intrinsically related to the structure of organic molecules and to crystallographic arrangements in minerals. Driven by the desire to chemically replicate natural dye molecules, scientists embarked on a journey that led to the discovery of synthetic dyes in 1856, when William Perkin accidentally synthesised the coloured molecule mauveine (or aniline purple). By the end of the 19th century, more than 400 synthetic dye formulations were patented, and the numbers grew exponentially during the first decades of the 20th century, as the invention of new molecules accompanied and partially drove the industrial revolution. However, it is not correct to say that synthetic molecules replaced the natural ones, as there is evidence that traditional methods and materials continued to be used and are still in use today.

The 15 articles selected for this Special Issue represent a window onto this very complex research scenario, for which materials knowledge is often not sufficient to obtain the desired results. This leads to highly inter-disciplinary studies, in which scientists, conservators, historians, art historians and archaeologists join forces to answer research questions on colour.

Citation: Tamburini, D. Colour Analysis: An Introduction to the Power of Studying Pigments and Dyes in Archaeological and Historical Objects. *Heritage* **2021**, *4*, 4366–4371. https://doi.org/10.3390/heritage4040241

Received: 5 November 2021
Accepted: 11 November 2021
Published: 15 November 2021

Sotiropoulou et al.'s review [1] on the use of molluscan purple (or "true" purple—in Greek, *porphyra* = πορφύρα) as a pigment in wall paintings from the Aegean area, summarises the research carried out on probably the most prestigious organic colourant in history, and integrates it with new scientific evidence. The article reports on the use of calcium carbonate from ground mollusc shells as the substrate on which to precipitate the organic dye, thus pointing to the recycling of these shells, which were considered as waste products from the "purple industry". Connections between textile dyeing and pigment production are also supported by archaeological evidence, as numerous loom weights were found in the sites under consideration. The wall paintings themselves, in which the purple pigment is used to depict several details in women's costumes, implicitly reinforce such a connection. The review ultimately showcases how science, archaeology and art history come together to offer new keys for the interpretation of a topic that has fascinated generations of researchers and scholars.

Pigments in wall paintings are also the main subject treated by Rampazzi et al. in their investigation on 16th-century Italian wall paintings [2]. In addition to summarising evidence of a traditional palette of inorganic pigments, such as red and yellow ochres, ultramarine blue, bianco Sangiovanni, cinnabar/vermilion and azurite, other pigments, such as clinochlore, Brunswick green and ultramarine yellow were detected in the upper layers of samples that also showed complex stratigraphy. This is reported as the first evidence for the use of such pigments in wall paintings. Furthermore, as these pigments were all synthesised in the 19th century, their presence provides dating information for the execution of re-painting campaigns. The authors further the investigation, by discussing the ubiquitous identification of aragonite as an indication of the use of shells as an aggregate in the mortar, a practice that was known from the Roman period, but that has never been attested in later wall paintings. These results, together with the report on the poor conservation state of the paintings, enable a fuller picture to be drawn of the undocumented history of the church that houses them and will hopefully foster conservation interventions.

Sixteenth-century artist materials are also the topic of Balbas et al.'s contribution [3]. The study brings the reader to Mexico, and focuses on a polychrome maize stem sculpture. The identification of the pigments is only one part of a detailed investigation aimed at reconstructing the production/conservation history of the sculpture. Traditional Mexican materials, such as maize, paper made of cotton fibres, and *colorín* wood are present alongside materials traditionally encountered in polychrome objects of European origin (gypsum and animal glue for the ground; vermillion, cochineal lake and lead white for the red shades), thus reflecting the convergence of indigenous and European artistic traditions in this type of objects. The historical research and stylistic analysis led to hypothesise the exact workshop in which the sculpture was produced. Computer tomography and stratigraphic analysis of the polychromy highlighted original elements and alterations. All the observations come together in a captivating discussion.

In the studies introduced so far, samples were taken and analysed to answer some of the research questions. However, in some cases, sampling is simply not an option, as the works of art under investigation are too precious or too fragile to allow for samples, even very tiny ones, to be removed from them. This is particularly true for artworks on parchment and paper, as highlighted in the contributions by Agostino et al. [4] and Dill et al. [5], respectively. The development of non-invasive and portable spectroscopic techniques is one of the most active research areas in heritage science. Agostino et al. show how an analytical workflow based on optical microscopy, fibre optic reflectance spectroscopy (FORS), fibre optic molecular fluorimetry (FOMF), X-ray fluorescence spectrometry (XRF) and micro-Raman spectroscopy can unveil the palette used to decorate Medieval illuminated manuscripts [4]. This information was then used to underline the fact that certain pigments and colourants were used in a hierarchical way, thus providing further insight into the modus operandi of the artist. On the other hand, Dill et al.'s contribution showcases the potential of non-invasive imaging techniques as opposed to point analysis [5]. Macro-XRF (MA-XRF), hyperspectral imaging (HSI), photometric stereo imaging and transmitted light imaging were used to investigate three hand-coloured prints showing the work of artist-naturalist Maria Sibylla Merian (1647–1717) and answer questions related to the attribution and dating of the prints. The observations enabled the prints and paper to be attributed to a posthumous edition of the artist's work published as a French translation in 1771, whereas the colouring was carried out after 1920, as shown by the presence of cadmium red and titanium white pigments, probably by an art dealer or collector trying to add value to the prints.

Modern pigments and colourants are the topic of several articles in this collection. These materials represent a challenge for scientists and conservators for many reasons, including the limited information available on their chemical composition and degradation. Hence, characterisation studies, such as the contribution by Pause et al. [6], are fundamental to building databases and enhancing our knowledge of these materials. In their study, the authors test the use of a handheld Raman device for the non-invasive identification of synthetic organic pigments in historic pre-1950 varnished paint-outs from manufacturer Royal Talens. Historic books and catalogues from colour manufacturers represent an important source of information, as highlighted in other studies as well [7]. In fact, the confusion around nomenclature and chemical formulae of synthetic organic pigments and dyes is a challenge in itself. In addition to reporting on the feasibility of the identification technique, the article contributes to a better understanding of artists' materials used between 1932 and 1950.

Building this foundation of knowledge is the first step towards more focused studies, such as the ones reported by Lizun et al. [8] and Martins et al. [9]. Lizun et al. take on the complex task of exploring the painting technique and materials adopted by Singapore artist Liu Kang during his period in Paris (1929–1932), by analysing 14 paintings with a wide array of non- and micro-invasive analytical techniques [8]. The detailed technical examination is coupled with historical information derived from archives and contemporary catalogues. The research leads the authors to reveal the use of a relatively restricted palette of pigments, such as ultramarine blue, viridian green, chrome yellow, iron oxides, organic reds, lead white, and bone black, with some additional pigments used sporadically, such as cobalt blue, Prussian blue, emerald green, cadmium yellow, cobalt yellow, and zinc white, while the painting technique was in line with the exploratory phase of the artist, who was developing his own style during this early stage of his career.

Martins et al. focus on an iconic work by Henri Matisse with the intention of identifying the pigments used in the gouaches of the illustrated book, "Jazz" [9]. Three different copies were investigated, and the results were consistent, revealing the use of 39 pigments, including both mineral and synthetic organic pigments. The article also discusses the use of multivariate statistical analysis for the treatment of large datasets and the limitations of certain non-invasive techniques for the identification of organic molecules. But most importantly, the research was able to inform on the condition of the prints, in order to support conservation strategies. In particular, similar colour shades were obtained using very different materials with different light sensitivities, thus impacting conservation decisions.

In Haddad et al.'s contribution, Alexander Calder's "Man-Eater with Pennants" is at the centre of an investigation to reveal the original paint colours hidden beneath layers of repainting [10]. In addition to showing an application on a sculpture rather than a two-dimensional object, the article reports on the difficulty of distinguishing original materials and subsequent applications in modern objects, and on the different degradation pathways occurring during outdoor exposure. The original paint layers were identified as containing Prussian blue, parachlor red and chrome yellow, whereas the many layers of overpaint contained titanium white, molybdate orange, several β-naphthol reds, red lead, and ultramarine blue. Interestingly, the initial intention to use a maquette model of the sculpture to confirm the original palette of the artist was discarded, as the analyses showed that the maquette was painted after "Man-Eater" was first installed. The results ultimately informed the restoration campaign that led to the sculpture once again being exhibited in the Museum of Modern Art's sculpture garden in New York City.

The implication of scientific investigation on conservation and display decision-making is also treated in Rayner et al.'s contribution [11], which focuses on an ancient Greek terracotta *krater* treated with selective in-painting with cadmium orange (CdSSe). Some of the restauration areas showed alteration after one year from the application of cadmium orange. These areas were associated with the presence of chlorine, and, upon recreation of the stratigraphy in mock-up samples, it was observed that the cadmium orange degrades in the presence of chlorine and light, forming selenium-rich structures. The research provides new information on the degradation of this pigment and led to the removal of the altered paint layer and re-treatment of the object with non-cadmium containing pigments.

Modern synthetic dyes are discussed in Ciccola et al.'s contribution [12], which reports on an interesting case study focusing on the analysis of two Indian leather puppets made in the 1970s. The spectroscopic analysis by reflectance spectroscopy and surface enhanced Raman spectroscopy (SERS) led to the identification of malachite green dye, crystal/methyl violet, rhodamine B and eosin Y, although some azo dyes and some complex mixtures were also present but difficult to disclose using only a spectroscopic approach. In fact, organic dyes often have complex molecular compositions that require the integration of a chromatographic approach, in order to separate the molecular components and accurately identify the dye source(s). Doherty et al. address this point in their contribution [13]. Following the recent structural elucidation of urolithin C, one of the main degradation products of the brazilwood dye [14], this article describes the suitability of SERS to identify this molecular marker in a variety of textile samples. The correlation of the data obtained by SERS with those obtained by high performance liquid chromatography (HPLC) with both a diode array detector (DAD) and a mass spectrometry detector (MS) enabled the results to be confirmed. Additionally, other dyes were identified in some of the samples, highlighting the possibility to detect brazilwood by SERS even in the presence of other dyes.

Weld, another well-known natural dye, is the focus of the contribution by Veneno et al. [15]. The authors made reconstructions of five 19th-century recipes of weld lake pigments discovered in the Windsor & Newton archive database, and then characterised them by FTIR and HPLC-DAD. The article discusses the steps necessary for the commercial preparation of the pigments, provides new information on FTIR bands specific for the identification of weld lake pigments, and shows the effects of the different matrices deriving from the different manufacture of the pigments.

Finally, specific attention to dyed textiles is given in the contributions by Campos Ayala et al. [16] and Łucejko et al. [17]. In their article, Campos Ayala et al. address the characterisation of the dyes used to produce primary (red, blue, and yellow) and secondary (purple, orange, and green) colours in Peruvian textiles spanning five major civilizations: the Paracas Necropolis, the Nazca, the Wari, the Chancay, and the Lambayeque [16]. With the intention to validate the use of ambient ionisation MS techniques, such as direct analysis in real time (DART-MS) and paper spray MS, the authors report on the results obtained from both reference samples and archaeological textiles, thus providing the reader with an important database. Comparative results obtained by HPLC-DAD integrate the results and support the conclusions. The article clearly shows the variety of dye sources available in the Andean region and the difficulties in identifying yellow dyes, once again underlining the complexity of conducting this type of research, as the biodiversity of the natural world becomes a challenge in itself. From South America to Northern Europe, the contribution by Łucejko et al. addresses the investigation of textile fragments, including rare silk embroideries, from the Viking Gokstad ship grave in Norway, dating to ca. 900 AD [17]. The severe fading of the colours made the analytical work challenging, but thanks to the application of HPLC-MS/MS, molecular markers for the presence of madder (probably from *Rubia tinctorum*) were detected in most samples. Additionally, hypotheses are made about the possible use of different dye recipes to obtain different colour shades (from yellowish to red) using only madder as dye source.

The overview provided by this selection of articles shows only a small fraction of the potential of colour analysis in heritage science. Analytical challenges related to the growing need for non-invasive analyses are further driving technological developments in this research field. However, obtaining accurate molecular information from samples is sometimes the only way to provide the desired answers. Through scientific analysis, our knowledge of pigments and dyes, as well as their degradation, grows year on year. As a result, researchers are offered new tools to address their questions in a stimulating and cross-disciplinary environment.

I hope that this collection of articles will inspire future research on pigments and dyes to go towards the direction of not just looking at colour in terms of wavelengths, electronic transitions, chemical components or molecular interactions, but to focus more on the artists and craftsmen who made and used colours both in ancient and more recent times. Switching the point of view from the colours themselves to the hands and minds that produced them, processed them, experimented with them, and applied them has the potential to unlock new answers, and certainly new questions, on this fascinating research topic.

Funding: This research received no external funding.

Data Availability Statement: Not applicable.

Conflicts of Interest: The author declares no conflict of interest.

References

1. Sotiropoulou, S.; Karapanagiotis, I.; Andrikopoulos, K.S.; Marketou, T.; Birtacha, K.; Marthari, M. Review and New Evidence on the Molluscan Purple Pigment Used in the Early Late Bronze Age Aegean Wall Paintings. *Heritage* **2021**, *4*, 171–187. [CrossRef]
2. Rampazzi, L.; Corti, C.; Geminiani, L.; Recchia, S. Unexpected Findings in 16th Century Wall Paintings: Identification of Aragonite and Unusual Pigments. *Heritage* **2021**, *4*, 2431–2448. [CrossRef]
3. Quintero Balbas, D.; Sánchez-Rodríguez, E.; Zárate Ramírez, Á. Holy Corn. Interdisciplinary Study of a Mexican 16th-Century Polychrome Maize Stem, Paper, and Colorín Wood Sculpture. *Heritage* **2021**, *4*, 1538–1553. [CrossRef]
4. Agostino, A.; Pellizzi, E.; Aceto, M.; Castronovo, S.; Saroni, G.; Gulmini, M. On the Hierarchical Use of Colourants in a 15th Century Book of Hours. *Heritage* **2021**, *4*, 1786–1806. [CrossRef]
5. Dill, O.; Vermeulen, M.; McGeachy, A.; Walton, M. Multi-Modal, Non-Invasive Investigation of Modern Colorants on Three Early Modern Prints by Maria Sibylla Merian. *Heritage* **2021**, *4*, 1590–1604. [CrossRef]
6. Pause, R.; van der Werf, I.D.; van den Berg, K.J. Identification of Pre-1950 Synthetic Organic Pigments in Artists' Paints. A Non-Invasive Approach Using Handheld Raman Spectroscopy. *Heritage* **2021**, *4*, 1348–1365. [CrossRef]
7. Tamburini, D.; Shimada, C.M.; McCarthy, B. The molecular characterization of early synthetic dyes in E. Knecht et al's textile sample book "A Manual of Dyeing" (1893) by high performance liquid chromatography—Diode array detector—Mass spectrometry (HPLC-DAD-MS). *Dye. Pigment.* **2021**, *190*, 109286. [CrossRef]
8. Lizun, D.; Kurkiewicz, T.; Szczupak, B. Exploring Liu Kang's Paris Practice (1929–1932): Insight into Painting Materials and Technique. *Heritage* **2021**, *4*, 828–863. [CrossRef]
9. Martins, A.; Prud'hom, A.C.; Duranton, M.; Haddad, A.; Daher, C.; Genachte-Le Bail, A.; Tang, T. Jazz Colors: Pigment Identification in the Gouaches Used by Henri Matisse. *Heritage* **2021**, *4*, 4205–4221. [CrossRef]
10. Haddad, A.; Randall, M.; Zycherman, L.; Martins, A. Reviving Alexander Calder's Man-Eater with Pennants: A Technical Examination of the Original Paint Palette. *Heritage* **2021**, *4*, 1920–1937. [CrossRef]
11. Rayner, G.; Costello, S.D.; McClelland, A.; Akey, A.; Eremin, K. Preliminary Investigations into the Alteration of Cadmium Orange Restoration Paint on an Ancient Greek Terracotta Krater. *Heritage* **2021**, *4*, 1497–1510. [CrossRef]
12. Ciccola, A.; Serafini, I.; D'Agostino, G.; Giambra, B.; Bosi, A.; Ripanti, F.; Nucara, A.; Postorino, P.; Curini, R.; Bruno, M.; et al. Dyes of a Shadow Theatre: Investigating Tholu Bommalu Indian Puppets through a Highly Sensitive Multi-Spectroscopic Approach. *Heritage* **2021**, *4*, 1807–1820. [CrossRef]
13. Doherty, B.; Degano, I.; Romani, A.; Higgitt, C.; Peggie, D.; Colombini, M.P.; Miliani, C. Identifying Brazilwood's Marker Component, Urolithin C, in Historical Textiles by Surface-Enhanced Raman Spectroscopy. *Heritage* **2021**, *4*, 1415–1428. [CrossRef]
14. Peggie, D.A.; Kirby, J.; Poulin, J.; Genuit, W.; Romanuka, J.; Wills, D.F.; De Simone, A.; Hulme, A.N. Historical mystery solved: A multi-analytical approach to the identification of a key marker for the historical use of brazilwood (*Caesalpinia* spp.) in paintings and textiles. *Anal. Methods* **2018**, *10*, 617–623. [CrossRef]
15. Veneno, M.; Nabais, P.; Otero, V.; Clemente, A.; Oliveira, M.C.; Melo, M.J. Yellow Lake Pigments from Weld in Art: Investigating the Winsor & Newton 19th Century Archive. *Heritage* **2021**, *4*, 422–436.

16. Campos Ayala, J.; Mahan, S.; Wilson, B.; Antúnez de Mayolo, K.; Jakes, K.; Stein, R.; Armitage, R.A. Characterizing the Dyes of Pre-Columbian Andean Textiles: Comparison of Ambient Ionization Mass Spectrometry and HPLC-DAD. *Heritage* **2021**, *4*, 1639–1659. [CrossRef]
17. Łucejko, J.J.; Vedeler, M.; Degano, I. Textile Dyes from Gokstad Viking Ship's Grave. *Heritage* **2021**, *4*, 2278–2286. [CrossRef]

heritage

MDPI

Article

Review and New Evidence on the Molluscan Purple Pigment Used in the Early Late Bronze Age Aegean Wall Paintings

Sophia Sotiropoulou [1,*], Ioannis Karapanagiotis [2], Konstantinos S. Andrikopoulos [3], Toula Marketou [4], Kiki Birtacha [5] and Marisa Marthari [6]

1 Foundation for Research and Technology-Hellas, Institute of Electronic Structure and Laser (FORTH-IESL), N.Plastira 100, 70013 Vassilika Vouton, Crete, Greece
2 University Ecclesiastical Academy of Thessaloniki, 54250 Thessaloniki, Greece; y.karapanagiotis@aeath.gr
3 Department of Physics, University of Patras, 26504 Rion, Greece; kandriko@upatras.gr
4 Ministry of Culture, 22nd Ephorate of Prehistoric and Classical Antiquities, Ippoton Street, 85100 Rhodes, Greece; kbepka@yahoo.gr
5 The Archaeological Society at Athens, Excavations at Akrotiri, Thera, Tholou 10, 10556 Athens, Greece; kikibir@otenet.gr
6 Ministry of Culture, 21st Ephorate of Prehistoric and Classical Antiquities, Epameinonda St., 10555 Athens, Greece; mmarthari@gmail.com
* Correspondence: sophiaso@iesl.forth.gr

Abstract: The production and use of the pigment extracted from the murex molluscs is discussed here in association with the purple textile dyeing industry in the Prehistoric Aegean. "True" purple has been identified in a number of archaeological finds dating from the early Late Bronze Age, found in old and recent excavations at three different but contemporary sites: Akrotiri and Raos on Thera, and Trianda on Rhodes. The chemical composition of the shellfish purple pigment either found in lump form or applied on wall paintings is discussed in relation to the archaeological context of several examined finds and with reference to Pliny's *purpurissum*. The results of a comprehensive methodology combining new data obtained with molecular spectroscopies (microRaman and FTIR) and already reported data obtained with high performance liquid chromatography coupled with a diode array detector (HPLC–DAD) applied to samples of the murex purple finds are discussed in comparison to published data relating to few other instances of analytically proven murex purple pigment found in the Aegean over the timespan of its documented exploitation.

Keywords: Late Cycladic I; brominated indigoids; Muricidae; murex; *purpurissum*; true purple; microRaman; FTIR; HPLC–DAD

Citation: Sotiropoulou, S.; Karapanagiotis, I.; Andrikopoulos, K.S.; Marketou, T.; Birtacha, K.; Marthari, M. Review and New Evidence on the Molluscan Purple Pigment Used in the Early Late Bronze Age Aegean Wall Paintings. *Heritage* **2021**, *4*, 171–187. https://doi.org/10.3390/heritage4010010

Received: 19 December 2020
Accepted: 11 January 2021
Published: 14 January 2021

1. Introduction

The exploitation of molluscs providing the fabulous purple dye (in Greek, porphyra = πορφύρα) in the Prehistoric Aegean, as evidenced by archaeological research, illuminates many aspects of the societies that developed it and bequeathed it to subsequent civilizations. The production at a commercial scale of "true purple" dye, obtained from the secretions of shellfish species native to the Mediterranean Sea, had a significant social and economic impact in the Aegean Bronze Age. Dyeing seems to have been an integral part of the production and distribution of textiles, whereas the extraction and the application process of the purple dye, demanding exceptional skills and much know-how, was presumably a preserved secret among the guild of specialized experts. The impact of the murex dye industry in the ancient Mediterranean societies was related to the value of dyed garments and colored textiles in general as symbols of social status and power, with the shellfish purple dye representing an added value and denoting the highest societal rank in many cultures around the Mediterranean basin during the Bronze Age and far later on [1,2].

Although the earliest record for the purple dye comes from inscriptions on the Mycenaean Linear B clay tablets dating to the 13th century BC, where the term "royal purple"

(Knossos X 976) was probably mentioned for the first time [3], important shell deposits of gastropods of the Muricidae family have been studied and confidently attributed to a non-dietary use of molluscs far earlier. Heavily fragmented shell debris in a primary production context or concentrations of crushed murex in a secondary use context are considered evidence of purple dye production activity at many Aegean sites, dating mainly from the Middle and Late Minoan periods, and in certain cases from the Early Minoan period [4]. The early archaeological evidence of the purple industry around the Mediterranean Basin is significant in both the Aegean and the Levantine littoral [4–9]. The question of whether the secret of the purple dyeing process and its exceptional properties was first discovered in the Aegean or the Levant has yet to be resolved, but given the established trading relations between the Aegean and the eastern Mediterranean coasts in the middle of the second millennium BC, it is reasonable to suggest that the process of extracting a commercial dye from murex species may have been practiced contemporaneously in many of the important Eastern Mediterranean centers, where familiarity with marine molluscs is evidenced by the shell middens, as well as by shells found in second uses (mortar, etc.) [10].

The chances of finding the proper product of the purple dyeing industry, a purple-dyed textile, decrease geometrically as we go back in time. However, residues of dyed textile, in which mollusc purple is identified, date back to the Late Bronze Age. In present evidence, the earliest cases belong to archaeological sites in Syria. Purple-dyed textile fragments were found first in a royal tomb complex in the Bronze Age palace of ancient Qatna, Tell Mishrife, Syria, discovered during excavations in 2002 [11]. It is estimated that the tomb was used for 300–400 years for the burial of kings, prior to the destruction of the palace by the invading Hittites in 1340 BC, a date that is extremely early given the fragility of the fiber substrate. Brominated and non-brominated indigoid compounds of a similar composition were identified in purple-colored extracts both from fragments of woven fabrics and samples in brown-stained areas of the layer of sediment covering the tomb floor, from which it was concluded that the colorants present in the sediment extracts were residues of the purple used to dye the royal fabrics [12]. Then, in a more recent excavation, shellfish purple-dyed textile residues dating to the 18th–16th century BC were identified in Chagar Bazar [13]. This earliest direct evidence of purple-dyed textiles confirm the early dating of the purple-dyeing industries at the coastal Levantine sites in Syria (Minet el-Beidha/Ras Shamra) and in Lebanon (Sarepta), so far documented only through the studied heaps of murex, as well as the identified purple substances on shreds of jar-type vessels with purple sediment in their interiors [7,10,14–17].

Another direct piece of evidence for the shellfish-purple dyeing industry raised from an excavated complex alluding to a dyeing installation is provided by containers that preserve purple traces, analytically proven to contain indigoid compounds [1,18]. One of the earliest instances of such purple-stained vessels with crushed shells, which dates back to the Middle Minoan period, MM IIB, was found at the site of Pefka, near Pacheia Ammos, on Crete [19,20].

A third, equally strong direct piece of evidence for purple production is the identification of the pigment prepared for use in painting; we are referring to the *purpurissum*, according to Pliny the Elder and his vat-dyeing recipe [21]. The murex purple dyestuff in the form of pigment with a calcium carbonate base and its application in painting was first identified and fully characterized in the context of the Xeste 3 wall paintings at Akrotiri on Thera [22–24], referring in fact to Pliny's vat-dyeing recipe for *purpurissum.*

Over the past 30 years, deep insights have been given into the production of murex purple, its coloring and medicinal properties, and the technology of its application, as a result of research findings in various disciplines [17,25–30]. The recent literature on the investigation of the chemical composition of purple, either dye or pigment, identified in archeological contexts, as well as on the factors that may affect and differentiate the color at all stages of production and the dyeing process, is extensive and yet far from being exhaustive [31,32].

This paper comparatively examines the results of the analysis of samples from three archaeological finds of the purple pigment in lump form, coming from the excavations of Akrotiri on Thera and Trianda on Rhodes, as well as of a representative sample from a purple paint detail on a wall painting at the recently excavated (2009–2012) Raos site, Thera, in which the valuable pigment was used. The chemical composition of the purple pigments and paints dating from the early Late Bronze Age (~17th century BC or earlier), identified at the three contemporary Aegean sites (Akrotiri and Raos on Thera, and Trianda on Rhodes), is discussed with reference to the archaeological context of the unique finds. Among the finds discussed, we included the lump of purple pigment found in Akrotiri, Complex D, room 12 (AKR-10891), which was the starting point and the motivation for the development of a comprehensive methodology that allows us today not only to recognize the nature the organic colorant but also to determine its exact composition in chromophores, as well as to identify the inorganic material base that makes it a pigment for painting on different substrates. Therefore, the methodology that has been progressively developed over the last 20 years and applied case-by-case to finds that came to light at different times is applied here in order to discuss the results for a set of samples that makes sense to study comparatively. New spectroscopic results, through microRaman and FTIR techniques applied on the investigated purple pigments and a purple wall painting detail, follow through previous data obtained on other true purple samples [23,24]. The spectroscopic data are combined with previously reported chromatographic studies of samples taken from the same archaeological finds to achieve their complete characterization [30]. The data collected with high performance liquid chromatography coupled with a diode array detector (HPLC–DAD) are presented here in detail and are interpreted, along with the spectroscopic results, with a fresh view from an interdisciplinary angle on the basis of new findings, allowing better understanding and interpretation of the results relating to their different excavation context as they come from different proper archaeological sites, which are, however, contemporary and belong to the same geopolitical context of the Late Bronze Age I in the Aegean.

2. Materials and Methods

2.1. The Archaeological Samples under Investigation

2.1.1. Samples Taken from Three Finds of Purple Pigment Excavated in Lump Form

- AKR-10891: A sample from a lump of purple pigment found in the Late Cycladic I/Late Minoan IA city at Akrotiri, Thera, in Complex D, room 12. A first non-invasive investigation of this pigment with X-ray fluorescence spectrometry was reported in [33]. A microsample taken from the same find with inv. no. 10891 was investigated in the past using microRaman, FTIR, and HPLC techniques and compared with samples taken from purple paint details in the wall paintings of Xeste 3, belonging to the iconographic ensemble of the "Saffron Gatherers" and the "Procession of Women with Bouquets" [22–24]. The new sample was analyzed again in the present study and compared with the rest of the samples, and the microRaman and FTIR spectroscopic data are presented here for the first time.
- AKR-10882: A sample from a very small quantity of purple pigment found in the new pillar shaft 53A, during the excavations conducted in the course of construction of the new shelter of the archaeological site at Akrotiri.
- TRI-PR 13: A sample from a quantity of purple pigment found in the excavation conducted in the Paraskeva plot, in the southwest sector of the Late Bronze Age settlement at Trianda (Ialysos), east of the Trianda stream and the modern street above it.

2.1.2. Microsample Taken from the Wall Painting Excavated at Raos, Thera

- RAOS-KAL1. A microsample taken from a rosette purple detail on a wall-painting fragment from the excavated site of Raos, Thera. The site of Raos is located on the south side of the caldera, with an excellent view of the Kamenes islets, Therasia, and north Thera. A tiny portion of the sample was embedded in epoxy resin (Struers CaldoFix-2) and prepared in a polished cross-section to be studied with microRaman spectroscopy. The rest of the sample was used for HPLC analysis, first reported in [30] and interpreted comparatively in this work, along with the microRaman spectroscopic results, presented here for the first time.

2.2. MicroRaman Spectroscopy Experimental Details

Analyses focusing, under microscope, on the purple particles were made with a Horiba Jobin-Yvon HR800 dispersive Raman spectrometer equipped with Edge filters and using the 458 nm excitation wavelength of an air-cooled (Melles Griot) Ar^+ laser. Laser power at the sample was settled on 100 μW in order to preserve the sample integrity. Sample visualization and Raman scattering collection were performed through a reverse microscope notably equipped with a 100× objective, which allows a spatial resolution close to 1 μm (Note: In the ideal case the spot is ~0.5 μm—the diffraction limit). The 600 groves/mm grating used gives a spectral resolution of about 3 cm^{-1} for the signal recorded with an air-cooled CCD detector. The spectra presented were baseline corrected to remove the intense fluorescence contribution exhibited by the measured materials.

2.3. FTIR Spectroscopy Experimental Details

A Biorad FTS 175 system was employed for the acquisition of the FTIR spectra covering the wavenumber range of 4000–400 cm^{-1}. A few particles (~1 mg) from each of the raw pigment samples were set aside and mixed with anhydrous KBr and then pressed into pellets. The spectra were collected in transmission mode and converted to absorbance mode with a spectral resolution of 4 cm^{-1} representing an average of 64 scans.

2.4. HPLC–DAD Experimental Details

Samples were treated with hot dimethyl sulfoxide (DMSO) to extract and solubilize the colorants [30]. The HPLC-DAD system (Thermoquest, Manchester, UK) consisted of a 4000 quaternary HPLC pump, a SCM 3000 vacuum degasser, an AS3000 auto sampler with column oven, a Rheodyne 7725i Injector with 20 μL sample loop, and a Diode Array Detector UV 6000LP. The HPLC was operated using (A) water + 0.1% (v/v) trifluoroacetic acid and (B) acetonitrile + 0.1% (v/v) trifluoroacetic acid [30]. Separation was achieved with an Altima C18 (Alltech Associates Inc., Chicago, IL, USA) column (5 μm particle size, 250 mm × 3.0 mm) at a stable temperature of 35 °C using the following gradient elution [30]: 0–0.5 min: 50% B isocratic; 0.5–12 min: linear gradient to 60% B; 12–12.5 min 60% B isocratic; 12.5–14 min linear gradient to 70% B; 14–17 min linear gradient to 85% B; 17–18 min 85% B isocratic; and 18–21 min linear gradient to 100% B. The amounts of the analyzed samples were in the order of 0.5 mg.

3. The Archaeological Context of the Finds under Investigation

3.1. The Purple Pigment Finds from Akrotiri, Thera

Two lumps of purple pigment, found free of any support or container in rooms of different buildings in the city, were identified as having a similar physical appearance.

The first lump (inv. no. 10891), with a volume of about 20 cm^3, was found in (Building Complex Delta, Romford, UK) room 12. We assume that the color was stored not in powder form after any treatment such as grinding, but rather in raw form, as a cohesive mass, because following its excavation and storage in a glass container for over 30 years it was still cohesive. The quantity, form, and context indicate that it was not an accidental find in the field.

We do not know if it was kept in a vessel or other container of perishable material (bag from plant fibers, basket, etc.). A triton shell was found next to the lump of purple. Particularly interesting is the fact that from the same place came a large amount of yellow ochre, which weighed about 8 kg and was probably stored in a large pithos [34]. The pithos was found lying next to the triton and the purple pigment. The presence in the same room of two lumps of pigments—the purple, which is valuable and rare, and the ochre, which is common but here in such great quantity—is extremely important. The pigments were kept there, probably to be distributed subsequently to artists and artisans.

The dimensions of the second purple find (inv. no. 10882) were comparably smaller (Figure 1). It was found in a closed area during the excavation of pillar shaft 53a for the new shelter of the excavation site. Pillar shaft 53a is located to the west of Xeste 3. The lump of pigment was very small and could have been an accidental find derived from the destruction layer. However, the finding in the same place of a clay vase containing dissolved masses of impure red ochre suggests that perhaps the purple pigment was kept there intentionally.

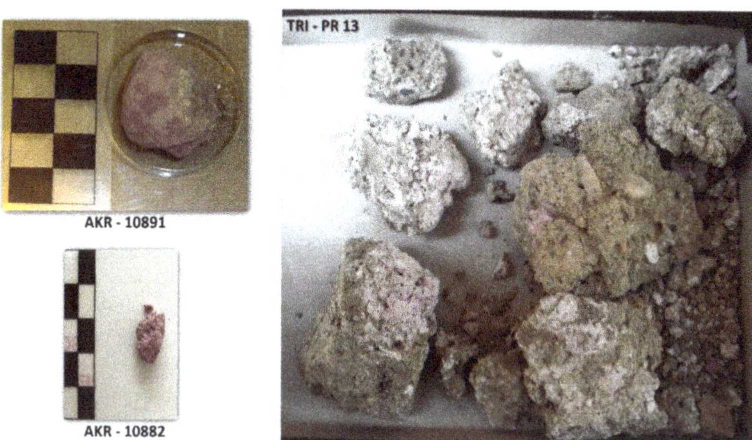

Figure 1. Photographs of the three purple pigment finds AKR-10891, AKR-10882, and TRI-PR13. (The photograph of pigment inv. no 10891 is of the part exhibited in the Museum of Prehistoric Thera, Thera, Greece).

3.2. The Purple Pigment Find from Trianda, Rhodes

A mass of purple pigment (TRI-PR13) was found at a depth of 3.40 m below the present surface in a layer of debris discarded in the northeast open space outside the monumental building located north of an E–W street crossing the southern zone of the prosperous Late Bronze Age IA city at Trianda [35]. The building (partially revealed in max. dimensions: 16.10 × 13.80 m) is a Xeste (of ashlar masonry), similar to the Xestai at Akrotiri, Thera [35,36]. The pigment was recovered from the earth filling of the debris bordering the west facade of the Xeste, partly covered with tephra from the Thera eruption. It was found just beneath a small fresco fragment (APX 2098) [37] depicting an elaborate black running spiral on a shiny whitish surface, among other fragmentary monochrome, polished dark grey, reddish, orange, or whitish frescoes; pieces of round tripod offering tables made of lime plaster; a reddish schist slab; and fragments of conical or some semiglobular cups and painted Late Bronze Age IA pottery, as well as animal teeth and bones. The pigment had the appearance of a cohesive dry mass, most probably once kept in a container of perishable material (see Figure 1).

The context comprised a secondary deposition of several materials fallen among the pre-eruption earthquake ruins, which were moved to outside the ruined building in order to clean it and prepare for its reconstruction. Similar contexts appear very often in the Late Bronze Age IA layers, mostly outside buildings, and they are indicative of the intensive cleaning activities of the earthquake ruins at Trianda before the tephra fall [36] (pp. 84–85), [38] (pp. 31–32), [39] (pp. 103–104, 107, 109). The excavation in the Paraskeva plot yielded abundant evidence of such intensive activities, since the exterior areas of the Xeste north of the E–W street was covered almost entirely with such secondary depositions containing many important materials [36] (pp. 83–85) and provided strong contextual evidence for the explanation of several aspects of the settlement and the activities of the people.

3.3. Purple Paint Details on a Wall Painting Excavated at Raos, Thera

The excavations at Raos brought to light a significant Late Cycladic I/Late Minoan IA complex atop a long crest consisting of a building surrounded by a spacious courtyard bounded by a robust stone-built enclosure wall [40–42]. Despite the poor state of preservation by Theran standards, the evidence shows that the Raos "villa" would have been quite impressive, with an upper story, high-quality wall paintings, and rather luxurious household equipment. It was situated on the outskirts of the Late Cycladic I/Late Minoan IA city at Akrotiri, less than 1 km from its northwest limit, on the natural land route leading from the coast to the hinterland of the Akrotiri Peninsula, where there are extensive tracts of arable land.

The fresco fragment from which the sample was taken derived from the north part of the southeast wing of the Raos residential building. It belonged to a heap of fresco fragments recovered on the floor of the ground floor. The heap covered an area of approximately 2 m^2. Most of the fresco fragments were found on entire or broken mud bricks and must have come from a collapsed mud-brick wall.

Near the heap of fresco fragments, two dressed tuff bases of a pier-and-door partition were found in situ. The evidence shows that a pier-and-door partition provided access to the southeast wing via a paved corridor at the heart of the building. Almost no moveable finds, neither in situ nor fallen from an upper story, were excavated in the north part of the southeast wing, i.e., in the space where the wall paintings were found. The exact plan and arrangement of the wing is difficult to grasp because of its almost complete destruction. However, both the particular architectural features (pier-and-door partition, frescoed mud-brick wall) and the absence of moveable finds indicate that this area was possibly used for rituals or formal receptions.

All the fresco fragments belonged to one composition, which we called the "Wall Painting with the Rosettes and Lilies." This is tripartite in structure. An upper narrow frieze of polychrome rosettes, some of which have purple petals, crowns a wide middle zone (Figure 2). In this main zone, pendent red triangular surfaces alternate with upright triangles, covered with scattered red lily flowers on a white ground. A lower zone of oblong panels embellished with colored wavy bands imitates a polychrome veined marble dado [41] (pp. 212–213), [42] (pp. 137–139).

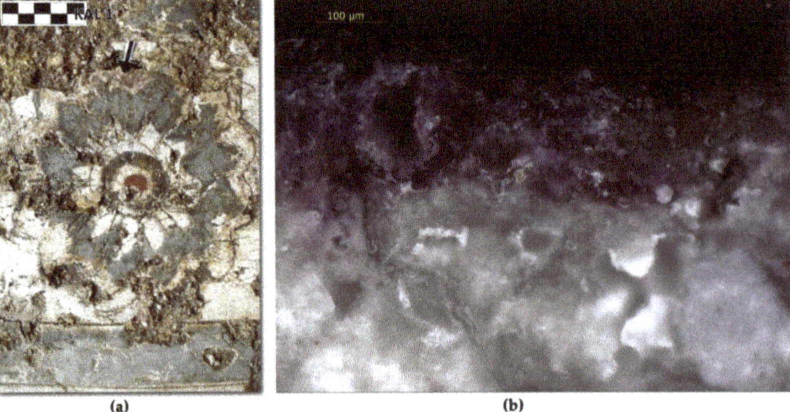

Figure 2. (**a**) Photograph of the wall-painting detail from Raos, with an indication of the sampling site on the purple petal of the rosette and (**b**) microphotograph of the cross-section of sample RAOS-KAL1.

4. Analytical Results and Discussion

4.1. Identification of the Brominated Indigoid Chromophores in the Purple Pigment

microRaman spectra were acquired in all the examined archaeological purple samples, on grains set aside from the three finds of purple pigment lumps (AKR-10882, AKR-10891, and TRI-PR13) and on the microsample from the purple paint detail in the Raos wall painting (RAOS-KAL1) prepared in a polished section, shown in Figure 2. The spectra are presented comparatively in (Figure 3); they present a perfect match of their peaks in both position and relative intensity for most of the characteristic vibrational bands of 6,6′-dibromoindigotin (DBI), the main chromophore compound of murex purple. For comparison, the spectrum taken from a reference 6,6′-DBI substance, synthesized according to previously described methods [43], is given in the same graph.

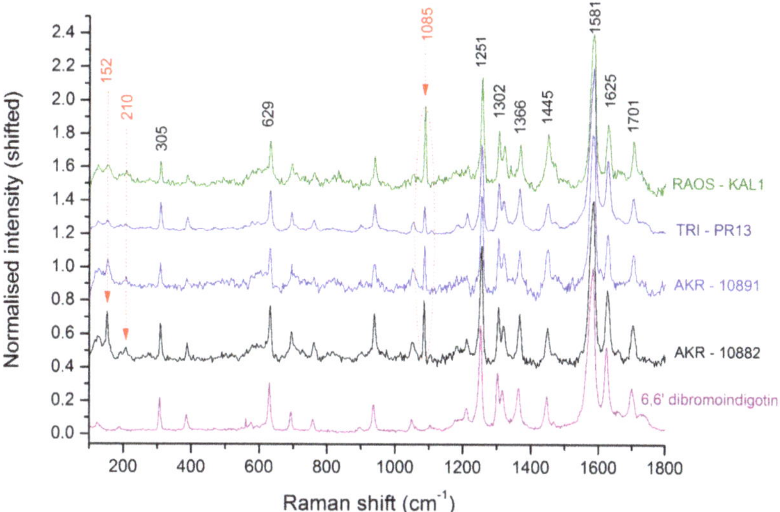

Figure 3. MicroRaman spectra acquired on samples from the purple pigment finds from Akrotiri and Trianda, (AKR-10882, AKR-10891, and TRI-PR13), as well as from the purple paint detail on the wall painting from Raos (RAOS-KAL1). The spectral signature of 6,6′-DBI is displayed in all the spectra acquired on the archaeological samples, shown in comparison with the spectrum taken on a synthetic 6,6′-DBI substance, [43].

The band positions and assignments to the vibrational modes for the 6,6′-DBI, the main chromophore compound of Tyrian purple, were previously published in comparison with indigo and its corresponding dihalogenated species 6,6′-DXI (where X: I, Br, F, Cl) [44]. A closer look at the spectra of the indigo and the dihalogenated indigoids, comparatively, allows for many similarities to be noticed, which is to be expected given that the majority of the detectable Raman bands are attributed to the indigoid structure. However, the exact position and the relative intensities of the major bands are proper to each halogenated indigoid compound, as the different halogeno-substituents induce differential changes (wavelength shift or relative intensity variance) in the characteristic vibrations that are in common, which makes the Raman spectrum a fingerprint for its identification. The specificity of each Raman spectrum extends even to differentiate two DBI positional isomers, as demonstrated in a recent paper comparing 5,5′-DBI to 6,6′-DBI and showing that the position of the bromine substituents affects many of the key bands of DBI and especially the wavelength position and the relative intensity of the band assigned to the ν(CBr) stretching vibration [45]. The 6,6′-DBI in particular presents the Raman bands corresponding to the benzene ring quadrant stretching vibrations ν(C=C) at ~1581 (s) cm^{-1} and ~1625 (ms) cm^{-1}, the band assigned to the carbonyl stretching ν(C=O) at ~1700 (m) cm^{-1}, the band corresponding to the bending δ(NH), δ(CH) modes at ~1445 (m) cm^{-1} and ~1360 (m) cm^{-1}, the stretching ν(C–N) modes at ~1300 (m) cm^{-1}, and a strong band at ~1250 (s) cm^{-1} corresponding to δ(CH), δ(C=O) bending modes. In addition, the band proper to the brominated indigoid compounds, corresponding to the stretching modes of the Br-C bonds, which for the 6,6′-DBI are at the symmetrical positions 6,6′, is strongly resolved at ~305 (s) cm^{-1}.

The identification of the 6,6′-DBI in the purple pigment through a well-resolved microRaman spectrum answers, satisfactorily in most cases, the primary archaeological question: the recognition of the murex identity of the dye, which automatically ascertains its added value. However, with Raman analysis, it is unlikely to detect any other than the DBI chromophore compound possibly present in the dye.

With the HPLC–DAD analysis, a more accurate determination of all chromophores present in the samples was possible. Three major peaks were detected in the HPLC profiles (Figure 4), which correspond to 6-bromoindigotin (MBI), 6,6′-dibromoindigotin (DBI), and 6,6′-dibromoindirubin (DBIR). Moreover, indigotin (IND), 6′-bromoindirubin (MBIR) and 6-bromoindirubin (MBIR) were detected in the chromatograms but corresponded to very small HPLC peaks, as illustrated in Figure 4 in the chromatogram of sample AKR-10882 shown in higher detail. Finally, indirubin (IR), which is usually found in molluscan extracts in small amounts [46–48], was not detected in any of the examined archaeological samples.

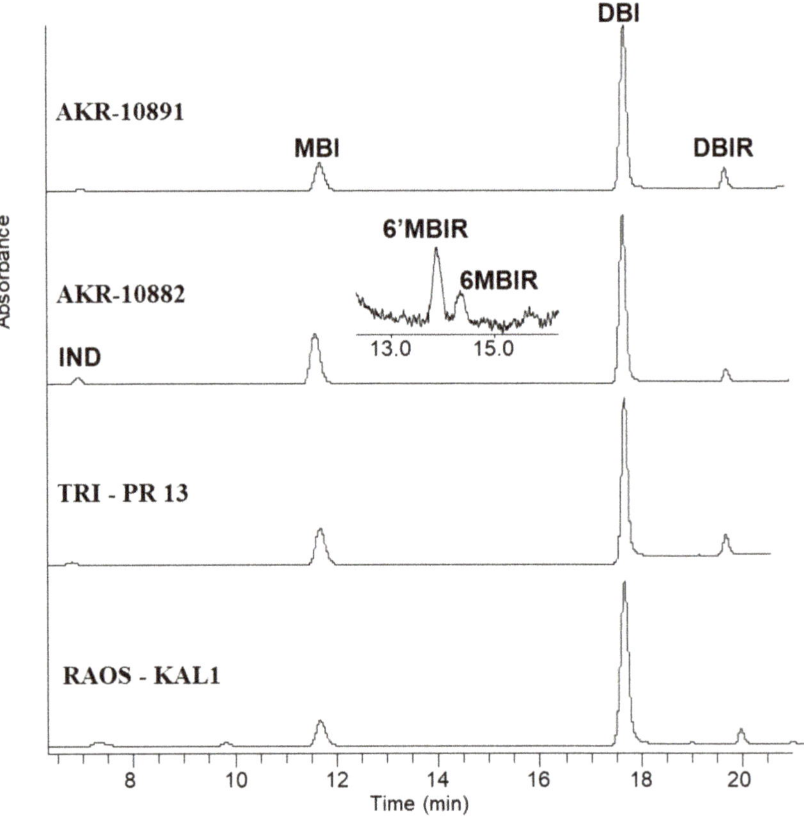

Figure 4. Chromatograms that were collected at 288 nm for the archaeological samples. The major HPLC peaks correspond to MBI, DBI and DBIR. Much smaller peaks were recorded for IND, 6′MBIR and 6MBIR as revealed for instance in the chromatogram of sample AKR-10882, shown in higher detail. The graphs for the samples AKR-10891 and AKR-10882 were adapted from [30] (with permission from Elsevier) and [32], respectively.

The aforementioned observations, which are visualized in the HPLC chromatograms, are clearly stated in the relative (%) values estimated from the integrated HPLC peak areas, as given in Table 1. The measurements were carried out using a monitoring wavelength of 288 nm [30]. Apparently, the given numbers are not the actual mass compositions of the samples. Based on the findings of a previous study it was to be expected that the % of DBI and IND reported in Table 1 would be an underestimation and overestimation, respectively, of the actual relative mass concentrations of the two compounds [48]. However, the semi-quantitative results, which are summarized in the table, are a good approximation of the actual concentrations and a very useful approach to compare the samples.

Table 1. Relative (%) integrated HPLC peak areas measured at 288 nm. The results are provided separately for the major HPLC peaks (MBI, DBI and DBIR) and as a sum for the minor HPLC peaks.

Sample	MBI	DBI	DBIR	IND + 6′MBIR + 6MBIR
AKR-10891	20.0	71.5	7.0	1.5
AKR-10882	26.9	65.5	3.8	3.8
TRI-PR13	24.2	65.9	6.9	3.1
RAOS-KAL1	15.5	77.3	5.6	1.6

The chromatographic analysis in Figure 4 and Table 1 confirmed the identification of molluscan purple in the four archaeological samples and revealed that the purple pigments and paints have similar compositions. DBI is the dominant compound in the four chromatograms of Figure 4, followed by MBI and DBIR. In the case it was detected, IND corresponded to an extremely small peak. These results suggest that, based on the identified composition of indigoid chromophores, the Late Cycladic purple pigment should have had a pure purple hue, with balanced proportions of red and blue components if not with a tendency to reddish purple. The actual color of the investigated samples, as appeared to the naked eye (see Figures 1 and 2a) and under microscope (see Figure 2b and Figure 6), as well as the measurements with VIS spectroscopy in the find (inv. no. 10891) [23], confirmed the HPLC-based assumption: There was no predominant bluish tint, which is often observed in molluscan pigments rich in blue IND [49].

As reported in previous published works, in the Mediterranean, there are three mollusc species that were used for dyeing and painting in antiquity: *Hexaplex trunculus* L. (*Murex trunculus*), *Bolinus brandaris* L. (*Murex brandaris*), and *Stramonita haemastoma* (*Thais haemastoma*). Studies on the significant amounts of shells found in several settlements on Crete (Palaikastro, Kouphonisi [5,50,51], Kommos [52], Chryssi, Pefka [53,54]), Kythera, Thera, Greece) and other insular and continental sites in Greece, as well as on the Eastern Mediterranean coasts, all agree on the almost exclusive majority of *M. trunculus*, with the exception of the Kythera case (a small settlement near Kastri), where *M. brandaris* is attested to be in abundance [55], whereas the incidence of *T. haemastoma* is always significantly lower.

The extraction of the Bronze Age Aegean purple principally from the *M. trunculus* as suggested by the archaeomalacological investigations is here supported by the HPLC findings with reference to related chemical studies [30]. These results about the biological source of the purple, although indicating the foremost exploitation of *M. trunculus* in purple dyeing, do not exclude the scenario of the addition of a regulated proportion of raw material from the *M. brandaris* species in order to obtain a unique, warmer shade of the expensive color.

4.2. Characterization of the Inorganic Substrate and Discussion of the Preparation Method of the Pigment as a Branch in the Operating Chain of the Purple Dyeing Industry

In the microRaman spectra, Figure 3, the characteristic vibrational band at 1085 cm^{-1} evidenced the presence of calcium carbonate in the composition of all the examined purple samples, either those from the pigment lumps (AKR-10891, AKR-10882, and TRI-PR 13) or the sample from the Raos wall paintings (RAOS-KAL1). This band, assigned to the ν_1

symmetric stretching of CO_3^{2-} in the calcium carbonate molecule, is rather similar for the two calcium carbonate allotropes (~1086.2 cm^{-1} for pure calcite and 1085.3 cm^{-1} for pure aragonite) [56]. This slight difference is smaller than the instrument's spectral resolution (~3 cm^{-1}) and therefore could not be used for the distinction of the two polymorphs in the archaeological samples. Given that the resolved band at 152 cm^{-1} (Figure 3) assigned to lattice vibrational mode is also common for calcite and aragonite, it is customary to focus on other additional, though weak, vibrational bands for the identification of the two allotropes. The most characteristic differences in the Raman spectra of calcite and aragonite are (a) the relative integrated intensities (with respect to the ν_1 band) of the bands at ~283.6 cm^{-1} for aragonite and ~281.2 cm^{-1} for calcite (for aragonite the integrated intensity of this band is at least 20 times weaker than the corresponding one of calcite, for which it is the second strongest peak in the spectrum) and (b) a band at 214.7 cm^{-1} in the aragonite Raman spectrum, which is specific and may as well be used for its identification [56]. If the 1085 cm^{-1} band in the spectra shown in Figure 3 was assigned to calcite, a corresponding band at ~280 cm^{-1} (characteristic also for calcite) should have been resolved in the Raman spectra, which was not the case. The latter indirectly suggests aragonite as the calcium carbonate allotrope identified here. Additionally, a weak band resolved at ~210 cm^{-1}, whose intensity increased proportionately to the respective one of 1085 cm^{-1}, agreed with the aforementioned suggestion.

Complementary data were collected with the corresponding FTIR spectra acquired from a few particles of each of the investigated samples. It should be mentioned that the FTIR spectra refer to a higher amount of sample, and thus are more representative of the bulk pigment than the Raman spectra acquired on single particles focused under the 100× objective of the Raman microscope, which allows a spatial resolution close to 1 μm. In the FTIR spectra shown in Figure 5a, it is possible to identify with certainty both aragonite and calcite in the inorganic substrate of the pigment based on the bands that are specific to one of the calcium carbonate polymorphs. In the case of the samples taken from wall-painting details, calcite may have derived also from the lime-based ground or may refer to chalk white as a pigment in the painting. However, the identification of both calcium carbonate allotropes and aragonite in particular in the purple pigment found in lump form puts in evidence the marine origin of the calcium carbonate substrate of the organic pigment. In the FTIR spectra, the well-resolved bands at 1475, 1081, 860, 711, and 700 cm^{-1} perfectly reproduce the respective characteristic bands of pure aragonite. Among the IR absorption bands of the carbonate ion, the split peaks at 711 and 700 cm^{-1} (ν_4 mode) and the peak at 1081 cm^{-1} (ν_1 mode) are specific to the aragonite structure whereas the peak assigned to the ν_2 mode appears with a difference of ~15 cm^{-1} for the two polymorphs, namely, at 860 cm^{-1} for aragonite versus 875 cm^{-1} for calcite [57,58]. Additionally, the stronger and broad absorption band assigned to the ν_3 mode, presenting a significant shift for calcite (at 1421 cm^{-1}) compared to aragonite (at 1475 cm^{-1}), further confirmed the presence of both calcium carbonate polymorphs in the composition of the purple pigments. For the identification of the DBI compound, only the two main bands at 1635 and 1612 cm^{-1} were barely resolved on the slope of the main broad carbonate band, Figure 5b, [26,43,45].

Examining a sample of the purple pigment from Trianda (TRI-PR13) under microscope, it was possible to observe the aragonite crystals with the characteristic needle-shaped form and clear appearance, as shown in Figure 6 [58,59]. In agreement with the presented FTIR data, in an older study of the purple lump (AKR-10891) with X-ray diffraction analysis [60], magnesium-rich calcite was also identified, which together with aragonite strengthened the assumption that the inorganic base of the pigment was possibly obtained from crushed and ground shells. The "recycling" of shells, of any species, either food residues or debris of purple production (*Muricidae* family), was a well-known practice not only for obtaining the base of the purple pigment but also for their use as coarse aggregate in the floor mortars or as raw material for the production of lime.

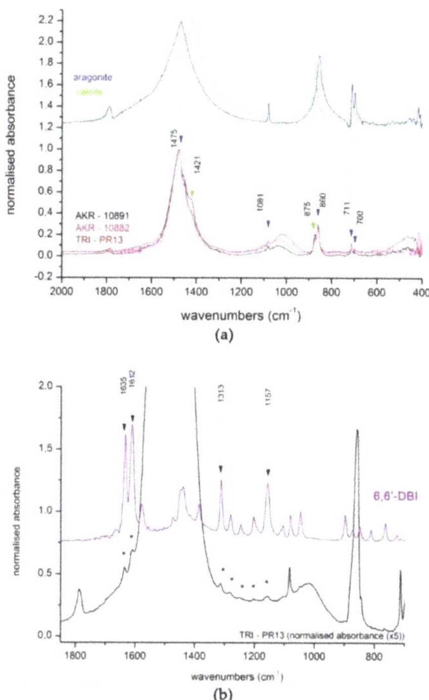

Figure 5. (**a**) FTIR spectra acquired in transmission mode on KBr pellets of samples from the purple pigments (AKR-10882, AKR-10891, and TRI-PR13) found free of any support. Both aragonite and calcite are identified in the inorganic substrate of the pigment. (**b**) Zoom in on FTIR Scheme 13. The major characteristic vibrational bands of the 6, 6′-DBI compound, namely, at 1635, 1612, 1313, and 1157 cm^{-1}, are barely resolved on the slopes of the main broad carbonate band.

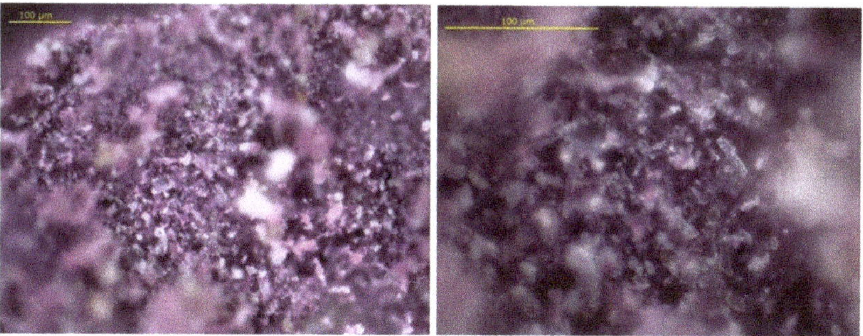

Figure 6. Microphotographs in different magnifications (original magnification 200×, 500×) of the purple pigment TRI-PR13 from Trianda in white light. The transparent longitudinal crystals of aragonite are evident in the purple matrix in the higher magnification, in particular.

The composition of the purple pigment consisting of a calcium carbonate substrate to which the purple dye was adsorbed, possibly by immersion in the alkaline juice as described by Pliny the Elder [21] in the procedure for preparing the pigment *purpurissum*, attests to its preparation in a vat, rendering improbable the application of a direct "primitive" technique. These data allow the establishment of vat-dyeing know-how to be traced at least back to the Late Bronze Age in the Aegean [23,61,62].

Reviewing the published results of analyses of shellfish purple pigment found in different archaeological contexts over time, in different applications and substrates, it appears that the inorganic base of the pigment can vary [32]. The choice of the inorganic base may have been related not only to the availability of the raw material but may also have been dictated by the substrate to which it was applied. For the wall paintings, the calcitic basis of the purple presents a perfect affinity with the lime-based ground and the calcium white used in a mixture with the rest of the pigments for rendering the light colors. An interesting alternative instance of shellfish purple was identified in the paint decoration of a funerary *kliné*, in Daskyleion (5th century BCE), with kaolinite being the main constituent of the inorganic base of the organic pigment, applied in that case on a white ground made of gypsum [63]. In the same paper, the HPLC data acquired from samples extracted from the purple paint detail on the *kliné* were discussed in comparison with the data acquired from an extract from a purple dyed-textile find, found within the same tomb. The perfect match of the HPLC profiles of the two extracts, from the purple pigment and the purple-dyed textile in the same context, constitutes a strong analytical demonstration of the argument suggesting the production of the purple pigment was integrant to the purple dyeing industry of textiles, according to Pliny's text.

Moving even further onward in time, and referring to a shellfish purple pigment identified in the paint decoration of Hellenistic terracotta figurines excavated from a rock-cut tomb in Chania, Crete (ca. 300 BC) [64], another white earth, huntite was identified in association with the purple pigment. Although in that case huntite was not directly attributed to the inorganic base of the purple pigment, this should be considered as a possibility. Interestingly, in a more recent investigation [65], huntite was also identified in association with the molluscan purple color revealed among bones and fragments of unidentified materials preserved in a gold larnax of the chamber of Tomb II, in the Macedonian royal tomb complex at Aegeae discovered by Prof. M. Andronikos. These data provide evidence for a wider range of white earth that apparently was known to be appropriate for use in the purple dyeing/pigment preparation operating chain.

4.3. Indirect Evidence for a Purple Dyeing Industry in the Aegean Sites of Thera and Rhodes

Although there is strong evidence for relating the purple color in the Theran and Rhodian contexts, with the Late Bronze Age industry of purple dyeing, there are no conclusive data for the location of the dyeing workshops, since there is not yet evidence of such installations in the excavated parts of the settlements. Instead, we present here only a number of strong arguments that support their possible existence outside the excavated sections, which correspond to residential areas.

The significant quantities of seashells found, most of them fragmented, in different buildings and streets at Akrotiri [66], corresponding to the stratigraphic and chronological sequence of the Late Cycladic I period, are considered food refuse, as the probability that dyeing installations existed near an inhabited area is highly unlikely. The largely attested second use of the shells in many instances as construction material for the floors of many rooms (Xeste 3, Pithoi Storeroom A2, B6 inter al.) [67] (pp. 195–196), (Figure 6), [68] (p. 124), as well as in mortar, as a filler in pottery production, etc., could more likely be associated with systematic recycling of waste products from the purple industry, with organized food scrap collection being less likely [66,69] (Figure 4), [70] (pp. 43, 44), (Figures 5 and 6), [5,71,72]. In another context, the evidence for the second use of murex shells, without indication of whether they are remains of foodstuffs or dyes, either in the production of lime plaster or in pieces visible on plastered surfaces probably to strengthen floor beddings, is discussed extensively in [73].

However, the murex finds (mostly *M. trunculus*), in fact crushed, whether food debris or used as construction material or as a filler in pottery production, etc., dating to the Late

Cycladic I/Late Minoan IA period, are proof of familiarity with the molluscs, implying the existence of a murex purple-dye industry at Akrotiri [5,72,74].

The lumps of mollusc purple found at Akrotiri together with lumps of other pigments in the same contexts could be intended for painting. The identification of mollusc purple pigment on murals at Akrotiri [22,75] and Raos [42] suggests its most probable application in monumental painting on Thera. The pigment was used selectively on the wall paintings. It was applied on certain petals of the rosettes in the "Rosette and Lily" wall painting from the Raos residential building. It was also used to render the petals of the crocus flowers, a necklace, and many details of the richly adorned female garments, and snoods in the wall paintings of the "Saffron Gatherers," the "Procession of Women with Bouquets," and the "Adorants" from Xeste 3 at Akrotiri [22].

The use of purple in the women's costumes implicitly refers to a possible second (or probably first) application of the purple color on Thera, the dyeing of textiles. In addition, textile production would have been quite a flourishing activity at the site, given the numerous loom weights found there [76–79]. All this evidence from Thera, both scientific (physical and chemical identification) and iconographic, reinforces the view that there was murex purple dye production on the island.

Excavations conducted since 1984 in the prehistoric settlement at Trianda on Rhodes have yielded evidence of a rather frequent use of *M. trunculus* in the settlement, not only by their sporadic presence along with other finds associated with food consumption, but also by the large quantities of crushed shells used either for reinforcing lime plaster or for constructing a strong substrate for plastered pavements, as in the case of the polythyron in the Markos plot [38]. On the other hand, many pigments, mostly red or pink and in some cases blue or white, were found in lumps or in powder form inside vases, mainly conical cups, or even without containers. Consequently, some of the above pigments might had been used by the painters, itinerant or local, who created the amazing frescoes that decorated the Late Bronze Age IA and even the earlier buildings at Trianda. Most recently, many lumps of pink pigment and pigments of other colors were found in the Middle Bronze Age levels of the settlement. Shiny red plasters decorated the houses of the settlement during this early period [36]. A deposit of thousands of murex shells, found in an open area close to a Middle Bronze Age polythyron, recently revealed during the excavation of the Atsiknoudas plot [80], south of the main Late Bronze Age city, provided further evidence of the early technologies and the activities of the society at Trianda, before the Minoan expansion. Moreover, the presence of many loom weights in an area east/southeast of the main settlement suggests large-scale textile production [36], which might be connected with dyeing as well. However, the only purple production workshop excavated thus far on the island of Rhodes is the Hellenistic installation found in the town of Rhodes, south of the Hellenistic fortification wall [81].

5. Conclusions

Lumps of mollusc purple pigment of similar physical appearance have been excavated at the Aegean sites of Akrotiri on Thera and Trianda on Rhodes dating from the Early Late Cycladic period. The same pigment was identified in purple details of certain wall paintings at Akrotiri and Raos, on Thera.

The chemical composition of the purple pigment was identified with molecular spectroscopies, microRaman and FTIR. It consisted of a calcium carbonate substrate to which the shellfish purple dye was adsorbed, possibly by immersion into a dye bath. The calcium carbonate substrate of the pigment, consisting mainly of aragonite and calcite, put in evidence its shell origin and referred to the "recycling" of shells of any species. The HPLC profiles collected from the archaeological samples were all similar, as they contained the same indigoid compounds in a comparatively similar relative content. MBI, DBI, and DBIR were recorded to be in the majority, whereas IND, 6′MBIR, and 6MBIR were present in much lower proportions.

Although there is not yet direct evidence for the location of purple dyeing installations in the Theran or Rhodian Late Bronze Age contexts, there are several arguments presented in this work that allow for the purple color identified in Akrotiri and Raos on Thera as well as in Trianda on Rhodes to be related with the Aegean Late Bronze Age industry of murex purple dyeing. The deposits of murex shells as well as the largely attested second use of the shells in many instances at Akrotiri and Trianda could more likely be associated with the systematic recycling of waste products from the purple industry than with an organized deposition of food debris. In addition, the large-scale textile production suggested by the numerous loom weights found in both sites might be connected with dyeing as well. Finally, the use of the purple pigment analyzed in this paper in several scenes in the Xeste 3 wall paintings at Akrotiri to render with naturalism not only the purple petals of the crocus flowers but also several details in women's woven costumes implicitly reinforces the probability of a local purple textile dyeing industry at the beginning of the Late Bronze Age.

Author Contributions: Conceptualization, S.S.; analytical methodology and investigation, S.S., I.K. and K.S.A.; archaeological context, T.M., K.B. and M.M.; writing—original draft preparation, review and editing, all authors. All authors have read and agreed to the published version of the manuscript.

Funding: This research received no external funding.

Institutional Review Board Statement: Not applicable.

Informed Consent Statement: Not applicable.

Data Availability Statement: Data available on request from the authors.

Acknowledgments: The authors would like to thank Christos Doumas, Director of the Excavations at Akrotiri, Thera, for permission to publish and support in studying the archaeological material from the settlement at Akrotiri. MicroRaman measurements were carried out in the Laboratory of Dynamics, Interactions and Reactivity LADIR, UMR 7075 CNRS/UPMC, Campus CNRS, Thiais. Special thanks are due to Ludovic Berlot-Gurlet for his valuable assistance. The main corpus of analytical data obtained with HPLC and FTIR techniques was produced and elaborated during previous affiliation of the first, second, and third authors to the Ormylia Foundation, Art Diagnosis Centre.

Conflicts of Interest: The authors declare no conflict of interest.

References

1. Marín-Aguilera, B.; Iacono, F.; Gleba, M. Colouring the mediterranean: Production and consumption of purple-dyed textiles in pre-roman times. *J. Mediterr. Archaeol.* **2018**, *31*, 127–154. [CrossRef]
2. Harris, S. From value to desirability: The allure of worldly things. *World Archaeol.* **2017**, *49*, 681–699. [CrossRef]
3. Ventris, M.G.F.; Chadwick, J. *Documents in Mycenaean Greece*, 2nd ed.; Cambridge University Press: Cambridge, UK, 1973; p. 573.
4. Reese, D.S. Industrial Exploitation of Murex Shells; Purple-dye and Lime Production at Sidi Khrebish, Benghazi(Berenice). *Libyan Stud.* **1980**, *11*, 79–93. [CrossRef]
5. Reese, D.S. Palaikastro Shells and Bronze Age Purple-Dye Production in the Mediterranean Basin. *Annu. Br. Sch. Athens* **1987**, *82*, 201–206. [CrossRef]
6. Reese, D.S. Iron age shell purple-dye production in the Aegean. In *Kommos IV, The Greek Sanctuary*; Part 1; Shaw, J.W., Shaw, M.C., Eds.; Princeton University Press: Princeton, NJ, USA, 2000; pp. 643–645.
7. Reese, D.S. Shells from Sarepta (Lebanon) and East Mediterranean Purple-Dye Production. *Mediterr. Archaeol. Archaeom.* **2010**, *10*, 1–31.
8. Brogan, T.M.; Betancourt, P.P.; Apostolakou, V. The purple dye industry on Eastern Crete. In *KOSMOS: Jewellery, Adornment and Textiles in the Aegean Bronze Age*; Lafineur, R., Nosch, M.-L., Eds.; Aegaeum 33; Peeters: Lueven-Liège, Belgium, 2012; pp. 187–193.
9. Carannante, A. Archaeomalacology and purple-dye: State of the art and new prospects of research. In *Moluscos y Púrpura en Contextos Arqueológicos Atlántico-Mediterráneos: Nuevos Datos y Reflexiones en Clave de Proceso Histórico*; Cantillo, J.J., Bernal, D., Ramos, J., Eds.; Universidad de Cádiz: Cadiz, Spain, 2014; pp. 273–282.
10. Macadam, H.I. Dilmun revisited. *Arab. Archaeol. Epigr.* **1990**, *1*, 49–87. [CrossRef]
11. Pfälzner, P. The world of the living and the world of the dead. *Ger. Res.* **2004**, *26*, 16–20. [CrossRef]
12. James, M.; Reifarth, N.; Mukherjee, A.; Crump, M.; Gates, P.; Sandor, P.; Robertson, F.; Pfälzner, P.; Evershed, R. High prestige Royal Purple dyed textiles from the Bronze Age royal tomb at Qatna, Syria. *Antiquity* **2009**, *83*, 1109–1118. [CrossRef]

13. Breniquet, C.; Desrosiers, S.; Nowik, W.; Rast-Eicher, A. Les textiles découverts dans les tombes de l'âge du Bronze moyen à Chagar Baza, Syrie. In *Chagar Bazar (Syrie) 8. Les Tombes Ordinaires de l'âge du Bronze Ancien et Moyen des Chantiers D, F, H, I (1999-2011)*; Tunca, O., Baghdo, A., Eds.; Peeters: Leuven-Liège, Belgium, 2018; pp. 11–32.
14. McGovern, P.E.; Michel, R.H. Royal Purple dye: Tracing the chemical origins of the industry. *Anal. Chem.* **1985**, *57*, 1514A–1522A.
15. McGovern, P.E.; Michel, R.H. Royal Purple dye: The chemical reconstruction of the ancient Mediterranean industry. *Acc. Chem. Res.* **1990**, *23*, 152–158. [CrossRef]
16. McGovern, P.E.; Michel, R.H. Royal Purple Dye: Its Identification by Complementary Physicochemical Techniques. *Masca Res. Pap. Sci. Archaeol.* **1990**, *7*, 69–76.
17. Koren, Z.C. Archaeo-chemical analysis of Royal Purple on a Darius I stone jar. *Microchim. Acta* **2008**, *162*, 381–392. [CrossRef]
18. Kalaitzaki, A.; Vafiadou, A.; Frony, A.; Reese, D.S.; Drivaliari, A.; Liritzis, I. Po-pu-re: Workshops, use and archaeometric analysis in pre-Roman central eastern Mediterranean. *Mediterr. Archaeol. Archaeom.* **2017**, *17*, 103–130. [CrossRef]
19. Apostolakou, S.; Betancourt, P.; Brogan, T.; Mylona, D. Chryssi and Pefka: The production and use of purple dye on Crete in the Middle and Late Bronze Age. In *Purpureae Vestes V. Textiles, Basketry and Dyes in the Ancient Mediterranean World*; Ortiz, J., Alfaro, C., Turell, L., Martínez, M.J., Eds.; Universitat de València: Valencia, Spain, 2016; pp. 199–208.
20. Koh, A.A.; Betancourt, P.P.; Pareja, M.N.; Brogan, T.M.; Apostolakou, V. Organic residue analysis of pottery from the dye workshop at Alatsomouri-Pefka, Crete. *J. Archaeol. Sci. Rep.* **2016**, *7*, 536–538. [CrossRef]
21. Pliny the Elder, The Natural History, XXXV, 26. Available online: http://data.perseus.org/citations/urn:cts:latinLit:phi0978.phi001.perseus-eng1:35.26 (accessed on 10 January 2021).
22. Chryssikopoulou, E.; Sotiropoulou, S. To *ιώδες στην παλέτα του Θηραίου ζωγράφου*. In *Αργοναύτης, Τιμητικός τόμος για τον καθηγητή Χρίστο Γ*; Ντούμα, A., Βλαχόπουλος, A., Μπίρταχα, K., Eds.; Η Καθημερινή: Athens, Greece, 2003; pp. 490–504.
23. Sotiropoulou, S.; Karapanagiotis, I. Conchylian purple investigation in prehistoric wall paintings of the Aegean area. In *Indirubin, the Red Shade of Indigo*; Meijer, L., Guyard, N., Skaltsounis, A.-L., Eisenbrand, G., Eds.; Life in Progress Editions: Roscoff, France, 2006; pp. 71–78.
24. Karapanagiotis, I.; Sotiropoulou, S.; Chryssikopoulou, E.; Magiatis, P.; Andrikopoulos, K.S.; Chryssoulakis, Y. Investigation of Tyrian purple occurring in prehistoric wall paintings of Thera. In *The Diversity of Dyes in History & Archaeology*; Kirby, J., Ed.; Archetype Publications: London, UK, 2017; pp. 82–89.
25. Wouters, J. A new method for the analysis of blue and purple dyes in textiles. *Dyes Hist. Archaeol.* **1992**, *10*, 17–21.
26. Cooksey, C.J. Tyrian Purple: 6,6′-Dibromoindigo and Related Compounds. *Molecules* **2001**, *6*, 736–769. [CrossRef]
27. Meijer, L.; Skaltsounis, A.L.; Magiatis, P.; Polychronopoulos, P.; Knockaert, M.; Leost, M.; Ryan, X.P.; Vonica, C.A.; Brivanlou, A.; Dajani, R.; et al. GSK-3-selective inhibitors derived from Tyrian purple indirubins. *Chem. Biol.* **2003**, *10*, 1255–1266. [CrossRef]
28. Meijer, L.; Guyard, N.; Skaltsounis, L.; Eisenbrand, G. (Eds.) *Indirubin, the Red Shade of Indigo*; Life in Progress Editions: Roscoff, France, 2006.
29. Haubrichs, R. L'etude de la pourpre: Histoire d'une couleur, chimie et experimentations. *Preist. Alp.* **2005**, *40*, 133–160.
30. Karapanagiotis, I.; Mantzouris, D.; Cooksey, C.; Mubarak, M.S.; Tsiamyrtzis, P. An improved HPLC method coupled to PCA for the identification of Tyrian Purple in archaeological and historical samples. *Microchem. J.* **2013**, *110*, 70–80. [CrossRef]
31. Cooksey, C. Tyrian purple: The first four thousand years. *Sci. Prog.* **2013**, *96*, 171–186. [CrossRef]
32. Karapanagiotis, I. A Review on the Archaeological Chemistry of Shellfish Purple. *Sustainability* **2019**, *11*, 3595. [CrossRef]
33. Aloupi, E.; Karydas, A.G.; Paradellis, T. Pigment analysis of wall paintings and ceramics from Greece and Cyprus. *X-ray Spectrom.* **2000**, *29*, 18–24. [CrossRef]
34. Sotiropoulou, S.; Perdikatsis, V.; Birtacha, K.; Apostolaki, C.; Devetzi, A. Physicochemical characterization and provenance of colouring materials from Akrotiri-Thera in relation to their archaeological context and application. *Archaeol. Anthr. Sci.* **2012**, *4*, 263–275. [CrossRef]
35. Marketou, T. Ιαλυσία-Τριάντα. Νότιος Τομέας Υστεροχαλκού οικισμού: Οδός Τριαντών (οικόπεδο Π. Παρασκευά, Κ.Μ. 668[A]). *ΑΔ* **1998**, *53*, 952–953.
36. Marketou, T. Ialysos and its neighbouring areas in the MBA and LB I periods: A chance for peace. In *The Minoans in the Central, Eastern and Northern Aegean—New Evidence. Acts of a Minoan Seminar 22–23 January 2005 in Collaboration with the Danish Institute at Athens and the German Archaeological Institute at Athens*; Macdonald, C.F., Hallager, E., Niemeier, W.-D., Eds.; Monographs of the Danish Institute at Athens: Athens, Greece, 2009; Volume 8, pp. 73–96.
37. Marketou, T. The art of the wall painting at Ialysos on Rhodes: From the early second millennium BC to the Eruption of the Thera volcano. In *ΧΡΩΣΤΗΡΕΣ/PAINTBRUSHES. Wall-Painting and Vase-Painting of the Second Millennium BC in Dialogue, Proceedings of the International Conference on Aegean Iconography, Akrotiri, Thera, Greece, 24–26 May 2013*; University of Ioannina/Hellenic Ministry of Culture and Sports—Archaeological Receipts Fund: Athens, Greece, 2018; pp. 261–275.
38. Marketou, T. New evidence on the topography and site history of prehistoric Ialysos. In *Archaeology in the Dodecanese*; Dietz, S., Papachristodoulou, I., Eds.; National Museum of Denmark Department of Near Eastern and Classical Antiquities: Copenhagen, Denmark, 1988; pp. 27–33.
39. Marketou, T. Santorini tephra from Rhodes and Kos: Some chronological remarks based on the stratigraphy. In *Thera and the Aegean World III*; Hardy, D., Renfrew, A.C., Eds.; Thera Foundation: London, UK, 1990; Volume 3, pp. 100–119.
40. Marthari, M. Ραός: Οικόπεδο Μ. Αλεφραγκή. *ΑΔ* **2012**, *56–59*, 105–106.

41. Marthari, M. The attraction of the pictorial" reconsidered: Pottery and wall-paintings, and the artistic environment on Late Cycladic I Thera in the light of the most recent research. In *ΧΡΩΣΤΗΡΕΣ/PAINTBRUSHES: Wall-Painting and Vase-Painting of the Second Millennium BC in Dialogue*; Vlachopoulos, A.G., Ed.; University of Ioannina/Hellenic Ministry of Culture and Sports—Archaeological Receipts Fund: Athens, Greece, 2018; pp. 205–221.
42. Marthari, M. Raos and Akrotiri: Memory and identity in LC I/LM IA Thera as reflected in settlement patterns and ceramic production. In *MNHMH/MNEME: Past and Memory in the Aegean Bonze Age*; Borgna, E., Caloi, I., Carinci, F.M., Laffineur, R., Eds.; Peeters: Leuven-Liège, Belgium, 2019; pp. 135–144.
43. Clark, R.J.H.; Cooksey, C.J. Monobromoindigos: A new general synthesis, the characterization of all four isomers and an investigation into the purple colour of 6,6′-dibromoindigo. *New J. Chem.* **1999**, *23*, 323–328. [CrossRef]
44. Karapanayiotis, T.; Villar, S.E.J.; Bowen, R.D.; Edwards, H.G.M. Raman spectroscopic and structural studies of indigo and its four 6,6A-dihalogeno analogues. *Analyst* **2004**, *129*, 613–618. [CrossRef]
45. Smith, G.D.; Chen, V.J.; Holden, A.; Keefe, M.H.; Shannon, L.G. Analytical characterization of 5,5′-dibromoindigo and its first discovery in a museum textile. *Herit. Sci.* **2019**, *7*, 62. [CrossRef]
46. Karapanagiotis, I.; de Villemereuil, V.; Magiatis, P.; Polychronopoulos, P.; Vougogiannopoulou, K.; Skaltsounis, A.-L. Identification of the Coloring Constituents of Four Natural Indigoid Dyes. *J. Liq. Chromatogr. Relat. Technol.* **2006**, *29*, 1491–1502. [CrossRef]
47. Mantzouris, D.; Karapanagiotis, I. Identification of indirubin and monobromoindirubins in Murex Brandaris. *Dyes Pigment.* **2014**, *104*, 194–196. [CrossRef]
48. Vasileiadou, A.; Karapanagiotis, I.; Zotou, A. Determination of Tyrian purple by high performance liquid chromatography with diode array detection. *J. Chromatogr. A* **2016**, *1448*, 67–72. [CrossRef] [PubMed]
49. Koren, Z.C.; Verhecken-Lammens, C. Microscopic and chromatographic analyses of molluskan purple yarns in a late Roman period textile. *E-Preserv. Sci.* **2013**, *10*, 27–34.
50. Bosanquet, R.C. Dicte and the Temples of Dictaean Zeus. *Ann. Br. School Athens* **1943**, *40*, 60–77. [CrossRef]
51. Barber, E.J.W. *Prehistoric Textiles: The Development of Cloth in the Neolithic and Bronze Age*; Princeton University Press: Princeton, NJ, USA, 1991.
52. Ruscillo, D. Faunal remains and murex dye production. In *Kommos V: The Monumental Minoan Buildings*; Shaw, J.W., Shaw, M.C., Eds.; Princeton University Press: Princeton, NJ, USA, 2006; pp. 776–840.
53. Apostolakou, S. A workshop for dyeing wool at Pefka near Pacheia Ammos. *Kentro* **2008**, *11*, 1–2.
54. Apostolakou, V.; Brogan, T.M.; Betancourt, P.P. The minoan settlement on chryssi and its murex industry. In *KOSMOS: Jewellery, Adornment and Textiles in the Aegean Bronze Age, Proceedings of the 13th International Aegean Conference, Copenhagen, Denmark, 21–26 April 2010*; Nosch, M.L., Laffineur, R., Eds.; Peeters: Lueven-Liège, Belgium, 2012; pp. 179–182.
55. Coldstream, J.N.; Huxley, G.L. *Kythera—Excavations and Studies*; Faber: London, UK, 1972.
56. De La Pierre, M.; Carteret, C.; Maschio, L.; André, E.; Orlando, R.; Dovesi, R. The Raman spectrum of $CaCO_3$ polymorphs calcite and aragonite: A combined experimental and computational study. *J. Chem. Phys.* **2014**, *140*, 1–12. [CrossRef] [PubMed]
57. Vagenas, N. Quantitative analysis of synthetic calcium carbonate polymorphs using FT-IR spectroscopy. *Talanta* **2003**, *59*, 831–836. [CrossRef]
58. Chang, R.; Choi, D.; Kim, M.H.; Park, Y. Tuning Crystal Polymorphisms and Structural Investigation of Precipitated Calcium Carbonates for CO2 Mineralization. *ACS Sustain. Chem. Eng.* **2017**, *5*, 1659–1667. [CrossRef]
59. Parker, J.E.; Thompson, S.P.; Lennie, A.R.; Potter, J.; Tang, C.C. A study of the aragonite-calcite transformation using Raman spectroscopy, synchrotron powder diffraction and scanning electron microscopy. *Cryst. Eng. Comm.* **2010**, *12*, 1590–1599. [CrossRef]
60. Perdikatsis, V. *XRD Analysis Report: Personal communication*; School of Mineral Resources Engineering, Technical University of Crete: Chania, Crete, Greece, 2003.
61. Cardon, D. Purple from molluscs. In *Natural Dyes*; Archetype Publications: London, UK, 2007; Chapter 11; pp. 559–562.
62. Doumet, J. *Etude sur la Couleur Pourpre Ancienne et Tentative de Reproduction du Procédé de Teinture de la Ville de Tyr Décrit par Pline l'Ancien*; Imprimerie Catholique: Beirut, Lebanon, 1980.
63. Papliaka, Z.E.; Konstanta, A.; Karapanagiotis, I.; Karadag, R.; Akyol, A.A.; Mantzouris, D.; Tsiamyrtzis, P. FTIR imaging and HPLC reveal ancient painting and dyeing techniques of molluskan purple. *Archaeol. Anthropol. Sci.* **2017**, *9*, 197–208. [CrossRef]
64. Maravelaki-Kalaitzaki, P.; Kallithrakas-Kontos, N. Pigment and terracotta analyses of Hellenistic figurines in Crete. *Anal. Chim. Acta* **2003**, *497*, 209–225. [CrossRef]
65. Maniatis, Y.; Arvaniti, T.; Antikas, T.G.; Wynn-Antikas, L.; Orsini, S.; Ribechini, E.; Colombini, M.P. Investigation of an unusual composite material found in the larnax with cremated bones in royal tomb II at Vergina. In Proceedings of the 40th International Symposium on Archaeometry, Los Angeles, CA, USA, 19–23 May 2014.
66. Karali, L. Το Μαλακολογικό υλικό του Ακρωτηρίου (Malacological Material of Akrotiri). In *One Day Conference on Akrotiri 19/12/87, Ημερίδα, Ακρωτήρι Θήρας, είκοσι χρόνια έρευνας (1967–1987) Συμπεράσματα-Προβλήματα-Προοπτικές*; Εν Αθήναις Αρχαιολογική Εταιρεία: Athens, Greece, 1992; Volume αρ.116, pp. 163–169.
67. Palyvou, C. *Ακρωτήρι Θήρας. Η οικοδομική τέχνη*; Εν Αθήναις Αρχαιολογική Εταιρεία: Athens, Greece, 1999.
68. Palyvou, C. *Akrotiri, Thera: An Architecture of Affluence 3500 Years Old*; Prehistory Monographs 15; INSTAP Academic Press: Philadelphia, NJ, USA, 2005.

69. Karali, L. Sea shells, land snails and other marine remains from Akrotiri. In *Thera and the Aegean World III, Proceedings of the Third International Congress, Santorini, (3-9/9/1989)*; Thera Foundation: London, UK, 1990; pp. 410–415.
70. Karali, L. *Shells in Aegean Prehistory*; British Archaeological Reports International Series 761; Archaeopress: Oxford, UK, 1999.
71. Doumas, C. *Thera: Pompeii of the Ancient Aegean: Excavations at Akrotiri, 1967–1979*; Thames and Hudson: London, UK, 1983.
72. Doumas, C. The excavations at Thera and the Aegean late bronze age. *Endeavour* **1983**, *7*, 144–149. [CrossRef]
73. Brysbaert, A. Murex uses in plaster features in the Aegean and eastern Mediterranean Bronze age. *Mediterr. Archaeol. Archaeom.* **2007**, *7*, 29–51.
74. Aloupi, E.; Maniatis, I.; Paradellis, T.; Karali, L. Analysis analysis of a purple material found at Akrotiri. In *Thera and the Aegean World III, Proceedings of the Third International Congress, Santorini, Greece, 3–9 September 1989*; Hardy, D.A., Doumas, C.G., Sakellarakis, J.A., Warren, P.M., Eds.; The Thera Foundation: London, UK, 1990; Volume 1, pp. 488–490. [CrossRef]
75. Chrysikopoulou, E. Use of murex purple in the Wall-paintings at Akrotiri, Thera. *ΑΛΣ* **2005**, *3*, 77–80.
76. Tzachili, I. All important yet elusive: Looking for evidence of cloth-making at Akrotiri, in 1990. In *Thera and the Aegean World III, Proceedings of the Third International Congress, Santorini, Greece, 3–9 September 1989*; Hardy, D.A., Doumas, C.G., Sakellarakis, J.A., Warren, P.M., Eds.; The Thera Foundation: London, UK, 1990; Volume 1, pp. 380–389.
77. Tzachili, I. Υφαντική και υφάντρες στο Προ ̈ι στορικό *Αιγαίο 2000–1000 π.Χ*; Πανεπιστημιακές Εκδόσεις Κρήτης: Heraklion, Greece, 1997.
78. Tzachili, I. Ποικίλα. In *Ακρωτήρι Θήρας. Δυτική Οικία: Τράπεζες, λίθινα, μετάλλινα, ποικίλα*; Doumas, C., Ed.; Η Εν Αθήναις Αρχαιολογική Εταιρεία: Athens, Greece, 2007; pp. 245–282.
79. Michaelidou, A. *Weight and Value in Pre-Coinage Societies, vol. II, Sidelights on Measurement from the Aegean and the Orient*; Μελετήματα 61; Research Centre for Greek and Roman Antiquity, National Hellenic Research Foundation: Athens, Greece, 2008.
80. Marketou, T. Time and space in the middle bronze age aegean world: Lalysos (Rhodes) a gateway to the eastern Mediterranean. In *Space and Time in Mediterranean Prehistory*; Souvatzi, S., Hadjii, A., Eds.; Routledge: London, UK; Taylor and Francis Group: New York, NY, USA, 2014; pp. 176–195.
81. Marketou, T. Εργαστήριο Πορφύρας στην πόλη της Ρόδου. In *Ρόδος 2400 Χρόνια: Η πόλη της Ρόδου από την Ίδρυσή της μέχρι την κατάληψη από τους Τούρκους (1523)*; Διεθνές Επιστημονικό Συνέδριο. Ρόδος, 24–29 Οκτωβρίου 1993. Πρακτικά Τόμος Α; Κυπραίου, Ε., Ζαφειροπούλου, Ν., Eds.; Υπουργείο Πολιτισμού: Athens, Greece, 1999; pp. 243–252.

 heritage

MDPI

Article

Unexpected Findings in 16th Century Wall Paintings: Identification of Aragonite and Unusual Pigments

Laura Rampazzi [1,2,*], Cristina Corti [3], Ludovico Geminiani [3] and Sandro Recchia [3]

1 Dipartimento di Scienze Umane e dell'Innovazione per il Territorio e Centro Speciale di Scienze e Simbolica dei Beni Culturali, Università degli Studi dell'Insubria, via Valleggio 9, 22100 Como, Italy
2 Istituto per le Scienze del Patrimonio Culturale, Consiglio Nazionale delle Ricerche (ISPC-CNR), via Cozzi 53, 20125 Milano, Italy
3 Dipartimento di Scienza e Alta Tecnologia, Università degli Studi dell'Insubria, via Valleggio 11, 22100 Como, Italy; cristina.corti@uninsubria.it (C.C.); lgeminiani@studenti.uninsubria.it (L.G.); sandro.recchia@uninsubria.it (S.R.)
* Correspondence: laura.rampazzi@uninsubria.it

Abstract: Sixteenth century wall paintings were analyzed from a church in an advanced state of decay in the Apennines of central Italy, now a remote area but once located along the salt routes from the Po Valley to the Ligurian Sea. Infrared spectroscopy (FTIR-ATR), X-ray diffraction (XRD) and scanning electron microscopy (SEM) with a microprobe were used to identify the painting materials, as input for possible future restoration. Together with the pigments traditionally used for wall painting, such as ochre, ultramarine blue, bianco di Sangiovanni, cinnabar/vermilion, azurite, some colors were also found to have only been used since the 18th century. This thus suggests that a series of decorative cycles occurred after the church was built, confirmed by the multilayer stratigraphy of the fragments. Some of these colors were also unusual, such as clinochlore, Brunswick green, and ultramarine yellow. The most notable result of the analytical campaign however, was the ubiquitous determination of aragonite, the mineralogical form of calcium carbonate, mainly of biogenic origin. Sources report its use in Roman times as an aggregate in mortars, and in the literature it has only been shown in Roman wall paintings. Its use in 16th century wall paintings is thus surprising.

Keywords: aragonite; Brunswick green; clinochlore; FTIR; mortars; SEM-EDX; ultramarine yellow; vermilion; wall paintings; XRD

Citation: Rampazzi, L.; Corti, C.; Geminiani, L.; Recchia, S. Unexpected Findings in 16th Century Wall Paintings: Identification of Aragonite and Unusual Pigments. *Heritage* **2021**, *4*, 2431–2448. https://doi.org/10.3390/heritage4030137

Academic Editor: Diego Tamburini

Received: 30 August 2021
Accepted: 12 September 2021
Published: 15 September 2021

1. Introduction

Conservation science helps preserve the memory of monuments which have been shown to be in such a declining state of neglect and decay that time is running out. Monuments that very probably cannot be restored as part of a conservation project risk disappearing without a trace. An analytical campaign of what is still possible to characterize can therefore reveal the materials, artistic techniques, decorations to future memory of a disappearing beauty. The church of Santo Stefano in Selva (Cerignale) (Figure 1), in the Apennines of Piacenza (central Italy), is in a declining state of neglect and decay. Given its location in a remote area, with no tourism, it is not a high priority for the national heritage. The local community pressed for an analytical campaign on the fragments of the few surviving paintings, in order to preserve their memory.

The area between the Po Valley and the Ligurian Sea was once much travelled, and there were also several historical pilgrimages, along important trade ridgeways, such as the salt routes, which involved the exchange of local goods. One of these routes, the "Strada del Cifalco", followed the natural path of the Trebbia Valley along the ridge between the Aveto and Trebbia rivers, connecting the north of what today is Liguria with Piacenza. It is thus not surprising that the Apennines, the mountainous system of this area, are rich in artistic testimonies, which unfortunately were neglected during the depopulation after the Second World War. This artistic history can be seen in the municipality of Cerignale too: the village of Ponte Organasco, with its medieval layout and many tower houses; the medieval castle of Cariseto, which hosted Emperor Frederick Barbarossa; and the church of Santo Stefano in Selva. The building is located in an area which was once a point of transit and rest for the muleteers travelling between the Trebbia valley and the Aveto valley, on their way to Liguria [1].

Figure 1. The remains of wall paintings in the church of S. Stefano.

Documentation on the history of the paintings is scarce. The building is listed in a church register which records the churches that in 1523 sent financial dues to the Papal State. The ruins still reveal the elaborate construction technique adopted to simulate a certain opulence, such as the vaulted ceiling, which looks like it is made of stone, but is actually a wooden structure with the intrados made of reeds covered with painted plaster. The facade of the single-nave church is relatively intact, and recalls the classical Greek temple structure. The lesenes support the very elaborate trabeation and triangular tympanum. Three transverse round arches have also remained intact. No records of the wall paintings have been found in the archives. We can thus only assume that the first cycle of paintings dates back to the construction of the building in the 16th century.

The church has almost completely lost the internal wall paintings (Figure 1). The decay originated from the partial settlement of the ground, around the middle of the last century, and more recently from a bolt of lightning that hit the stone bell tower causing a fire that almost completely destroyed the roof.

The present work reports the analytical campaign conducted on fragments from the few wall paintings still in place. The samples were analyzed with optical microscopy and scanning electron microscopy equipped with an energy-dispersive spectrometer, also on polished cross sections, and with X-ray diffraction and infrared spectroscopy on powders. The analyses obtained important information on the pictorial materials, pigments and binders used for the decoration of the walls. Traditional and unusual pigments were revealed together with the widespread use of aragonite probably from shells utilized as a building material, which, to the best of our knowledge, has never before been found in 16th century wall paintings.

We believe that the results could contribute to maintain the memory of the history of the church and help support a material study as part of future conservation work.

2. Materials and Methods

Table 1 reports the samples. Since most of the samples had clearly different overlapping color backgrounds, when possible without risk of contamination, the individual layers were separated with a scalpel and analyzed individually. Fragments of the eight most complex samples were analyzed as polished cross sections. From each fragment, the support mortar layer was taken and characterized.

Table 1. Description of the samples, identified pigments and employed analytical techniques (OM: optical microscopy; FTIR-ATR: Fourier-transform infrared spectroscopy in attenuated total reflection; XRD: X-ray diffraction; SEM-EDX: scanning electron microscopy-energy dispersive X-ray spectroscopy; PCS: polished cross section).

Sample	Images	Colors	Identified Pigments	Analytical Techniques
ST1		aqua green	blanc fixe ($BaSO_4$), clinochlore ($(Mg_4Fe_2Al(Si_3Al) O_{10}(OH)_8)$), ultramarine blue ($Na_8Al_6Si_6O_{24}S_4$)	OM, FTIR-ATR, XRD, SEM-EDX
ST2		aqua green	blanc fixe ($BaSO_4$), ultramarine blue ($Na_8Al_6Si_6O_{24}S_4$)	FTIR-ATR, XRD, SEM-EDX, PCS
		blue	ultramarine blue ($Na_8Al_6Si_6O_{24}S_4$)	
		red	red ochre (Fe_2O_3, clay)	
		white	blanc fixe ($BaSO_4$)	
		white	bianco di Sangiovanni ($CaCO_3$)	
		yellow	yellow ochre (FeOOH, clay)	
ST3		aqua green	ultramarine blue ($Na_8Al_6Si_6O_{24}S_4$), ultramarine yellow ($BaCrO_4$)	FTIR-ATR, SEM-EDX, PCS
		blue	ultramarine blue ($Na_8Al_6Si_6O_{24}S_4$)	
		white	bianco di Sangiovanni ($CaCO_3$)	
		yellow	yellow ochre (FeOOH, clay)	
ST4		blue	ultramarine blue ($Na_8Al_6Si_6O_{24}S_4$)	FTIR-ATR, XRD, SEM-EDX

Table 1. *Cont.*

Sample	Images	Colors	Identified Pigments	Analytical Techniques
ST5		blue	ultramarine blue ($Na_8Al_6Si_6O_{24}S_4$)	OM, FTIR-ATR, XRD, SEM-EDX, PCS
		red	red ochre (Fe_2O_3, clay)	
		white	bianco di Sangiovanni ($CaCO_3$)	
ST6	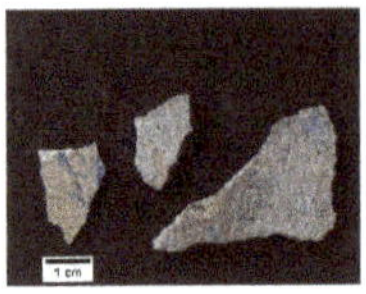	blue	ultramarine blue ($Na_8Al_6Si_6O_{24}S_4$)	FTIR-ATR, XRD, SEM-EDX
		red	red ochre (Fe_2O_3, clay)	
		yellow	cinnabar/vermilion (HgS), yellow ochre (FeOOH, clay)	
ST7		black	red ochre (Fe_2O_3, clay)	OM, FTIR-ATR, SEM-EDX
		blue	blanc fixe ($BaSO_4$), ultramarine blue ($Na_8Al_6Si_6O_{24}S_4$)	
		light blue	blanc fixe ($BaSO_4$), ultramarine blue ($Na_8Al_6Si_6O_{24}S_4$), white lead ($2PbCO_3$ $Pb(OH)_2$)	
		grey	manganese black (MnO_2), yellow ochre (FeOOH, clay)	
		dark ochre	manganese black (MnO_2), yellow ochre (FeOOH, clay)	
		yellow	yellow ochre (FeOOH, clay)	
ST8		green	Brunswick green ($(Cu,Zn)_2(OH)_3Cl$)	OM, FTIR-ATR, XRD, SEM-EDX
		orange	yellow ochre (FeOOH, clay)	
		yellow	chrome yellow ($PbCrO_4$)	
ST9		ochre	yellow ochre (FeOOH, clay)	FTIR-ATR, SEM-EDX
ST10		red	red ochre (Fe_2O_3, clay)	OM, FTIR-ATR, SEM-EDX
		yellow	yellow ochre (FeOOH, clay)	
ST11		green	Brunswick green ($(Cu,Zn)_2(OH)_3Cl$)	OM, FTIR-ATR, XRD, SEM-EDX
		yellow	chrome yellow ($PbCrO_4$), yellow ochre (FeOOH, clay)	

Table 1. *Cont.*

Sample	Images	Colors	Identified Pigments	Analytical Techniques
ST12		red	cinnabar/vermilion (HgS)	FTIR-ATR, SEM-EDX
		yellow	chrome yellow ($PbCrO_4$), yellow ochre (FeOOH, clay)	
ST13		red	cinnabar/vermilion (HgS), yellow ochre (FeOOH, clay)	FTIR-ATR, SEM-EDX
		yellow	chrome yellow ($PbCrO_4$), ultramarine yellow ($BaCrO_4$)	
ST14		blue	ultramarine blue ($Na_8Al_6Si_6O_{24}S_4$)	FTIR-ATR, SEM-EDX, PCS
		red	cinnabar/vermilion (HgS), red ochre (Fe_2O_3, clay)	
		white	bianco di Sangiovanni ($CaCO_3$)	
		yellow	chrome yellow ($PbCrO_4$), yellow ochre (FeOOH, clay)	
ST15		red	cinnabar/vermilion (HgS), red ochre (Fe_2O_3, clay)	OM, FTIR-ATR, SEM-EDX
		light red	blanc fixe ($BaSO_4$), red ochre (Fe_2O_3, clay)	
		white	bianco di Sangiovanni ($CaCO_3$)	
		yellow	chrome yellow ($PbCrO_4$), yellow ochre (FeOOH, clay)	
ST16		light blue	ultramarine blue ($Na_8Al_6Si_6O_{24}S_4$)	OM, FTIR-ATR, SEM-EDX
		red	cinnabar/vermilion (HgS), red ochre (Fe_2O_3, clay)	
		light red	red ochre (Fe_2O_3, clay)	
		yellow	chrome yellow ($PbCrO_4$)	
ST17		blue	ultramarine blue ($Na_8Al_6Si_6O_{24}S_4$)	OM, FTIR-ATR, SEM-EDX, PCS
		dark red	red ochre (Fe_2O_3, clay)	
		light red	red ochre (Fe_2O_3, clay)	
		white	bianco di Sangiovanni ($CaCO_3$)	
ST18		blue	ultramarine blue ($Na_8Al_6Si_6O_{24}S_4$)	OM, FTIR-ATR, XRD, SEM-EDX, PCS
		light blue	ultramarine blue ($Na_8Al_6Si_6O_{24}S_4$)	
		red	red ochre (Fe_2O_3, clay)	
		yellow	yellow ochre (FeOOH, clay)	

Table 1. *Cont.*

Sample	Images	Colors	Identified Pigments	Analytical Techniques
ST19		blue	ultramarine blue ($Na_8Al_6Si_6O_{24}S_4$)	OM, FTIR-ATR, SEM-EDX, PCS
		light blue	azurite ($2CuCO_3$ $Cu(OH)_2$	
		dark ochre	yellow ochre (FeOOH, clay)	
		red	red ochre (Fe_2O_3, clay)	
ST20		blue	ultramarine blue ($Na_8Al_6Si_6O_{24}S_4$)	OM, FTIR-ATR, SEM-EDX, PCS
		red	cinnabar/vermilion (HgS), red ochre (Fe_2O_3, clay)	
		white	bianco di Sangiovanni ($CaCO_3$)	
		yellow	chrome yellow ($PbCrO_4$), yellow ochre (FeOOH, clay)	
ST21		light blue	ultramarine blue ($Na_8Al_6Si_6O_{24}S_4$)	OM, FTIR-ATR, SEM-EDX
		red	red ochre (Fe_2O_3, clay)	
		white	bianco di Sangiovanni ($CaCO_3$)	
		yellow	yellow ochre (FeOOH, clay)	
ST22		light blue	ultramarine blue ($Na_8Al_6Si_6O_{24}S_4$)	OM, FTIR-ATR, SEM-EDX
		red	red ochre (Fe_2O_3, clay)	
		yellow	yellow ochre (FeOOH, clay)	
ST23		red	yellow ochre (FeOOH, clay)	OM, FTIR-ATR, SEM-EDX
		yellow	red ochre (Fe_2O_3, clay)	

2.1. Sampling

Samples were collected from detached fragments of wall paintings and stored in LDPE bags or containers.

2.2. Infrared Spectroscopy

FTIR-ATR spectra were acquired by means of a Thermo Scientific Nicolet iS10 instrument, in the range between 4000 and 600 cm^{-1}, 4 cm^{-1} resolution, 32 scans. The background was periodically registered. Spectra were interpreted by comparison with a homemade reference database and with the literature.

2.3. X-ray Diffraction

A selection of samples was analyzed as fine ground powders. X-ray diffraction analyses were performed using a Rigaku Miniflex 300 diffractometer (30 kV, 10 mA, Cu-K_α radiation (λ = 1.5418 Å), 5–55° Theta/2-Theta, step scan 0.02°, scan speed 3°/min). PDXL2 software supporting ICDD (The International Centre for Diffraction Data) PDF2 databases were used to identify the phases.

2.4. Scanning Electron Microscopy and Energy Dispersive X-ray Spectroscopy

The samples were observed without any pre-treatment with a FEI/Philips XL30 ESEM (low vacuum mode—1 torr, 20 kV, BSE detector). The possibility of working in "low vacuum mode" made it possible not to cover the samples with a conductive layer of carbon or gold. Depending on the size of the collected fragments, they were either observed whole or broken up to obtain a significant portion that was compatible with the size of the sample holder. The elemental analyses were carried out using an X-ray energy dispersive spectrometer, EDAX AMETEK Element, coupled to SEM.

A selection of fragments was also embedded in an epoxy resin, cross-cut with a diamond saw and then mechanically polished. Polished cross sections were then analyzed by SEM-EDX, too.

2.5. Optical Microscopy

Samples were observed with an optical microscope Nikon Eclipse LV150, equipped with a Nikon DS-FI1 digital image acquisition system. Images were acquired and elaborated using the NIS-elements F software. Polished cross sections were observed with a portable digital microscope MAOZUA USB001, and images were acquired using the software MicroCapture Plus.

3. Results

The fragments analyzed had different colored layers that were identified by IR spectra and XRD analysis of the powders or suggested by EDX elemental spectra and maps (Table 1).

Yellow and red ochre were identified based on XRD and FTIR patterns, and then confirmed by iron associated with silicon and aluminum in EDX maps. For example, the FTIR spectrum (Figure 2a) of the yellow areas of sample ST7 showed peaks at 3150 and 798 cm^{-1}, which could be attributed to OH stretching and out-of-plane deformation of goethite, respectively; the pattern presented signals at 3691, 3649 and 3619 and 912 cm^{-1} too, which resemble the outer and inner OH group stretching and deformation of kaolinite, respectively [2–4]. The signals at 1027 (Si-O-Si), 1004 (Si-O-Al) and 938 cm^{-1} (Al-O-H) confirm our findings. Kaolinite is normally associated with the presence of ochre [5]. The absorbance at 1162 cm^{-1} (Si-O asymmetric stretching) is due to the presence of quartz, as well as the peak at 691 cm^{-1} which is Si-O symmetric stretching mode [2,3].

The powders of sample ST1, which is colored in aqua green, were analyzed with XRD (Figure 3). A comparison with the instrument database revealed the presence of clinochlore ($Mg_3(Mg_2Al)((Si_3Al)O_{10})$, which is a member of the chlorite group, with a color that varies between yellowish green, olive green, blackish green and bluish green [6]. The characteristic peaks of calcite ($CaCO_3$), gypsum ($CaSO_4{\cdot}2H_2O$), bassanite ($CaSO_4{\cdot}0.5H_2O$), quartz (SiO_2) and barite ($BaSO_4$) were also observed.

The pattern of IR bands (Figure 4a) at around 1410 cm^{-1} (C-O asymmetric stretching) and 1793 cm^{-1} (combination band), 873 and 712 cm^{-1} (C-O out-of-plane and in-plane bending vibrations, respectively) is present in all the spectra, which suggests the presence of calcium carbonate as calcite from the binder fraction of the mortar substrate [7–12]. The spectra of the white areas present only the peaks at around 1793, 1410, 873 and 712 cm^{-1}, which, as just discussed, refer to calcium carbonate, which constitutes the pigment bianco di Sangiovanni [6].

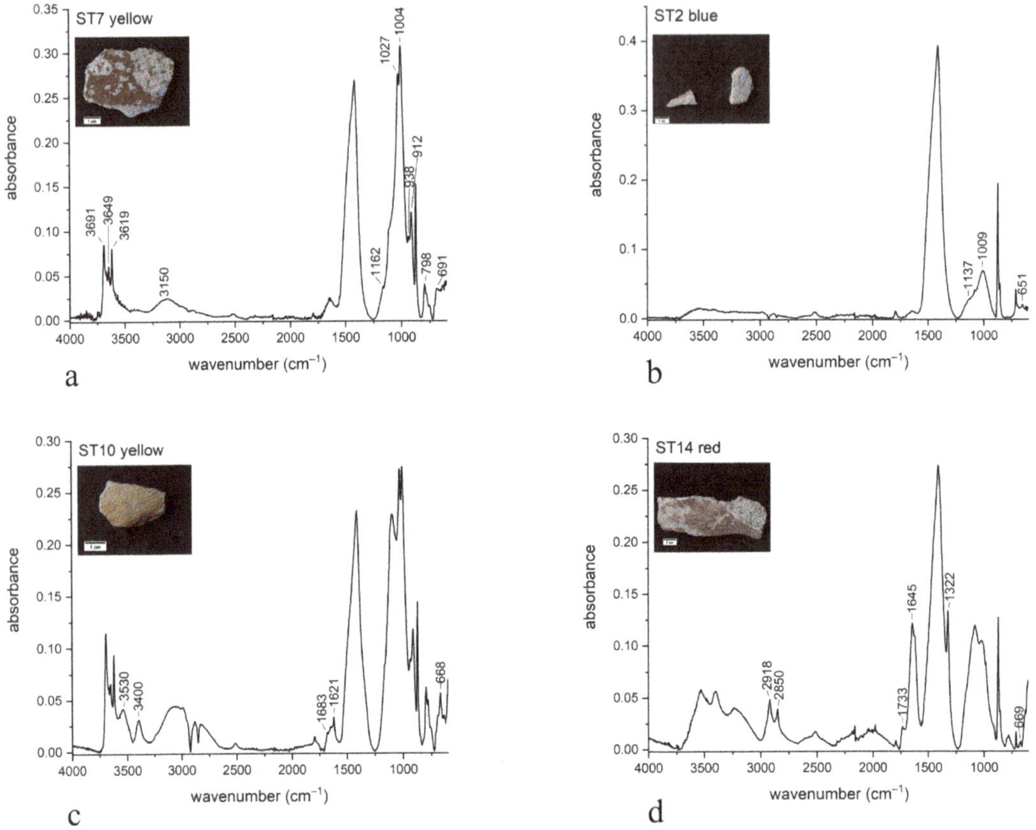

Figure 2. ATR spectra of: (**a**) yellow area of sample ST7, showing the presence of kaolinite, goethite and calcite; (**b**) blue area of sample ST2, showing the presence of ultramarine blue, calcite, aragonite; (**c**) yellow area of sample ST10, showing the presence of gypsum, calcite, ochre, traces of organic compounds; (**d**) red area of sample ST14, showing the presence of calcium oxalate, calcite, ochre, organic compounds.

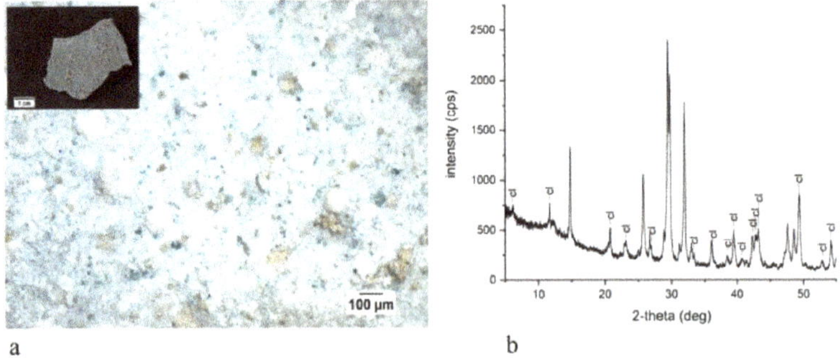

Figure 3. Aqua green area of sample ST1: digital microscope image of the surface (**a**) and XRD analysis on powders (**b**) showing the main signals of clinochlore (cl).

The specific peaks at around 1083, 855 and 702 cm^{-1} found in all the samples suggest the presence of calcium carbonate in the mineralogical form of aragonite, both in the support and mixed with the pigments. The 855 and 702 cm^{-1} signals are due to ν_2 and ν_4 CO_3^{2-} vibrations, respectively, and the absorbance at 1083 cm^{-1} could be attributed to ν_1 symmetric CO_3^{2-} stretching [13,14]. Figure 4a shows a representative spectrum of aragonite, identified in sample ST6, where the peaks are clearly distinguished from those of calcite. SEM images of sample ST19 (Figure 4b) clearly showed the typical sharp, needle-like crystals of aragonite [15]. Aragonite was always corroborated by the XRD pattern, which also distinguished between the two mineralogical forms of calcium carbonate (Figure 4c).

In some red samples, the EDX investigations showed the distribution of mercury and sulfur in the same area, suggesting the use of cinnabar/vermilion, as in sample ST15 (Figure 5). Similarly, the co-presence of chrome and lead in EDX maps indicated the presence of chrome yellow, as in the yellow area in sample ST11, where the pigment was confirmed by the signals of crocoite ($PbCrO_4$) present in XRD results (Figure 6).

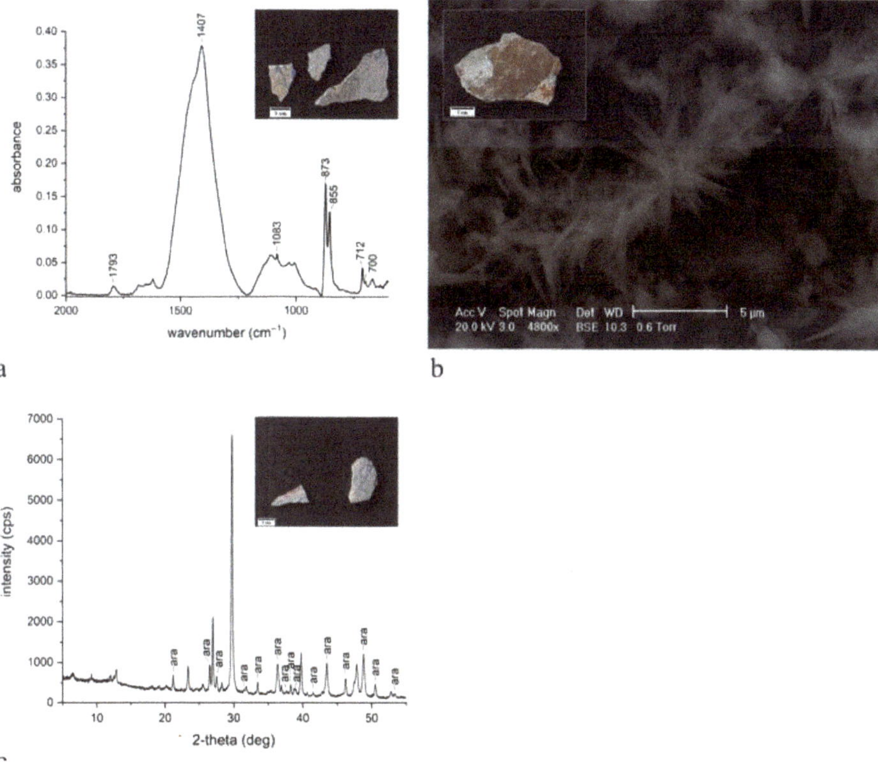

Figure 4. (**a**) ATR spectra of the red area of sample ST6, showing the peaks of calcite, aragonite, ochre; (**b**) SEM image in BSE mode of the surface of sample ST19: the typical sharp crystals of aragonite are shown; (**c**) XRD analysis on powders of mortar support of sample ST2, showing the main signals of aragonite (ara).

The spectrum of the blue areas showed peaks at 1137, 1009 and 651 cm^{-1}, which could be attributed to Si-O-Si and Si-O-Al stretching modes of ultramarine blue ($Na_8Al_6Si_6O_{24}S_4$), respectively [16]. Figure 2b shows a spectrum of ultramarine, as found in sample ST2.

Blanc fixe (barium sulphate $BaSO_4$) was found in some samples due to the presence of IR peaks at 3606, 1089 and 659 cm^{-1}, which can be attributed to OH (surface-adsorbed water molecules), SO_4^{2-} and Ba-S-O stretching modes, respectively [17]. EDX maps of sulfur and barium and XRD patterns of barite recorded in the same areas corroborated the hypothesis.

Gypsum was present in most samples, as highlighted by the XRD results and by the IR signals (Figure 2c) at around 3530 and 3400 (hydroxyl stretching bands), 1683 and 1621 (hydroxyl bending vibrations), 1113 (SO asymmetric stretching), and 668 cm^{-1} (SO asymmetric bending mode) [18].

Figure 5. Red area of sample ST15: (**a**) digital microscope image of the surface (scale bar: 100 µm); (**b**) SEM image in BSE mode of the painted layer (scale bar: 50 µm); (**c**) EDX elemental distribution map of mercury (scale bar: 50 µm); (**d**) EDX elemental distribution maps of sulfur (scale bar: 50 µm).

Some samples presented IR patterns (Figure 2d) showing absorbances at around 1645 and 1322 (C-O stretching vibration) and 668 cm^{-1} (O-C-O bending vibrations), which were ascribed to calcium oxalate (CaC_2O_4 nH_2O) [18]. The spectra also show the signals at around 2918, 2850 and 1733 cm^{-1}, which suggest the CH and C=O stretching absorption of organic compounds, respectively, but were too weak and few for a reliable identification [7].

In a few green areas, zinc, chlorine and copper appeared to be associated in the EDX maps, as observed in samples ST8 and ST11 (Figures 6 and 7). The combination of these elements suggests the compound ($(Cu,Zn)_2(OH)_3Cl$), known as Brunswick green. XRD analyses of the same samples suggested the presence of the natural equivalent of the pigment, i.e., paratacamite (Figure 6). In sample ST11, Brunswick green was identified in green areas close to yellow areas painted with chrome yellow, as clearly shown in Figure 6.

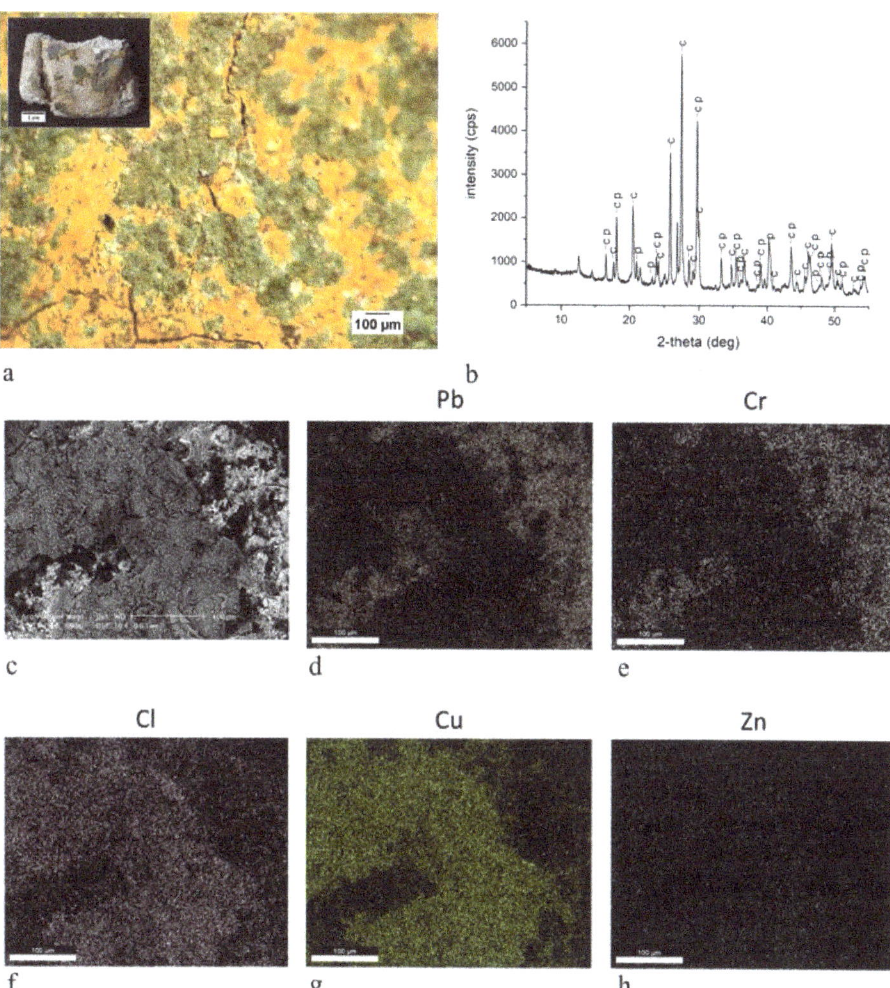

a b c d e f g h

Figure 6. Yellow and green areas of sample ST11: (**a**) digital microscope image of the surface (scale bar: 100 µm); (**b**) XRD analysis on powders, showing the main signals of crocoite (c) and paratacamite (p); (**c**) SEM image in BSE mode of the painted layer (scale bar: 100 µm); (**d**) EDX elemental distribution maps of lead (scale bar: 100 µm); (**e**) EDX elemental distribution maps of chromium (scale bar: 100 µm); (**f**) EDX elemental distribution maps of chlorine (scale bar: 100 µm); (**g**) EDX elemental distribution maps of copper (scale bar: 100 µm); (**h**) EDX elemental distribution maps of zinc (scale bar: 100 µm).

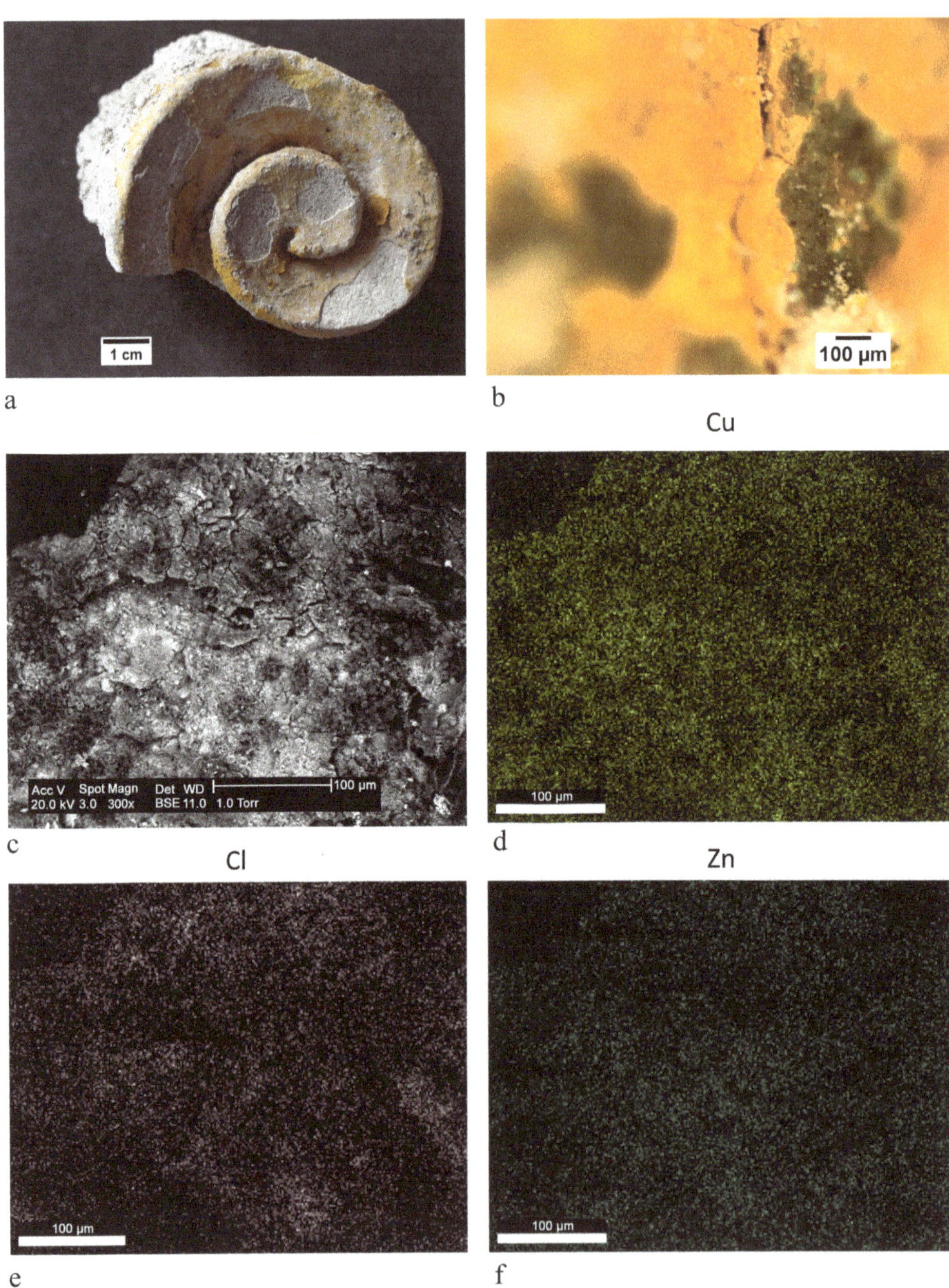

Figure 7. Green area of sample ST8: (**a**) image of the sample; (**b**) digital microscope image of details of the green areas of the painted surface (scale bar: 100 µm); (**c**) SEM image in BSE mode of the painted layer (scale bar: 100 µm); (**d**) EDX elemental distribution map of copper (scale bar: 100 µm); (**e**) EDX elemental distribution maps of chlorine (scale bar: 100 µm); (**f**) EDX elemental distribution maps of zinc (scale bar: 100 µm).

With regards to stratigraphical studies, optical and electronic microscopy revealed a number of colored layers ranging from three to eight. A combination of the EDX elemental maps recorded along the stratigraphy and the IR analysis of the single layers taken, where possible, with micro scalpel, suggested the pigments present in the samples (Table 1). Ochre is the most common pigment, generally alternated with bianco di Sangiovanni white and ultramarine blue. In samples ST3 and ST13, the EDX maps recorded the signals of barium and chrome in the same areas (Figure 8). The combination of these two elements resembles the pigment barium chromate called ultramarine yellow or lemon yellow [6]. Analysis of the polished cross sections revealed strange combinations of pigments. Figure 9 shows the stratigraphy of sample ST20. The inner red layer seems to consist of the overlapping of a layer of cinnabar/vermilion and ochre, due to the presence of an accumulation of mercury and iron, respectively. On the other hand, the alternation of only ochre and azurite was observed in sample ST19.

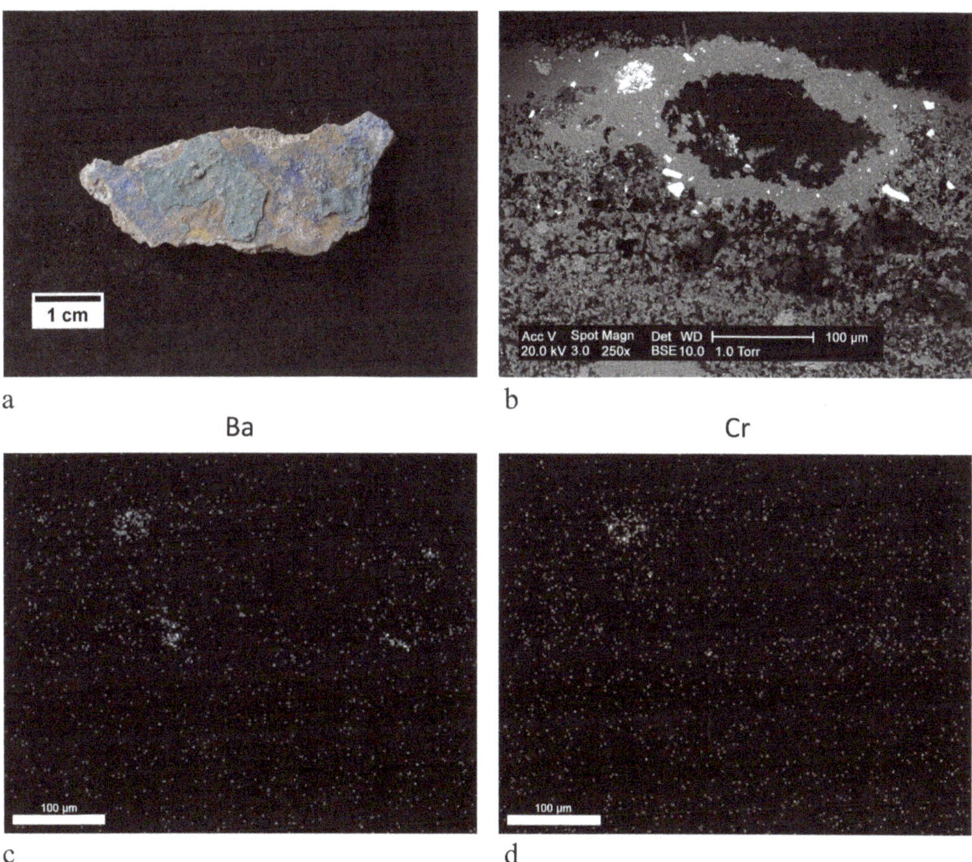

Figure 8. Yellow area of sample ST3: (**a**) image of the sample; (**b**) SEM image in BSE mode of the painted layer (scale bar: 100 µm); (**c**) EDX elemental distribution map of barium (scale bar: 100 µm); (**d**) EDX elemental distribution maps of chromium (scale bar: 100 µm).

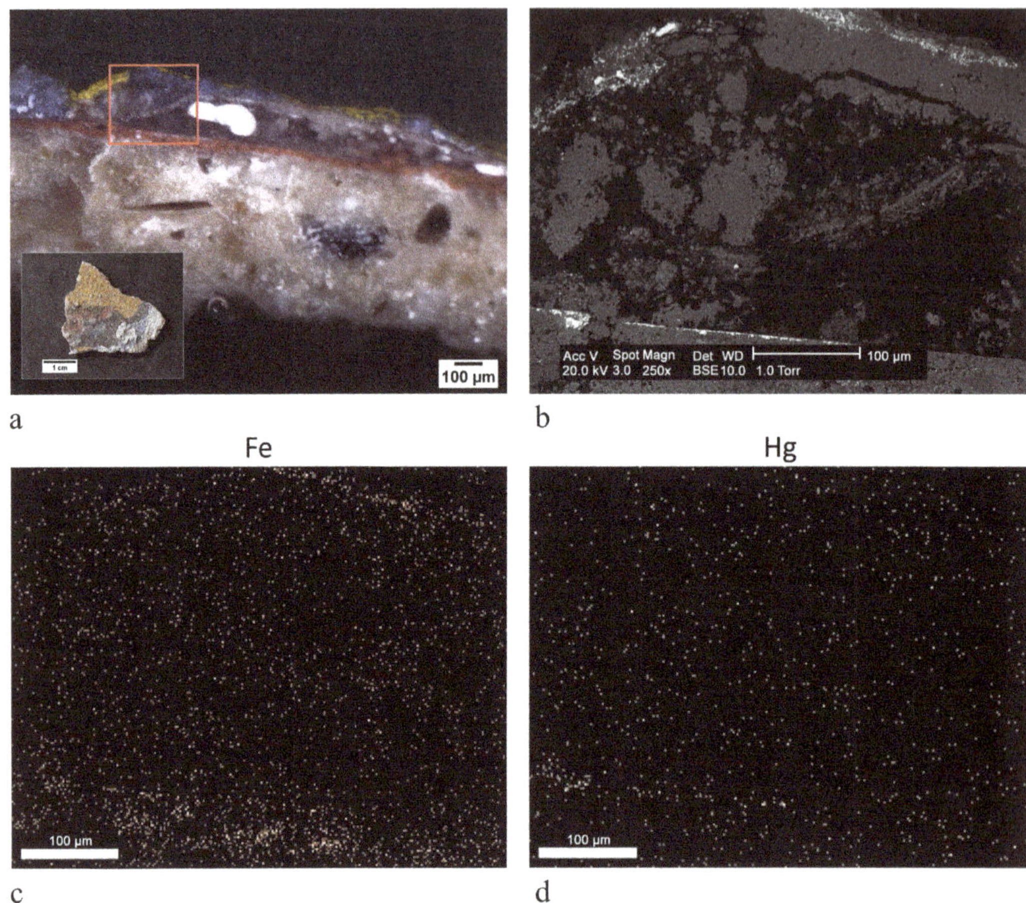

Figure 9. Polished cross section of sample ST20: (**a**) digital microscope image of the surface (scale bar: 100 µm); (**b**) SEM image in BSE mode of the painted layer (scale bar: 100 µm). The analyzed area roughly corresponds to the area evidenced in red in the figure (**a**); (**c**) EDX elemental distribution map of iron (scale bar: 100 µm); (**d**) EDX elemental distribution maps of mercury (scale bar: 100 µm).

4. Discussion

The pigments identified are both traditional and modern colors, such as chrome yellow and blanc fixe, above all as an extender, that were first used at the end of the 18th century [16].

Clinochlore was identified in the aqua green area of one sample. This mineral belongs to the chlorite family, which is one of the clay minerals that make up green earths, along with glauconite and celadonite [6]. The presence of only clinochlore for the green color is rare but possible, as documented in two articles [19,20].

In two samples, Brunswick green was found, which is another uncommon pigment. This color was first prepared in the town of Braunschweig ("Brunswick") in 1764 [6] and was one of the modern pigments that replaced the expensive malachite [21]. It consisted primarily of atacamite; however, a variation of the main recipe included the use of zinc salts to form paratacamite. The occurrence of copper and chlorine also indicates the presence of atacamite, which is also a possible pigment, or maybe the result of the decay of azurite. However, zinc was found in the same areas of the pigment, thus suggesting the presence of Brunswick green. We ruled out the presence of zinc due to zinc white. In fact, zinc was not found in other areas of the wall paintings. From the beginning of the 19th century, various recipes were developed to produce Brunswick green. However unfortunately the term "Brunswick green" was also applied to colors that referred to other compositions, such as emerald green, verdigris, Prussian blue mixed with chrome yellow and various yellow, brown and black pigments. According to Eastaugh et al. [6], the abundance of meanings suggests that, rather than a specific pigment, Brunswick green was just part of color terminology. The ambiguity of the term and the composition perhaps explain the scarce bibliography of findings of the color in paintings [22,23]. When Brunswick green is mentioned, in fact, the true pigment is misrepresented by a mixture of chrome yellow and Prussian blue. In some cases, the compound may be due to decay phenomena, and thus it is not easy to assess whether the pigment was used as a raw material or whether it was a decay product. In any case, the presence of this pigment also suggests a decorative intervention after the beginning of the 19th century, and therefore following the execution of the first mural paintings.

With regards to the detection of barium chromate, this pigment was commercially available and known throughout the 19th century as ultramarine yellow, lemon yellow, barium yellow, or baryta yellow [6]. It belongs to the family of yellow chromate pigments and has been used since the end of the 19th century, as verified by the finding in paintings of that period [24]. Among chromate pigments, it is the most lightfast and turns green if exposed to light for a long time. Findings of this pigment are also scarce [24–30].

The observation of the stratigraphic sections seems only able to suggest the painting techniques used by the artists. The issue, as emerges also from the literature [31–35], is still very controversial, and, to date, there are no reliable and univocal guidelines. Considering this, microscopic observation seems to suggest a fresco technique for the first decorative cycle, while the finding of organic substances in the outer layers indicates a secco technique for the subsequent interventions.

The presence of gypsum in the majority of samples is likely due to the sulphation of the lime binder fraction, given the poor conditions of the building and the lack of protection from precipitation. The chemical origin of calcium oxalate found in some fragments may be due to the mineralization of organic compounds [36], although films may have been formed as a by-product of surface microflora. SEM morphological investigations did not indicate the presence of either past or present bio-colonization, and traces of organic substances were found in samples containing calcium oxalate. This compound could have formed through the oxidative degradation of the organic binder and/or treatment applied onto the wall painting as part of conservation works.

Our most notable finding was the ubiquitous determination of aragonite, as a component of the supporting mortar. Aragonite is one of the mineralogical forms of calcium carbonate, being of biogenic origin (shells, coral, pearls). When found in the mortar, it is ascribed to the aggregate fraction, as the production of lime, also from the calcination of shells, facilitates the formation of calcite. The most common source of aragonite is from mollusk shells and endoskeleton corals. In some geological formations, it may be associated with specific rocks; however, these rocks are not present in the area of Cerignale.

Dilaria reconstructed in detail the use of aragonite in building technologies [37]. In ancient times, people used shells for food, rituals, decoration, fishing, and for lighting lamps. The shells of some species of Murex were also the by-products of the production of purple, an expensive pigment. Once crushed, shells were used as building material, mainly for floors and as an aggregate in mortar. Shells may have been added intentionally in the mixtures to save on the raw materials, and were used as a strong material but, at the same time, were light and therefore easy to transport. Unintentional use cannot be ruled out; for example, when river sands that were not well sieved were used as aggregate. In any case, the use of shells has been described in Roman sources in the 4th century C.E., when it seemed that they improved the set-up of the mixture. Dilaria reports numerous cases of findings in the Mediterranean area, e.g., Greece, Lebanon, Egypt, Italy, Tunisia, dating from the Bronze Era to the Byzantine Era [37]. Shells have mainly been found in the preparation of floors, wall coverings, foundations of buildings, mosaics, cisterns, but rarely for the mortar support of frescoes. In Italy, shells were used in the north-eastern and central areas, always along the marine coasts and always as building materials for floors and walls. In some cases, it has been possible to trace whether the use is associated with the use of non-sieved sands, in others with the production of purple from the Murex species. It is clear from Dilaria's study that, in every historical period, the use of shells was limited to coastal, lagoon or island territories, since the properties they conferred to mortars were not considered valuable enough to justify their trade.

In most findings on building materials reported in archaeological sources, identification is visual because the shells can still be seen, although partially crushed [37]. Regarding scientific articles on the analytical determination of shells in mortars, aragonite has been revealed by FTIR spectroscopy and XRD in Roman mortars from Caesarea Maritima [38] and in wall paintings exclusively from the Roman and Hellenistic eras [39–44]. Some of these authors have also suggested that aragonite may have been used as a pigment too, alone in white areas or mixed with other pigments, as described in treatises [6].

5. Conclusions

The analysis of the wall paintings of the church of S. Stefano attest to the use of a rich palette of colors and reveal a palimpsest of various painting cycles, as highlighted by the presence of multilayered stratigraphies. Together with common colors, such as ochre, ultramarine blue, bianco di Sangiovanni, cinnabar/vermilion, azurite, a few unusual pigments were identified, specifically clinochlore, Brunswick green, ultramarine yellow, which have rarely been found in painted works of art and never in wall paintings. The presence of these pigments datable to the 19th century, indicate a terminus post quem for the execution of the external pictorial layers. Under the microscope, the underlying layers do not appear to be particularly decayed and therefore did not need to be hidden with other colors. This suggests that the building was painted regularly following whatever tastes were current at the time, using synthetic pigments available on the market from the 19th century. Indeed, there are no records on the most recent restoration works and pictorial cycles. Thus, the identification of modern pigments contributes to shedding light on the non-documented history of the monument.

The results unfortunately also reveal, as expected, the dramatic past and present state of conservation of the paintings. Most have come away from the support; however, those that still survive have been covered on the surface by gypsum and calcium oxalate.

Finally, the ubiquitous presence of aragonite due to the intentional use of shells as an aggregate in mortar is particularly interesting. What is surprising is that the literature shows that this is the only case of this finding in wall paintings that are not from the Roman era. This therefore reveals the use of a building technology from the past in this area of the Apennines, perhaps due to its proximity to the Trebbia River and the Ligurian Sea. It is worth highlighting the importance of this area in trade between the Po Valley and Liguria. It was therefore a very frequented area, probably with the exchange of building practices that could have entailed the use of shells in mortars. It would be interesting to investigate other historical buildings in the area of Cerignale, as the discovery elsewhere of aragonite in mortars would corroborate our hypothesis. Shells were probably used as light aggregate and were therefore easy to transport to the church of S. Stefano, which is located far from the main roads and at an altitude of 700 m. The shells may have originated from the Trebbia River, in the valley below, or in the Ligurian sea, to which the Trebbia valley was connected.

The analyses carried out provide evidence of the unexpected richness and variety of the paintings in the church of S. Stefano, which is at real risk of fading. It is thus important to underline the fundamental role of conservation science, which, by studying the techniques and materials used by the artists of the past, may encourage future restoration projects and enhances and preserves the disappearing beauty.

Author Contributions: Conceptualization, L.R.; investigation, C.C., L.G. and L.R.; writing—original draft preparation, C.C. and L.R.; writing—review and editing, C.C., L.G., L.R. and S.R. All authors have read and agreed to the published version of the manuscript.

Funding: This research received no specific grant from any funding agency in the public, commercial or not-for-profit sectors.

Institutional Review Board Statement: Not applicable.

Informed Consent Statement: Not applicable.

Data Availability Statement: The data presented in this study are available on request from the corresponding author.

Acknowledgments: The authors would like to thank Chiara Zannin, Giorgio Castignoli, Giorgio Villa, The Curia Vescovile di Piacenza-Bobbio, The Soprintendenza Archeologia, Belle Arti e Paesaggio per le province di Parma e Piacenza, The Municipality of Cerignale and its mayor Massimo Castelli, The Pro Loco of Cerignale, The Fondazione Banca del Monte di Lombardia and Norberto Masciocchi for support.

Conflicts of Interest: The authors declare no conflict of interest.

References

1. Pearce, M. Reconstructing past transapennine routes: The Trebbia valley. *Reconstr. Past Transapennine Routes Trebbia Val.* **2002**, *6*, 181–183.
2. Wilson, M.J. *Clay Mineralogy: Spectroscopic and Chemical Determinative Methods*; Chapman & Hall: London, UK, 1994; ISBN 9780412533808.
3. Bikiaris, D.; Daniilia, S.; Sotiropoulou, S.; Katsimbiri, O.; Pavlidou, E.; Moutsatsou, A.P.; Chryssoulakis, Y. Ochre-differentiation through micro-Raman and micro-FTIR spectroscopies: Application on wall paintings at Meteora and Mount Athos, Greece. *Spectrochim. Acta Part A Mol. Biomol. Spectrosc.* **2000**, *56*, 3–18. [CrossRef]
4. Bugini, R.; Corti, C.; Folli, L.; Rampazzi, L. Unveiling the Use of Creta in Roman Plasters: Analysis of Clay Wall Paintings from Brixia (Italy). *Archaeometry* **2017**, *59*, 84–95. [CrossRef]
5. Crupi, V.; La Russa, M.F.; Venuti, V.; Ruffolo, S.; Ricca, M.; Paladini, G.; Albini, R.; Macchia, A.; Denaro, L.; Birarda, G.; et al. A combined SR-based Raman and InfraRed investigation of pigmenting matter used in wall paintings: The San Gennaro and San Gaudioso Catacombs (Naples, Italy) case. *Eur. Phys. J. Plus* **2018**, *133*, 369. [CrossRef]
6. Eastaugh, N.; Walsh, V.; Chaplin, T.; Siddall, R. *Pigment Compendium: A Dictionary and Optical Microscopy of Historical Pigments*; Pigment Compendium: A Dictionary and Optical Microscopy of Historical Pigments; Butterworth-Heinemann: Oxford, UK, 2008; ISBN 9780750689809.
7. Derrick, M.R.; Stulik, D.; Landry, J.M. *Infrared Spectroscopy in Conservation Science*; The Getty Conservation Institute: Los Angeles, CA, USA, 1999; ISBN 0892364696.
8. Bugini, R.; Corti, C.; Folli, L.; Rampazzi, L. Roman Wall Paintings: Characterisation of Plaster Coats Made of Clay Mud. *Heritage* **2021**, *4*, 48. [CrossRef]
9. Montoya, C.; Lanas, J.; Arandigoyen, M.; Navarro, I.; García Casado, P.J.; Alvarez, J.I. Study of ancient dolomitic mortars of the church of Santa Maria de Zamarce in Navarra (Spain): Comparison with simulated standards. *Thermochim. Acta* **2003**, *398*, 107–122. [CrossRef]
10. Igea, J.; Lapuente, P.; Martínez-Ramírez, S.; Blanco-Varela, M.T. Characterization of mudejar mortars from St. Gil Abbot church (Zaragoza, Spain): Investigation of the manufacturing technology of ancient gypsum mortars. *Mater. Constr.* **2012**, *62*, 515–529. [CrossRef]
11. Biscontin, G.; Pellizon Birelli, M.; Zendri, E. Characterization of binders employed in the manufacture of Venetian historical mortars. *J. Cult. Herit.* **2002**, *3*, 31–37. [CrossRef]
12. Crupi, V.; D'Amico, S.; Denaro, L.; Donato, P.; Majolino, D.; Paladini, G.; Persico, R.; Saccone, M.; Sansotta, C.; Spagnolo, G.V.; et al. Mobile Spectroscopy in Archaeometry: Some Case Study. *J. Spectrosc.* **2018**, *2018*, 1–11. [CrossRef]
13. Zhang, K.; Grimoldi, A.; Rampazzi, L.; Sansonetti, A.; Corti, C. Contribution of thermal analysis in the characterization of lime-based mortars with oxblood addition. *Thermochim. Acta* **2019**, *678*, 178303. [CrossRef]
14. Zhang, K.; Corti, C.; Grimoldi, A.; Rampazzi, L.; Sansonetti, A. Application of Different Fourier Transform Infrared (FT-IR) Methods in the Characterization of Lime-Based Mortars with Oxblood. *Appl. Spectrosc.* **2019**, *73*, 479–491. [CrossRef]
15. Žigovečki Gobac, Ž.; Posilović, H.; Bermanec, V. Identification of biogenetic calcite and aragonite using SEM. *Geol. Croat.* **2009**, *62*, 201–206. [CrossRef]
16. Roy, A. *Artists' Pigments: A Handbook of Their History and Characteristics*; Oxford University Press: London, UK, 1993; Volume 2, ISBN 9780894681899.
17. Aroke, U.O.; Abdulkarim, A.; Ogubunka, R.O. Fourier-transform Infrared Characterization of Kaolin, Granite, Bentonite and Barite. *ATBU J. Environ. Technol.* **2013**, *6*, 42–53.

18. Ranalli, G.; Bosch-Roig, P.; Crudele, S.; Rampazzi, L.; Corti, C.; Zanardini, E. Dry biocleaning of artwork: An innovative methodology for Cultural Heritage recovery? *Microb. Cell* **2021**, *8*, 91–105. [CrossRef]
19. Edreira, M.C.; Feliu, M.J.; Fernández-Lorenzo, C.; Martín, J. Roman wall paintings characterization from Cripta del Museo and Alcazaba in Mérida (Spain): Chromatic, energy dispersive X-ray fluorescence spectroscopic, X-ray diffraction and Fourier transform infrared spectroscopic analysis. *Anal. Chim. Acta* **2001**, *434*, 331–345. [CrossRef]
20. Gebremariam, K.F.; Kvittingen, L.; Banica, F.-G. Physico-Chemical Characterization of Pigments and Binders of Murals in a Church in Ethiopia. *Archaeometry* **2016**, *58*, 271–283. [CrossRef]
21. Scott, D.A. *Copper and Bronze in Art: Corrosion, Colorants, Conservation*; Getty Conservation Institute: Los Angeles, CA, USA, 2002; ISBN 0892366389.
22. Setti, M.; Lanfranchi, A.; Cultrone, G.; Marinoni, L. Archaeometric investigation and evaluation of the decay of ceramic materials from the church of Santa Maria del Carmine in Pavia, Italy. *Mater. Constr.* **2012**, *62*, 79–98. [CrossRef]
23. Gabrielli, N. Technical-scientific investigations to detect the temporal vicissitudes of the funeral monument of Innocent VIII (Giovanbattista Cibo, 1484–1492), compared with that of Sixtus IV (Francesco della Rovere 1471–1484), both made by Antonio del Pollaiolo. *Nat. Prod. Res.* **2019**, *33*, 926–936. [CrossRef]
24. Otero, V.; Campos, M.F.; Pinto, J.V.; Vilarigues, M.; Carlyle, L.; Melo, M.J. Barium, zinc and strontium yellows in late 19th–early 20th century oil paintings. *Herit. Sci.* **2017**, *5*, 46. [CrossRef]
25. Fulton, E.L.; Newman, R.; Woodward, J.; Wright, J. The Methods and Materials of Martin Johnson Heade. *J. Am. Inst. Conserv.* **2002**, *41*, 155. [CrossRef]
26. Klyachkovskaya, E.V.; Kozhukh, N.M.; Rozantsev, V.A.; Gaponenko, S.V. Layer-by-Layer Laser Spectrum Microanalysis of Easel-Painting Materials. *J. Appl. Spectrosc.* **2005**, *72*, 371–375. [CrossRef]
27. Vetter, W.; Schreiner, M. A Fiber Optic Reflection-UV/Vis/NIR-System for Non-Destructive Analysis of Art Objects. *Adv. Chem. Sci.* **2014**, *3*, 7–14.
28. Grifoni, E.; Briganti, L.; Marras, L.; Orsini, S.; Colombini, M.P.; Legnaioli, S.; Lezzerini, M.; Lorenzetti, G.; Pagnotta, S.; Palleschi, V. The chemical-physical knowledge before the restoration: The case of "The Plague in Lucca", a masterpiece of Lorenzo Viani (1882–1936). *Herit. Sci.* **2015**, *3*, 26. [CrossRef]
29. Romano, F.P.; Caliri, C.; Nicotra, P.; Di Martino, S.; Pappalardo, L.; Rizzo, F.; Santos, H.C. Real-time elemental imaging of large dimension paintings with a novel mobile macro X-ray fluorescence (MA-XRF) scanning technique. *J. Anal. At. Spectrom.* **2017**, *32*, 773–781. [CrossRef]
30. Cristea-Stan, D.; Constantinescu, B. Studies on pigments of religious mural paintings using a portable X-ray Fluorescence spectrometer—The cases of Urechesti-Cicanesti Arges and Icoanei Bucuresti churches. *Proc. Rom. Acad. Ser. A* **2019**, *20*, 347–352.
31. Regazzoni, L.; Cavallo, G.; Biondelli, D.; Gilardi, J. Microscopic Analysis of Wall Painting Techniques: Laboratory Replicas and Romanesque Case Studies in Southern Switzerland. *Stud. Conserv.* **2018**, *63*, 326–341. [CrossRef]
32. Kriznar, A.; Ruiz-Conde, A.; Sánchez-Soto, P.J. Microanalysis of Gothic mural paintings (15th century) in Slovenia: Investigation of the technique used by the Masters. *X-ray Spectrom.* **2008**, *37*, 360–369. [CrossRef]
33. Piovesan, R.; Mazzoli, C.; Maritan, L.; Cornale, P. Fresco and lime-paint: An experimental study and objective criteria for distinguishing between these painting techniques. *Archaeometry* **2012**, *54*, 723–736. [CrossRef]
34. Mugnaini, S.; Bagnoli, A.; Bensi, P.; Droghini, F.; Scala, A.; Guasparri, G. Thirteenth century wall paintings under the Siena Cathedral (Italy). Mineralogical and petrographic study of materials, painting techniques and state of conservation. *J. Cult. Herit.* **2006**, *7*, 171–185. [CrossRef]
35. Mora, P.; Mora, L.; Philippot, P. *La Conservazione delle Pitture Murali*; Editrice Compositori: Bologna, Italy, 1999; ISBN 8877941839.
36. Rampazzi, L. Calcium oxalate films on works of art: A review. *J. Cult. Herit.* **2019**, *40*, 195–214. [CrossRef]
37. Dilaria, S. Costruire ingegnosamente riutilizzando materiali poveri. L'impiego di conchiglie a fini edilizi ad Aquileia tra età repubblicana e tarda antichità. *REUDAR. Eur. J. Rom. Archit.* **2017**, *1*, 25. [CrossRef]
38. Asscher, Y.; van Zuiden, A.; Elimelech, C.; Gendelman, P.; Ad, U.; Sharvit, J.; Secco, M.; Ricci, G.; Artioli, G. Prescreening Hydraulic Lime-Binders for Disordered Calcite in Caesarea Maritima: Characterizing the Chemical Environment Using FTIR. *Radiocarbon* **2020**, *62*, 527–543. [CrossRef]
39. Brysbaert, A. Lapis Lazuli in an Enigmatic 'Purple' Pigment from a Thirteenth-Century BC Greek Wall Painting. *Stud. Conserv.* **2006**, *51*, 252–266. [CrossRef]
40. Mazzocchin, G.; Del Favero, M.; Tasca, G. Analysis of Pigments from Roman Wall Paintings Found in the "Agro Centuriato" of Julia Concordia (Italy). *Ann. Chim.* **2007**, *97*, 905–913. [CrossRef]
41. Mazzocchin, G.A.; Vianello, A.; Minghelli, S.; Rudello, D. Analysis of roman wall paintings from the thermae of 'Iulia Concordia'. *Archaeometry* **2010**, *52*, 644–655. [CrossRef]
42. Duran, A.; Jimenez De Haro, M.C.; Perez-Rodriguez, J.L.; Franquelo, M.L.; Herrera, L.K.; Justo, A. Determination of pigments and binders in pompeian wall paintings using synchrotron radiation—High-resolution X-ray powder diffraction and conventional spectroscopy—Chromatography. *Archaeometry* **2010**, *52*, 286–307. [CrossRef]
43. Corti, C.; Rampazzi, L.; Visoná, P. Hellenistic mortar and plaster from Contrada Mella near Oppido Mamertina (Calabria, Italy). *Int. J. Conserv. Sci.* **2016**, *7*, 57–70.
44. Garofano, I.; Perez-Rodriguez, J.L.; Robador, M.D.; Duran, A. An innovative combination of non-invasive UV–Visible-FORS, XRD and XRF techniques to study Roman wall paintings from Seville, Spain. *J. Cult. Herit.* **2016**, *22*, 1028–1039. [CrossRef]

 heritage

MDPI

Article

Holy Corn. Interdisciplinary Study of a Mexican 16th-Century Polychrome Maize Stem, Paper, and *Colorín* Wood Sculpture

Diego Quintero Balbas [1,*], Esteban Sánchez-Rodríguez [2] and Álvaro Zárate Ramírez [3]

[1] National Institute of Optics, CNR (CNR-INO), Largo Enrico Fermi, 6, 50125 Florence, Italy
[2] Laboratorio de Análisis y Diagnóstico del Patrimonio-COLMICH, Calle Cerro de Nahuatzen 85, Fraccionamiento Jardines del Cerro Grande, La Piedad 59370, Mexico; esanchez@colmich.edu.mx
[3] Seminario-Taller de Restauración de Escultura Policromada, Escuela de Conservación y Restauración de Occidente, Calle Analco 285, Analco, Guadalajara 44450, Mexico; j.zarate@ecro.edu.mx
* Correspondence: diegoivan.quinterobalbas@ino.cnr.it

Abstract: Maize stem sculptures, produced during the 16th and 17th centuries in New Spain (today, Mexico) are a clear example of the convergence of the artistic traditions from the American indigenous populations and European influence. This typology of sculptures is not limited to the Americas, as the examples found in European countries have shown. Therefore, a detailed technological investigation is required to correctly classify them. This work presents the interdisciplinary and multianalytical investigation of a 16th-century sculpture made with a maize stem preserved in Guadalajara city, Mexico. We used a set of techniques, such as CT, SEM-EDX, μ-FTIR, and μ-Raman, to study, from a macro to a micro level, the structure, the polychromy, and the modification of the sculpture. The results showed the use of maize stems, paper, and wood in the construction of the sculpture and the use of the traditional polychromy, as well as the numerous modifications that changed its appearance considerably resulting in its misclassification. We were able to associate the statue with the Cortés workshop (Mexico City region), probably produced in the decade of 1580, and track its liturgical use and historical development through the centuries.

Keywords: maize stem; New Spain; computerized tomography; FTIR; Raman; SEM-EDX; Mexican sculpture; polychrome sculpture

Citation: Quintero Balbas, D.; Sánchez-Rodríguez, E.; Zárate Ramírez, Á. Holy Corn. Interdisciplinary Study of a Mexican 16th-Century Polychrome Maize Stem, Paper, and *Colorín* Wood Sculpture. *Heritage* **2021**, *4*, 1538–1553. https://doi.org/10.3390/heritage4030085

Academic Editor: Diego Tamburini

Received: 24 June 2021
Accepted: 26 July 2021
Published: 31 July 2021

1. Introduction

The production of corn (*Zea mays* L.) stem light sculptures in New Spain (today Mexico) during the 16th and 17th centuries is a clear example of the convergence of two artistic traditions. On one hand, there are the Mesoamerican sculptural materials such as maize stems, agave inflorescence (*quiote*), and *colorín* wood (*Erythrina coralloides* DC.). On the other there are the European artistic traditions and iconographic models [1,2]. This technique is still used today in the Michoacán state (Western Mexico). Corn, the most characteristic material of this sculpture technique, was, and still is today, an important element in the Mexican culture. For example, according to the Mayan tradition, humankind was made of corn [3]. However, this type of sculpture was not exclusive to the New Spain viceroyalty. Bruquetas [4] reported examples of similar statues, made with glued fabrics and agave inflorescence, from the Andean region (previously the Peru viceroy). There are also historical records of the use of maize stem for sculpture production in the Kingdom of Guatemala [5].

It is possible to track the artistic tradition of American low weight sculptures through the Spanish *papelón* [6] until the Italian *cartapesta* [7] used at least since the 14th century as a profitable activity for artists thanks to an early serial production system of sculptures sold at lower prices due to the "poor" materials used. Because of the international trade routes of materials, such as gold, silver, and cochineal [8,9], as well as artwork during the Spanish Empire (16th–19th centuries) [10,11], some American light weight sculptures arrived in European countries such as Spain [12–14] and Croatia (previously the Republic of Ragusa) [10].

The chronicles written by the friars in charge of the evangelization of the indigenous populations reported the technical procedures to use maize stem as sculpture material. According to the historical records, the production of maize stem sculptures started with the harvest of the adult maize plant after the production of the corn, followed by the cleaning of the stems, debarked or not according to their use, by boiling them. Finally, they were let to dry slowly. In some cases, the sculptors added poisonous plants to the boiling bath to increase the resistance to biological attack [14]. The debarked stems were ground and bound with a natural adhesive to prepare the paste for modeling the volumes. The historical references indicate the use of a polysaccharide gum named *tatzingueni* or *tzauhtli* extracted from orchid bulbs [15], but so far only animal glue has been identified in the sculptures analyzed [16].

Unfortunately, today, we know the name of very few artists that used maize stems and other local materials such as *colorín* wood. The Cerda family, Matías and his son, Luis, active during the 16th and 17th centuries in the Michoacán region (Western Mexico), are the most famous sculptors of which we have historical records [1,17,18]. Additionally, the sculpture of the *Virgen del Pueblito* (today in Querétaro state in central Mexico) is attributed to the friar Sebastián Gallegos (active during the first half of the 17th century), who probably learnt the technique in Michoacán [19]. There are also records about the use of maize stem by the friar Félix de Mata in Guatemala [5].

Despite the reduced records about artists, the technical and formal characteristics of the sculptures allow their classification into "workshops" linked to specific regions. Amador Marrero [13,20] proposed nine workshops associated with three geographical areas: the western area of Mexico, specifically, what is today the Michoacán state, the central area concentrated in Mexico City, which was the capital of the New Spain viceroyalty, and Oaxaca in the southwest region (Figure S1).

In addition to their formal characteristics, in general, the internal structure of the sculptures allows their classification; the examples from the Michoacán region have a core made of maize stem which is attached using natural fibers or light wood (e.g., *colorín*) covered with debarked maize stem and maize stem paste. On the other hand, the sculptures produced in the central area are hollowed. Artists used two-half casts to obtain the initial shape, made of paper (similar to the *papelón* or *cartapesta* techniques) and, in some cases, codices [14,21]. To facilitate the removal of the paper from the mold, the artist applied gypsum or diatomaceous earth [22]. This system allowed a semi-serial production of the sculpture [14]. The artists achieved the final volumes by modeling debarked or maize stem paste over the paper or wooden core.

The polychrome sculpture production was a collaborative work as dictated by the regulations of the guilds; after the sculptor finished the volumes, a polychromer (a specialized painter) prepared the polychromy [23–25]. Very few records report on the polychromy technique used in the maize stem sculptures. Bonavit identified the use of a preparation layer called *ticatlali* or *tízar* (most probably gypsum), which was also reported by Carrillo y Gariel [21]. Regarding the pigments, Bonavit suggested the painters used carbon black and cochineal lake mixed with a siccative oil as binder [26].

The scientific investigation of maize stem sculptures is paramount for their correct identification, classification, and correct conservation. The variety of materials and the complexity of the structure make it necessary to use a multianalytic approach to understand the technology of these sculptures. Radiography [12,27] and computerized tomography [10,28,29] are the preferred technique for the non-invasive investigation of the internal structure. Endoscopy also offers this kind of information, but it is applicable only to hollowed sculptures that, due to their condition, allow the insertion of the camera [30]. Microscopic and spectroscopic methods enable the study of the polychrome surface [27,31,32]. The research published so far indicates that the artists followed the European tradition: gypsum and animal glue for the ground layer and pigments (e.g., lead white, azurite, vermilion, minium) mixed with siccative oils, and in a very few cases, animal glue, as a binder. There are also reports of the use of metallic leaves applied over red bole to decorate the loincloth [13,19,22].

Most recently, studies have focused on gaining a better understanding of the degradation mechanisms of maize stem as a sculpture material, particularly its physical properties linked to their preparation, by studying the mechanical properties and the resistance to biological damage according to the part of the stem used and the method of preparation [33].

This work presents an interdisciplinary investigation that covers the historical research, the stylistic analysis, and the technological study of a 16th-century maize stem sculpture, *el Señor del Santo Entierro* (Lord of the Holy Sepulcher). Before our study, scholars misclassified the sculpture and suggested it was a substitution of the original. We completed historical research together with a multianalytic examination, combining non-invasive and micro-invasive methods, using computerized tomography (CT) to understand the internal structure, optical and scanning electron microscopy (SEM-EDX), micro attenuated total reflection-Fourier transform Infrared (μ-ATR-FTIR) and micro-Raman (μ-Raman) spectroscopies for the study of the original polychrome surface and its modifications. We aimed to recover the statue history and study its technology to classify it correctly and help in its conservation.

2. Materials and Methods

2.1. The Sculpture of Señor del Santo Entierro

We studied the sculpture entitled *Señor del Santo Entierro* (Figure 1) currently at the *Nuestra Señora de la Soledad* church in Guadalajara city, in Jalisco state (Western Mexico).

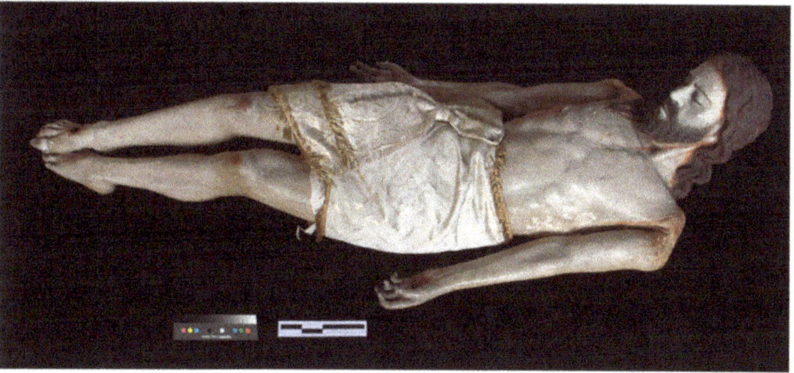

Figure 1. Sculpture of *Señor del Santo Entierro*, ca. 1580, Cortés workshop, Mexico.

2.2. Computerized Tomography

The computerized tomography (CT) was performed using a two-slide Siemens Emotion Duo 1 scanner with a standard protocol for lung tissue diagnosis (113 mA and 130 kV). The tomograms were acquired in slices of 3 mm with a separation between every measure of 0.5 mm and processed in the transverse, sagittal, and coronal planes.

2.3. Sample Preparation

Table 1 summarizes the samples analyzed and the analyses performed. The sampling areas are reported in Figure S2. Cross-section samples were imbibed in Nic Tone® transparent acrylic resin and polished until the sample surface was exposed. One of the cross-sections from the polychromy was imbibed initially with KBr and then in Implex® polyester resin. It was polished until the sample surface was exposed [34].

The fibers from the paper samples were partially separated under the stereomicroscope and then introduced into a test tube with distilled water and completely separated. A drop of the fibers sample was put over a glass slide and covered for observation.

The maize stem sample was boiled and cut using a razor blade to obtain a transverse view of the cellular structures. The slides were added to a sample holder with carbon adhesive tape.

Table 1. List of samples analyzed and a summary of the analysis performed.

Sample ID	Type	Analysis
SSE-1	Cross-section	Optical microscopy, SEM-EDX, μ-FTIR, μ-Raman
SSE-2	Cross-section	Optical microscopy, SEM-EDX, μ-Raman
SSE-5	Cross-section	Optical microscopy, SEM-EDX
SSE-6	Maize stem paste	SEM
SSE-8	Paper	Optical microscopy

2.4. Stereo and Optical Microscopy

The samples were observed and documented with a stereomicroscope Leica model EZ4HD and an optical microscope Leica model DM 4000 M with a Leica digital camera model DFC 450 C using 20× and 50× magnifications. We processed the images with the Leica Application Suite 4.0 software.

2.5. Scanning Electron Microscopy

The cross-section samples imbibed in polyester resin were completely covered with graphite adhesive tape leaving the sample uncovered. We applied two strips of aluminum adhesive tape on both sides of the sample to increase its conductivity.

Sample images were captured with a SEM Jeol model JSM-6390LV under low vacuum using 20 kV. Small areas according to the size of each layer were analyzed using EDX, the time of analysis was 100 s. Maps of the cross-section samples were obtained with 3 million counts. The data was processed with the INCA Suite 4.08 software.

2.6. Attenuated Total Reflection-Fourier Transformed Infrared Spectroscopy

Micro-ATR-FTIR (μ-ATR-FTIR) analyses were performed using a Thermo Scientific Nicolet iN10MX spectrometer in attenuated total reflection (ATR) mode with a Ge crystal. The spectra were recorded in the range between 4000 to 675 cm^{-1} with an optical aperture of 200 × 200 μm, corresponding to an effective investigated area of 50 × 50 μm, and a spectral resolution of 4 cm^{-1} and 64 scans. Maps were recorded with an optical aperture of 40 × 40 μm (effective investigated area of 10 × 10 μm) and a step size of 8 μm. The data was processed with OmnicPicta and Omic32 software.

2.7. Micro Raman Spectroscopy

μ-Raman analyses were performed with a Bruker Senterra Raman Microscope coupled to an Olympus BX 40 microscope equipped with a CCD camera. The spectra were recorded in the 100–3200 cm^{-1} range using a 785 nm He–Ne laser source and with excitation powers of 9 mW and 1 mW and a spectral resolution of 3–5 cm^{-1}. Acquisition time was 5–10 s and 5–10 accumulations to maximize the signal-to-noise ratio. Spectra were processed with OPUS and Origin Pro software.

3. Results

3.1. Historical Research

Unfortunately, little historical documentation about the sculpture has survived. Therefore, a partial history has been reconstructed by us using the documentation regarding where the sculpture has been located and examining the constant changes through the centuries.

The *Nuestra Señora de la Soledad y el Santo Entierro de Cristo* confraternity—constituted only by Spaniards—was founded in 1590 in the capital of the Kingdom of New Galicia (an autonomous kingdom that was part of the New Spain viceroyalty and today is the Guadalajara city in the Jalisco state) in the San Miguel Royal Hospital; its principal function was to organize the representations and processions for the Holy Friday that, according to historical documentation, was done for the first time in 1595 [35]. The confraternity commissioned the sculpture of the *Señor del Santo Entierro*, together with the sculpture of the *Señora de la Soledad*, for the liturgical representations of the Holy Friday.

A historical description of the confraternity procession indicates that the *Señor del Santo Entierro* sculpture was descended from the cross—the reason why the arms of the sculpture were articulated—and placed in a richly decorated urn, which was followed by angel statues holding the *Arma Christi* [35], which were also a visual didactic element for the evangelization. The last procession performed by the confraternity was in 1866 [35]. These religious performances were very popular and useful from a didactic and evangelical perspective in the early years of the Spanish colonization [36]. The processional uses of the images are directly linked to their materiality; the maize stem sculptures were famous because of their low weight. The mural paintings of Huejotzingo (Puebla state) and Teitipac (Oaxaca state) monasteries depict similar examples of the Holy Friday processions done in New Spain [36], which are still performed nowadays in Tzintzuntzan (Michoacan state) where another maize stem statue, *Señor del Rescate*, is used to represent the descendant of dead Christ's body, followed by a procession [37].

The *Nuestra Señora de la Soledad y el Santo Entierro de Cristo* confraternity was associated with the Archbasilica of Saint John Lateran in Rome in 1598 [38], the same year in which the sculptures were moved from the hospital and placed in the Shrine chapel inside the cathedral. Because of the great devotion towards the two sculptures, in 1599, a specific chapel was built for them, and by 1658, their own church was built in a park next to the cathedral [39,40].

Because of the "law of tolerance of sects" enacted in 1926 and the subsequent Cristero War [41], the church gradually reduced its activity and finally closed on 2 October 1933 by presidential decree [42], passing to the administration of the Federal Finance offices, and in 1935, it became an archive. Finally, in 1949, was destroyed as part of an urban modification program planed by Ignacio Díaz Morales and promoted by governor Jesús González Gallo [43]. After the closure of the church, the sculptures were placed initially in the cathedral and later in different private houses until the construction of the current church of Our Lady of Solitude, designed by Pedro Castellanos de Lambley. Its construction started in 1950, but today it remains unfinished [44].

3.2. *Stylistic Analysis and Workshop Adscription*

To ascribe the sculpture to a specific workshop, we performed a stylistic analysis of the sculpture. It is of natural size (170 cm) and the representation is anatomically correct. Two main elements allow the adscription of the sculpture to a specific workshop: the loincloth and the head. The loincloth is simple and short; it represents a single white cloth wrapped around the loin from right to left. The Spanish sculptor Alonso Berruguete (1490–1561) used this type of loincloth in his wooden sculptures [45]. Amador Marrero [14] identified similar loincloth designs in the early maize stem sculptures (around 1570) ascribed to the Cortés workshop. However, in the *Señor del Santo Entierro* sculpture, the loincloth is simpler; we suppose it is an intermediate stage between the early sculptures from the 1570s and the sculptures made in 1580s that have a simpler loincloth design (Figure 2). Regarding the facial representation, the beard is the only element we can consider, since the hair is a modification: it is short and forked and the mustache surrounds the mouth.

Based on these elements, we proposed the ascription of this sculpture to the Cortés workshop, associated with the central region, particularly Mexico City. It is probably closer to the production of the 1580s, in agreement with the historical documentation of the confraternity.

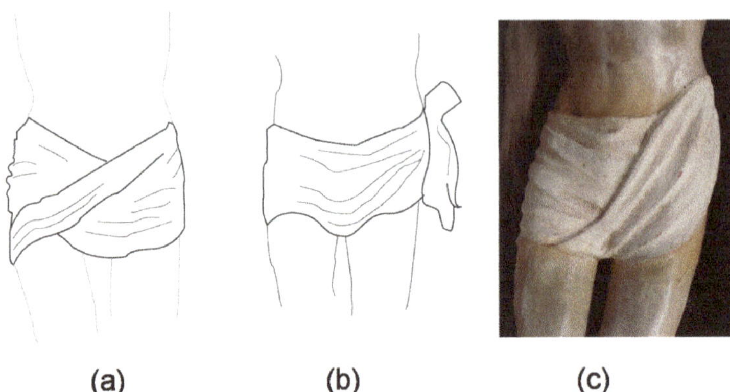

Figure 2. Comparison between different loincloth designs. (**a**) Loincloth general design identified in the Cortés workshop sculptures from the 1570s; (**b**) loincloth found in the Cortés workshop sculptures from the 1580s; (**c**) detail of the loincloth from the *Señor del Santo Entierro* sculpture.

3.3. Computerized Tomography

The tomographic images showed the original structure of the sculpture and some modifications (Figure 3). The body is hollowed and was constructed initially with a two-half cast (Figure 3c) to obtain a general shape. The artist modeled the volumes over the paper silhouette, first with debarked maize stems and later with maize stem paste.

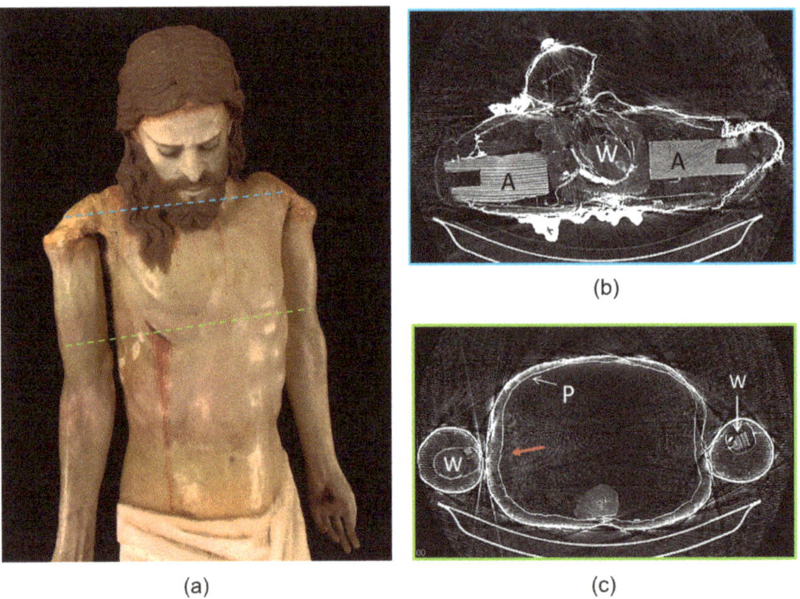

Figure 3. (**a**) Detail of the sculpture, the dashed lines in color indicate the different areas of the tomograms; (**b**) Axial tomogram of the shoulder areas, the image shows the articulation system (A) and the wooden elements used to insert the head of the sculpture (W); (**c**) Axial tomogram shows the hollowed structure, the paper layer (P) obtained using a two-half mold; the red arrow shows the union between the two halves obtained from the mold. The wooden elements used to reinforce the arms and insert the wooden hands are marked with W.

The upper head part (i.e., the cranial vault) was modified by adding three wooden fragments pasted together (Figure S3b,c). Additionally, the space between the head and hair (which was modeled using gypsum as suggested by SEM-EDX analysis, data not showed) suggests it is a modification. Small metallic nails (approximately 60) fixed it to the sculpture head (Figure S3a).

The tomographic study also allowed us to understand the shoulders articulation system (Figure 3b), it is a "*galleta*" or "*gozne de paleta*" system (similar to a hinge), done using a woodblock cut in the external extremity to create a space; another wooden element from the inside of the arm was inserted in the middle of the woodblock and fixed with a linchpin (Figure S3a). This particular articulation system was already used in Spanish sculptures in the 16th century [46]; however, previous studies on New Spain sculptures suggest that the "*galleta*" system was used after the 18th century, while the most common articulation system during the 16th century was spherical [47]. In light of this, the articulation system was probably substituted, which is reasonable considering the possible damage produced by the constant use.

Particularly in the loincloth area, we observed a hidden perforation on the left side (Figure 4b) that was probably used to insert a ribbon (today lost) when the sculpture was placed on the cross, as several paintings show, for example, *El Señor de Santa Teresa* (1760) by Francisco Antonio Vallejo (1722–1785) or the *El Señor de Chalma* (1719) by José de Mora (1642–1724), both today at the Museo Nacional de Arte in Mexico City.

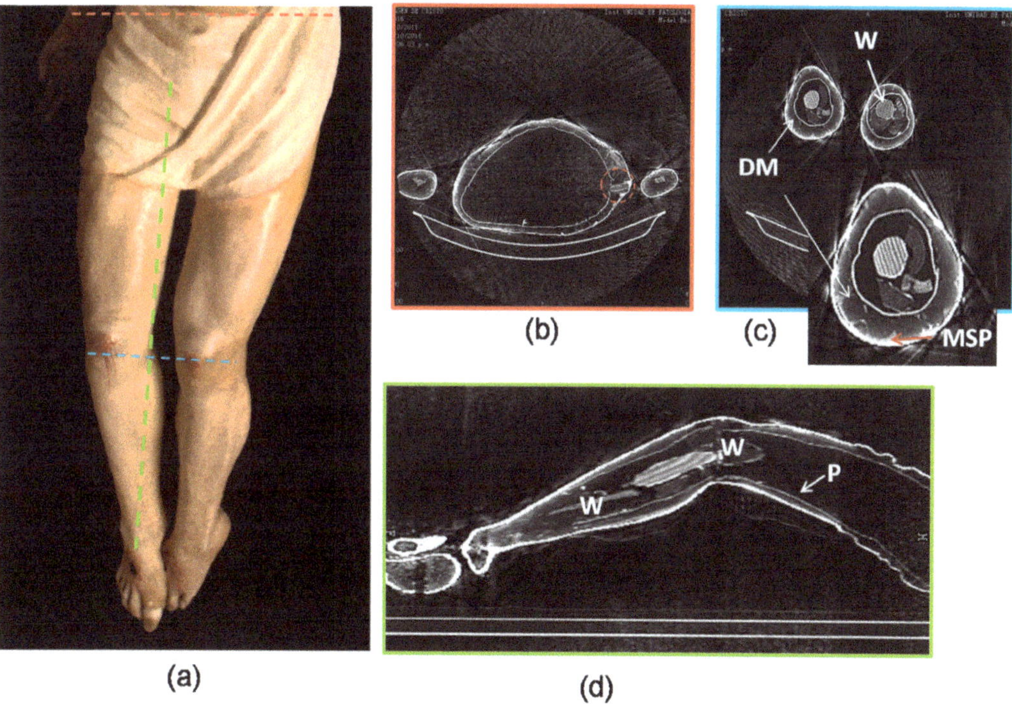

Figure 4. (**a**) Detail of the lower part of the sculpture, the dashed lines in color indicate the different areas of the tomograms; (**b**) Axial tomogram of the loincloth, the image shows a possible hold to insert a removable ribbon (marked with a red dashed circle), in the same region the separation of the two halves of the paper layer is shown; (**c**) The axial tomogram shows the wooden elements that reinforce the knees, the inset shows the circumference of the debarked maize (DM) stems and the maize stem paste (MSP) used to obtain the volumes; (**d**) Sagittal tomogram that shows the hollowed structure, the paper layer (P), and the wooden elements (W) used to reinforce the knees and to insert the feet.

In addition, the CT images showed the technical knowledge of the artist who used resistant materials in the areas where mechanical stress concentrates. The hands and feet were carved in wood (probably *colorín* to avoid increasing the weight) and some wooden elements to reinforce the neck, the arms (Figure 3b,d), and the legs (Figure 4d), as well as for fixing the hands and feet were inserted (Figure 4d). Several wooden elements reinforced the knees (Figure 4c,d).

3.4. Maize Stem and Paper Identification

Sample SSE-6, obtained from a fracture in the beard of the sculpture, confirmed the use of maize stem paste for modeling the volumes. The SEM images showed the characteristic cellular structures of the maize (*Zea mays* L.) vascular bundle (Figure 5a,b) [48]. The internal structure of the sculpture (sample SSE-8) was made with a thick cardboard-like paper made of long, yellowish fibers. Under the microscope, we identified cotton (*Gossypium hirsutum* L.) fibers (Figure 5c), which was the most common textile fiber used by the indigenous populations [49].

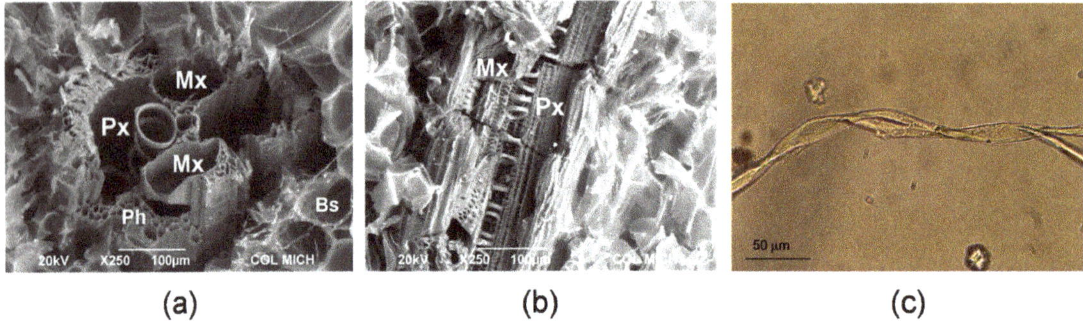

Figure 5. (**a**) SEM backscattering image of the maize stem sample (SSE-6) shows the maize vascular bundle, Px: protoxylem, Mx: metaxylem vessels, Ph: phloem, Bs: bundle sheath; (**b**) Longitudinal view of some structures of the maize vascular bundle, Px: protoxylem, Mx: metaxylem vessel; (**c**) Cotton fiber from the paper layer (sample SSE-8) under the optical microscope (50×), average diameter 10 µm.

3.5. Polychromy Analysis

3.5.1. Original Polychromy

The polychromy was done following the European artistic tradition. Over the final layer of maize stem paste, the artist applied a preparation layer made of *gesso grosso*, containing mainly anhydrite ($CaSO_4$), identified thanks to the 1095 cm^{-1} band from the νS=O in the µ-FTIR spectra (Figure 6f). Anhydrite, obtained by roasting gypsum, was traditionally used for ground layers in Southern Europe until the 16th century [50]. A particular characteristic of the ground layer of this sculpture is the presence of a cellulosic material (probably maize stem fragments) used as a filler (Figure 6a,d). These fragments are present in the different cross-section samples. The weak bands at 1656 cm^{-1} and 1548 cm^{-1} (data not showed) arising from ν(C=O) and NH, respectively [50], suggest the used of animal glue as a binder for the ground layer.

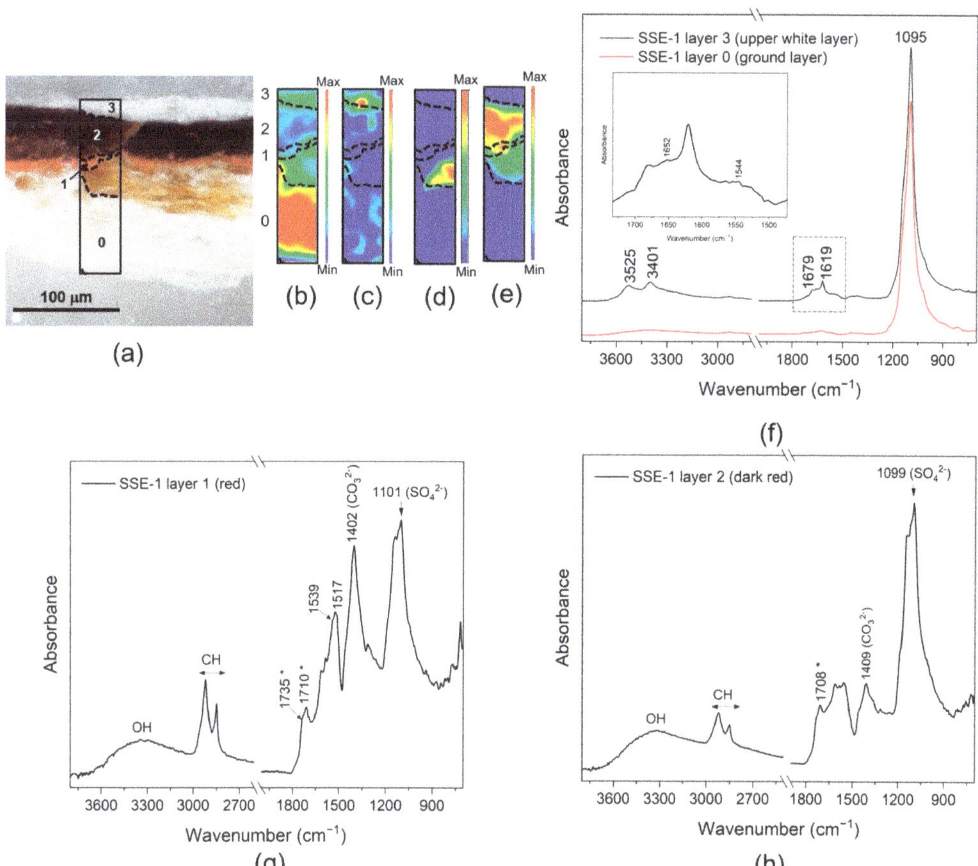

Figure 6. (**a**) SSE-1 cross-section, 20x; (**b**) Map of the band 1095 cm^{-1} from the $SO_4{}^{2-}$; (**c**) Map of the band at 3526 cm^{-1} from OH of gypsum; (**d**) Map obtained integrating the area of the band at 1029 cm^{-1} from the νCO at C_6 [51] of cellulosic material used as filler; (**e**) Map of the band at 1735 cm^{-1} from the C=O of the oil binder; (**f**) FTIR spectra from layers 0 (anhydrite) and 3 (gypsum) of sample SSE-1. The inset shows the bands attributed to animal glue as a binder; (**g**) FTIR spectrum obtained from layer 1, the bands associated with the binder marked with *; (**h**) FTIR spectrum of the dark red layer (layer 2) where the band at 1409 cm^{-1} indicates the addition of lead white and the band at 1099 cm^{-1} suggests the presence of sulfates (probably calcium sulfate).

The cross-section analysis suggested that the original loincloth decoration was simple as the artist used the white color of the ground layer. Additionally, we identified the remains of the painted blood in the samples SSE-1 and SSE-2. The SEM-EDX (Table 2) and µ-Raman (Figure 7b) results suggested that the artist used two different red pigments applied in two layers; the first one contains vermilion (HgS), which has an intense red color, while the second was made using a red lake, most probably a cochineal lake, to achieve a darker hue to represent coagulated blood. The SEM-EDX results (Table 2) indicated the presence of Al, K, and Si in the dark red layer, probably associated with the lake substrate (alum $[KAl(SO_4)_2 \cdot 12H_2O]$) and traces of Cu, previously related to the use of insects for the lake preparation [52]. Lead white was also added to the mixture, as suggested by the SEM-EDX (Table 2) and µ-FTIR results (Figure 6g,h), probably as a drier [53].

Table 2. SEM-EDX and spectroscopic results of the three cross-section samples analyzed.

Sample	No.	Layer	Main Elements	Trace Elements	µ-Raman/µ-FTIR
SSE-1	3	White layer	Ca, S	-	Gypsum ($CaSO_4 \cdot 2H_2O$), animal glue
	2	Dark red layer (blood)	Ca, S, Si, K, Pb, Al	Cu	Drying oil, Pb carboxylates
	1	Red layer (blood)	Hg, S, Ca	Si	Vermilion (HgS)
	0	Ground layer	Ca, S	-	Anhydrite ($CaSO_4$), cellulosic material, animal glue
SSE-2	6	White layer (flesh-tones)	Zn, Ba, S, Pb	Fe, Ca, Cr, Si	-
	5	White layer (flesh-tones)	Pb	-	-
	4	Yellowish layer (flesh-tones)	Pb, Ca,	Cl	-
	3	Layer dark red (blood)	Ca, Si, Al, K, Pb, S	Cu	-
	2	Layer red (blood)	Hg, S, Pb	Cu, Ca	Vermilion (HgS)
	1	Original flesh-tones	Ca, Pb	-	Cerussite ($PbCO_3$)
SSE-5	6	White layer (flesh-tones)	Zn, Ba, S, Pb	Fe, Ca, Cr, Si	-
	5	White layer (flesh-tones)	Pb	Ca, Cl	-
	4	Yellowish layer (flesh-tones)	Pb	Ca, Si, Cl	-
	0	Ground layer	Ca, S	-	-

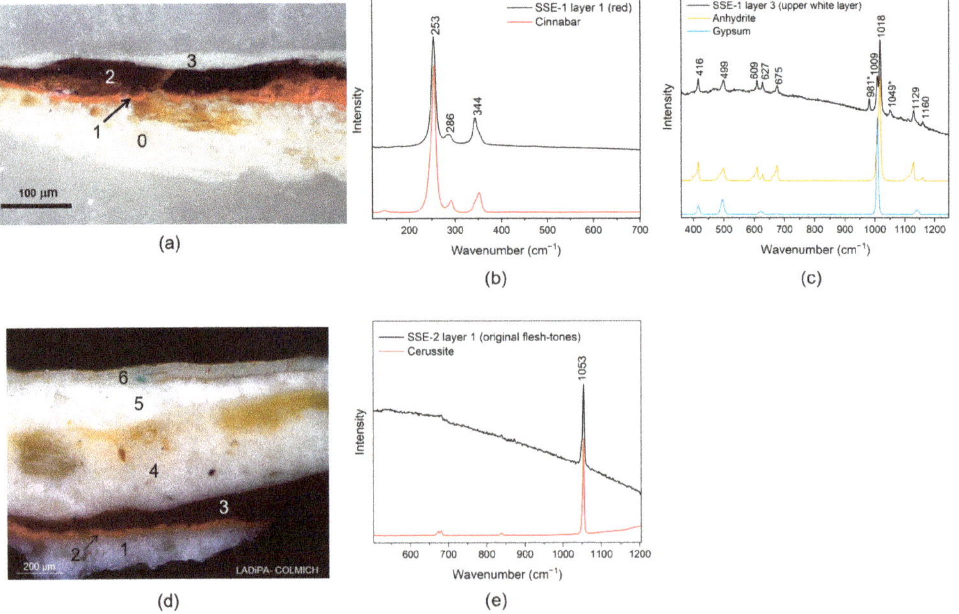

Figure 7. (**a**) SSE-1 cross-section, 20×; (**b**) Raman spectrum of the red layer (layer 1) of sample SSE-1, compared to that of cinnabar; (**c**) Raman spectrum obtained from the upper white layer (layer 3) of SSE-1 compared to the spectra of anhydrite and gypsum; (**d**) SEE-2 cross-section, polarized light, 50×; (**e**) Raman spectrum of the original flesh-tones (layer 1) compared to that of cerussite.

The original flesh-tones were rendered using lead pigments, as suggested by the SEM-EDX results (Table 2) of sample SSE-2. In the μ-Rama spectrum (Figure 7e), the band at 1053 cm^{-1}, arising from the $\nu_1(a'_1)$ CO_3^{2-}, suggested the presence of lead white [54], mainly constitute by cerussite ($PbCO_3$), which indicates an "inversed" proportion between cerussite (generally the minor component in the mixture) and hydrocerussite ($2 Pb_3(CO_3)_2(OH)_2$) produced by acidic processing of lead white pigment by washing or grinding with vinegar as suggested by Gonzalez [55].

The μ-FTIR analysis of sample SSE-1 (Figure 6e,g,h) suggests that the artist used a drying oil as a binder due to the band at 1735 cm^{-1} from the ν(C=O) of the esters of the oil. Additionally, the spectra showed the characteristic bands of degradation products from the binder, such as free fatty acids (suggested by the band at 1710 cm^{-1} [56]) and Pb carboxylates (associated with the band at 1517 cm^{-1} and a shoulder at 1539 cm^{-1} [56,57]).

3.5.2. Polychromy Modifications

The cross-section analyses allowed us to identify the modification to the original polychromy. The loincloth was covered with gypsum ($CaSO_4 \cdot 2H_2O$), identified by the two bands at 3528 cm^{-1} and 3398 cm^{-1} from the νO-H, and the band at 1095 cm^{-1} from the νS=O [50] identified in the μ-FTIR spectra (Figure 6f). Furthermore, the μ-Raman analysis suggested that a small amount of anhydrite is also present (Figure 7c). Additionally, the bands at 1651 cm^{-1} and 1544 cm^{-1} indicate the use of animal glue as a binder (inset of Figure 6f).

Regarding the flesh-tones, we identified two over paintings; the first composed of a double layer made of Pb-based pigments, mainly lead white, as suggested by the μ-Raman and SEM-EDX results (Table 2 and Figure S4). In these two layers, the SEM-EDX detected Cl, which can be associated with the presence of laurionite ($PbCl_2OH^-$) as a contaminant from the Pb ore deposits [58] or by the use of the Dutch method—developed mainly after the 16th century in the Netherlands—for the synthesis of lead white as suggested by Noun and colleagues [55,59].

We supposed that the second over painting of flesh-tones was performed using lithopone (ZnS, $BaSO_4$, and traces of ZnO), as suggested by the presence of Zn, Ba, and S. To obtain the required hue, other pigments containing Fe and Cr were added. The presence of lithopone in this layer allowed us to date this modification to after 1874 [60].

4. Discussion

Before our study, some scholars, particularly Orozco [44], who was also involved in the construction of the current *Nuestra Señora de la Soledad* church, suggested that the sculpture was probably a substitution of the original because its characteristics, mainly the polychromy and the hair, were atypical for the 16th-century maize stem images.

The multianalytical analysis allowed us to understand the complex structure of the sculpture and confirm that it is the original (Figure 8). The artist considered the function of the sculpture during the Holy Friday and the need for reinforcement in the areas subjected to higher mechanical stress, such as the hands, feet, knees, arms, and the neck. In particular, the wooden elements used to reinforce the neck and the modification of the hair, and the upper part of the head may suggest that, originally, the head could move for the representation of Christ and that, probably, a wig and metallic crown were applied, as was usual for these kind of sculptures during the liturgical representations. However, a more detailed investigation is required to confirm this hypothesis.

In addition to understanding the sculptural technology, based on the formal and technical examination, we were able to associate the sculpture to the Cortés workshop, particularly to the production in the 1580s decade, and identify the changes it has suffered over time. Indeed, the hollowed structure obtained using a two-half mold is similar to the structure of other previously studied sculptures attributed the Cortés workshop and can be associated to artists with a stronger influence of the European traditions (*papelón* or *cartapesta*) concentrated in the capital of the New Spain viceroyalty [2,10,12].

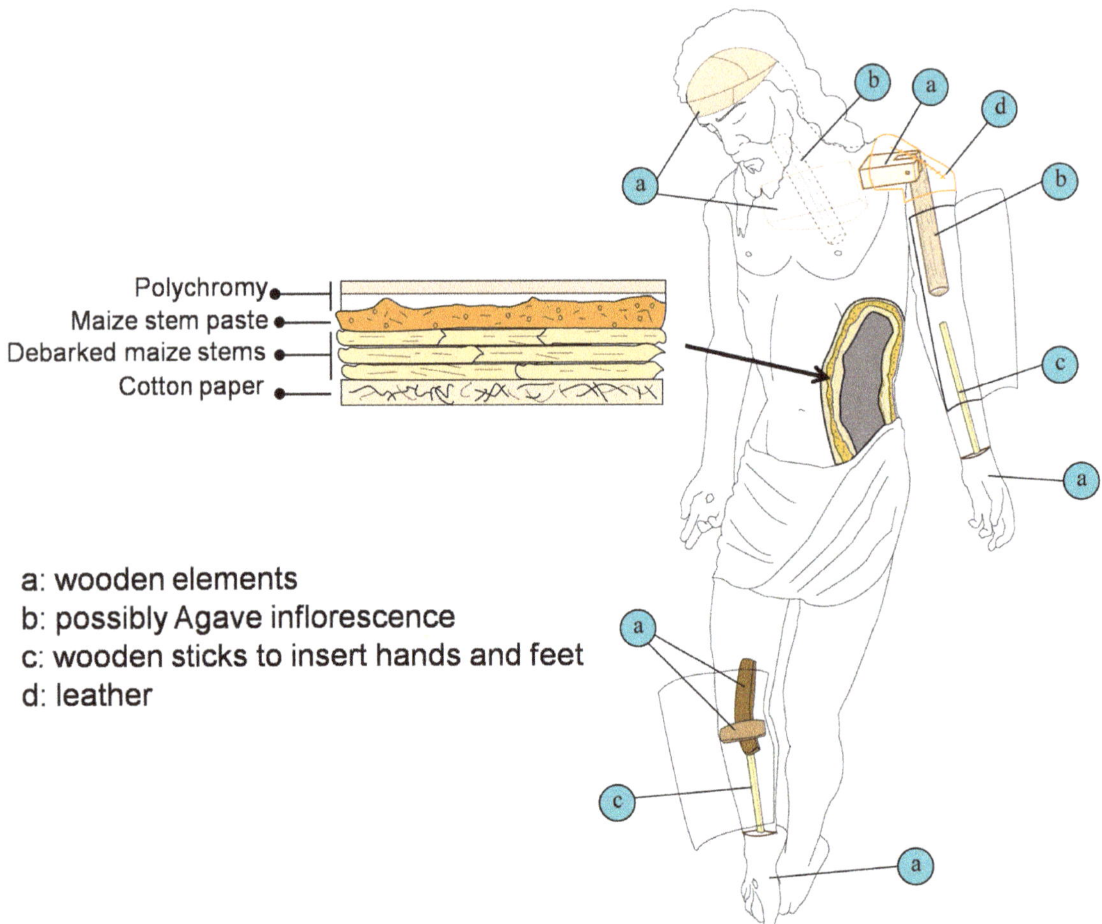

Figure 8. General scheme of the *Señor del Santo Entierro* sculpture.

The polychromy modifications can be associated with changes in preference and taste, particularly the flesh-tones. As the cross-section samples show, the profuse blooding represented in the original polychromy, which is a characteristic of the Christ images of the 16th and 17th centuries, was covered and substituted with more decorous polychromy for the tastes of the 19th century. Indeed, the Guadalajara cathedral, where the sculpture was placed for several years, suffered a complete renovation during the 19th century to adapt it to the neoclassical taste [61].

Finally, the presence in Guadalajara city of a sculpture made close to Mexico City despite the vicinity of the Michoacán region, suggests a particular link between the confraternity and the capital of New Spain. The acquisition of a statue produced in a specific region can be linked to a particular interest on certain workshops. Further studies are required to understand the trade routes of this typology of sculptures.

5. Conclusions

Our study allowed us to recover the history, understand the technical characteristic of both the structure and the polychromy, and correctly classify the *Señor del Santo Entierro* sculpture, as well as to understand its changes through the centuries. We were able to ascribe the sculpture to the Cortés workshop made around 1580. The analytical techniques showed that the statue suffered several modifications related to the damage produced by the constant use during the liturgical representations of the Holy Friday and to adapt it to the modern aesthetic taste.

As this study has shown, the technological complexity of the low weight maize stem sculptures requires a multianalytical approach, combining non-invasive and micro-invasive methods, to understand the structure and the nature of the materials. These sculpture techniques, which are not confined to Mexico, still require recognition around the world for their correct classification and the understanding of the technical variations according to the regions of productions. The historical investigations shed light on possible trade routes, which required further studies both inside the Mexican territory and in other parts of the Americas and Europe.

Supplementary Materials: The following are available online at https://www.mdpi.com/article/10.3390/heritage4030085/s1, Figure S1: Location of some maize stem sculptures register in Mexico; Figure S2: Sampling scheme; Figure S3: Topogram and tomograms of the upper part of the sculpture; Figure S4: SEM-EDX mapping of sample SSE-1, SSE-2, and SSE-5.

Author Contributions: Data curation, D.Q.B.; investigation, D.Q.B. and E.S.-R.; methodology, D.Q.B.; supervision, Á.Z.R.; writing—original draft, D.Q.B.; writing—review and editing, D.Q.B. and Á.Z.R. All authors have read and agreed to the published version of the manuscript.

Funding: Not applicable.

Institutional Review Board Statement: Not applicable.

Informed Consent Statement: Not applicable.

Data Availability Statement: The data presented in this study are available on request from the corresponding author.

Acknowledgments: We would like to express our gratitude to Genaro Santoscoy and Silvia Serratos for the access to the CT and Nelly Sigaut and Mirta Insaurralde from the Laboratorio de Análisis y Diagnóstico del Patrimonio (LADiPA-COLMICH) for the access to the microscopy techniques. We would also like to thank Rocco Mazzeo from the M2AD laboratory from the University of Bologna for the access to μ-FTIR analysis and Mariangela Vandini and Flavia Fiorillo from the Art Diagnostic laboratory of the University of Bologna for the access to the Raman instrumentation. Thanks to Antonio García Rangel and Tomás de Hijar Ornelas for the access to the sculpture.

Conflicts of Interest: The authors declare no conflict of interest.

References

1. Brito Benítez, E.L. Symbolism and Use of Maize in Pre-Hispanic and Colonial Religious Imagery in Mexico. *E-Conserv. J.* **2014**, *2*, 116–127. [CrossRef]
2. Amador Marrero, P.F. Singulares aportaciones desde la restauración para el conocimiento de la escultura ligera novohispana. El caso del Señor de la Ascensión (Cristo Resucitado) de la Catedral de Tlalnepantla, México, y su adscripción al Taller de Cortés. *Intervención* **2019**, *10*, 15–24. [CrossRef]
3. Huff, L.A. Sacred Sustenance: Maize, Storytelling, and a Maya Sense of Plac. *J. Lat. Am. Geogr.* **2006**, *5*, 79–96. [CrossRef]
4. Bruquetas, R. Un Conjunto de Esculturas en Maguey de la Iglesia de San Miguel de Mamara, en el Apurímac Andino. In *Escultura Ligera*; Ajuntament de València: Valencia, Spain, 2017; pp. 85–98.
5. Ávila Figueroa, E. Técnicas y Materiales de la Escultura Ligera Novohispana con Caña de Maíz: Una Aproximación historiográfica. Master's Thesis, Universidad Nacional Autónoma de México, Mexico City, Mexico, 2011.
6. Arias Martínez, M. Vestidas y de Pasta: Testimonios Documentales Sobre Escultura Procesional para la Cofradía de la Soledad de Madrid en el siglo XVI. In *Escultura Ligera*; Ajuntament de València: Valencia, Spain, 2018; pp. 119–133.

7. Casciaro, R. Cartapesta e Scultura Polimaterica nell'Italia del Rinascimento. In *La Scultura in Cartapesta, Sansovino, Bernini e i Maestri Leccesi Tra Tecnica e Artificio*; Silvana Editoriale: Milan, Italy, 2008; pp. 51–80.
8. Marichal Salinas, C. Mexican Cochineal, Local Technologies and the Rise of Global Trade from the Sixteenth to the Nineteenth Centurie. In *Global History and New Polycentric Approaches*; Perez Garcia, M., De Sousa, L., Eds.; Palgrave Macmillan: London, UK, 2018; pp. 255–273.
9. Álvarez, M.-T.; Villaseñor Black, C. Introduction: Art and Trade in the Age of Global Encounters, 1492–1800. *J. Interdiscip. Hist.* **2014**, *45*, 267–275. [CrossRef]
10. Amador Marrero, P.F.; Decroly, M.; Glaude, C.; Indekeu, C.; Marinković, A.; Marušić, M.M.; Matijević, J.; Portsteffen, H.; Sanyova, J. The Crucified Christ of Lopud, Croatia. A Unique Early Import of Mexican Polychromed Sculpture Made of Maize Stalks. *Z. Für Kunsttechnol. Konserv.* **2018**, *2*, 183–200.
11. Pérez de Castro, R.; Amador Marrero, P.F. La Recepción de Crucificados Ligeros Novohispanos en Castilla y León: Nuevos Ejemplos y Perspectivas. In *Torna Viaje. Tránsito Artístico Entre Los Virreinatos Americanos y La Metrópolis*, 1st ed.; Quiles, F., Amador Marrero, P.F., Fernández, M., Eds.; Andavira Editora S.L.: Santiago de Compostela, Spain, 2020; pp. 623–668.
12. Valverde Larrosa, C.; Martín García, J.C. Estudios radiográficos de tres de los grandes Cristos de caña de maíz identificados en España: El Cristo crucificado de Lerma (Burgos), el Cristo de Santa María de Vitoria-Gasteiz (Álava) y el Cristo de la buena muerte de Gran Canaria (Gran Canaria). *Intervención* **2015**, *6*, 39–52. [CrossRef]
13. Amador Marrero, P.F. *Traza Española, Ropaje Indiano. El Cristo de Telde y la Imagineria en Caña de Maíz*; Ayuntamiento de Telde: Telde, Spain, 2002.
14. Amador Marrero, P.F. Imaginería Ligera Novohispana en el Arte Español de los Siglos XVI y XVII. Ph.D. Thesis, Universidad de Las Palmas de Gran Canaria, Las Palmas, Spain, 2012.
15. Peña, A.; Capella, S.; González, C. Characterization and identification of the mucilage extracted from orchid bulbs (Bletia campanulata) by high temperature capillary gas chromatography (HT-CGC). *J. High Resolut. Chromatogr.* **1995**, *18*, 713–717. [CrossRef]
16. De Sahagún, B. *Historia General de las Cosas de la Nueva España*; Porrúa: Mexico city, Mexico, 1975.
17. Velarde Cruz, S.I. *Imaginería Michoacana en Caña de Maíz. Estudio Histórico y Catálogo de Imágenes en Morelia, Tupátaro, Pátzcuaro, Tzintzuntzan, Quiroga y Santa Fe de la Laguna, Mich. Siglos XVI-XVII*; Conaculta: Mexico City, Mexico, 2003.
18. De la Rea, A. *Crónicas de Michoacán*; UNAM: Mexico City, Mexico, 1991.
19. García-Abásolo, A.F.; García Lascurain, G. *Imaginería Indígena Mexicana. Una Catequesis en Caña de Maíz*; Publicaciones de la Obra Social y Cultural Cajasur: Córdoba, Spain, 2001.
20. Amador Marrero, P.F. Imaginería ligera en Oaxaca. El Taller de los grandes Cristos. *Bol. Monum. Hist.* **2009**, *3*, 45–60.
21. Carrillo y Gariel, A. *El Cristo de Mexicaltzingo. Técnica de las Esculturas en Caña*; Dirección de Monumentos Coloniales: Mexico City, Mexico, 1949.
22. Araujo Suárez, R.; Huerta Carrillo, A.; Guerrero Bolan, S. *Esculturas de Papel Amate y Caña de Maíz*; Fideicomiso Cultural Franz Mayer: Mexico City, Mexico, 1989.
23. Cennini, C. *Il Libro Dell'arte*, 10th ed.; Neri Pozza Editore: Vicenza, Italy, 2017.
24. Bassett, J.; Alvarez, M.-T. Process and Collaboration in a Seventeenth-Century Polychrome Sculpture: Luisa Roldán and Tomás de los Arcos. *Getty Res. J.* **2011**, *3*, 15–32. [CrossRef]
25. De Boodt, R. The Shop Floor of the Brussels Sculptor. In *Borman. A Family of Northern Renaissance Sculptors*; Debaene, M., Borman, A., Eds.; Harvey Miller Publishers: Turnhout, Belgium, 2019; pp. 34–45.
26. Bonavit, J. *Esculturas Tarascas de Caña de Maíz y Orquídeas, Fabricadas Bajo la Dirección del Ilustrísimo Señor don vasco de Quiroga*; Filmas Publicistas: Morelia, Mexico, 1947.
27. Buccolieri, A.; Buccolieri, G.; Castellano, A.; Quarta Colosso, P.; Miotto, L. Non-destructive techniques used during the restoration of the relief "Madonna and Child" by Jacopo Sansovino. *Appl. Phys. A* **2015**, *120*, 447–453. [CrossRef]
28. Sarrió Martín, M.F.; Juanes Barber, D.; Ferrazza, L. La imagen de San Miguel Arcángel del Ayuntamiento de Valencia. Análisis del Sistema Constructivo Mediante el Estudio con Tomografía Computarizada (TC). In *Escultura Ligera*; Ajuntament de València: Valencia, Spain, 2017; pp. 9–24.
29. Amador Marrero, P.F. El Santo Cristo Reliquia Muy Milagrosa. Análisis Interdisciplinario de una Imagen Novohispana de Papelón. In *Garachico y Sus Fiestas del Cristo. Apuntes Históricos y Crónicas de Prensa, Gobierno de Canarias, Ayuntamiento de la Villa y Puerto de Garachico*; Velázquez Ramos, C., Ed.; Centro de Iniciativas Turísticas: Las Caletillas, Spain, 2010; p. 43.
30. Labastida Vargas, L. El empleo de la videoscopia en el estudio de la imaginería ligera o de pasta de caña. *An. Inst. Investig. Estéticas* **2005**, *27*, 199–207. [CrossRef]
31. Freitas, R.P.; Ribeiro, I.M.; Calza, C.; Oliveira, A.L.; Felix, V.S.; Ferreira, D.S.; Pimenta, A.R.; Pereira, R.V.; Pereira, M.O.; Lopes, R.T. Analysis of a Brazilian baroque sculpture using Raman spectroscopy and FT-IR. *Spectrochim. Acta Part A Mol. Biomol. Spectrosc.* **2016**, *154*, 67–71. [CrossRef]
32. Casciaro, R.; Cassiano, A. *Sculture di età Barocca tra Terra d'Otranto, Napoli e Spagna. Catalogo Della Mostra (Lecce, 16 Dicembre 2007–28 Maggio 2008)*; De Luca Editore d'Arte: Rome, Italy, 2008.

33. Ortega-Ordaz, A.A.; Sánchez-Rodríguez, E.; Rojas-Abarca, L.; Bojórquez-Quintal, J.E.; Ku-González, A.; Cruz-Cárdenas, C.I.; Quintero-Balbás, D.I. Chemical and Morphological Decay in Maize Stem Sculptures. In *Science and Digital Technology for Cultural Heritage. Interdisciplinary Approach to Diagnosis, Vulnerability, Risk Assessment and Graphic Information Models, Proceedings of the 4th International Congress Science and Technology for the Conservation of Cultural Heritage, Seville, Spain, 26–30 March 2019*; Calderón, P.O., Puertio, F.P., Verhagen, P., Prieto, A.J., Eds.; CRC Press: Boca Raton, FL, USA, 2019.
34. Prati, S.; Rosi, F.; Sciutto, G.; Mazzeo, R.; Magrini, D.; Sotiropoulou, S.; Van Bos, M. Evaluation of the effect of six different paint cross section preparation methods on the performances of Fourier Transformed Infrared microscopy in attenuated total reflection mode. *Microchem. J.* **2012**, *103*, 79–89. [CrossRef]
35. De la Mota Padilla, M. *Historia del Reino de Nueva Galicia en la América Septentrional*; Universidad de Guadalajara: Guadajlara, Mexico, 1973.
36. Verdi Webster, S. Art, Ritual, and Confraternities in Sixteenth-Century New Spain. Penitential Imagery at the Monastery of San Miguel, Huejotzingo. *An. Inst. Investig. Estéticas* **1997**, *19*, 5–43. [CrossRef]
37. Cecilia Espinoza, M.; Ruiz Ángel, G. *El Mestizaje de las Artes en la Semana Santa Hispanoamericana y Española In Torna Viaje. Tránsito Artístico Entre los Virreinatos Americanos y la Metrópolis*; Quiles, F., Amador Marrero, P.F., Fernández, M., Eds.; Andavira Editora S.L.: Santiago de Compostela, Spain, 2020; pp. 547–564.
38. *Certificación de la Agregación de la Cofradía a la de Roma*; The Historical Archive of the Archdiocese of Guadalajara: Guadalajara, Mexico, 1598.
39. Alfaro Anguiano, C.G. La Colonia y la Época Clerical. In *Los Moradores de la Rotonda*; Anguiano, A., Gabriel, C., Eds.; Ayuntamiento de Guadalajara: Guadalajara, Mexico, 1992; p. 1.
40. Hernández Larrañaga, J. *Guadalajara: Identidad Perdida. Transformación Urbana en el Siglo XX*; Agreta: Guadalajara, Mexico, 2001.
41. Meyer, J. *La Cristiada*; Siglo Veintiuno Editores: Mexico City, Mexico, 1979.
42. Informe del estado del templo, 1933, Section Gobinerno, Serie Parroquias/Templo de la Soledad, Folder La Soledad Gdl (Catedral), The Historical Archive of the Archdiocese of Guadalajara, Mexico.
43. Kasis Ariceaga, A. *Monografías de Arquitectos del Siglo XX, Ignacio Díaz Morales, Gobierno de Jalisco*; Universidad Jesuita de Guadalajara: Guadalajara, Mexico, 2004.
44. Orozco, E.L. *Los Cristos de Caña de Maíz y Otras Venerables Imágenes de Nuestro Señor Jesucristo*; Amate: Mexico City, Mexico, 1970.
45. Gómez García, C. Disposición del Paño de Pureza en la Escultura del Cristo Crucificado Entre los Siglos XVI y XVII. Ph.D. Thesis, Universidad Complutense de Madrid, Madrid, Spain, 2007.
46. Fernández González, R. Sistemas de Articulación en Cristos del Descendimiento. Master's Thesis, Universidad Politécnica de Valencia, Valencia, Spain, 2012.
47. Calzada Martínez, H. La Escultura Articulada en el Distrito Federal: Arte, Ingenio y Movimiento. Master's Thesis, Universidad Nacional Autónoma de México, Mexico City, Mexico, 2011.
48. Shane, M.W.; McCully, M.E.; Canny, M.J. The Vascular System of Maize Stems Revisited: Implications for Water Transport and Xylem Safety. *Ann. Bot.* **2000**, *86*, 245–258. [CrossRef]
49. Stark, B.L. Long-term economic change: Craft extensification in the Mesoamerican cotton textile industry. *J. Anthropol. Archaeol.* **2020**, *59*, 101194. [CrossRef]
50. Melo, H.P.; Cruz, A.J.; Candeias, A.; Mirão, J.; Cardoso, A.M.; Oliveira, M.J.; Valadas, S. Problems of analysis by FTIR of Calcium Sulphate-based preparatory layers: The case of a group of 16th-century protuguese paintings. *Archaeom.* **2014**, *56*, 513–526. [CrossRef]
51. Cichosz, S.; Masek, A. IR Study on Cellulose with the Varied Moisture Contents: Insight into the Supramolecular Structure. *Materials* **2020**, *13*, 4573. [CrossRef] [PubMed]
52. Kirby, J.; Spring, M.; Higgitt, C. The Technology of Red Lake Pigment Manufacture: Study of the Dyestuff Substrate. *Natl. Gallery Tech. Bull.* **2005**, *26*, 71–87.
53. Matteini, M.; Moles, A. *La Chimica nel Restauro. I Materiali Dell'arte Pittorica*, 13th ed.; Nardini Editore: Firenze, Italy, 2010.
54. Bell, I.M.; Clark, R.J.H.; Gibbs, P.J. Raman spectroscopic library of natural and synthetic pigments (pre-~1850 AD). *Spectrochim. Acta Part A* **1997**, *53*, 2159–2179. [CrossRef]
55. Gonzalez, V.; Calligaro, T.; Wallez, G.; Eveno, M.; Toussaint, K.; Menu, M. Composition and microstructure of the lead white pigment in Masters paintings using HR Synchrotron XRD. *Michrochim. Acta* **2016**, *125*, 43–49. [CrossRef]
56. Mazzeo, R.; Prati, S.; Quaranta, M.; Joseph, E.; Kendix, E.; Galeotti, M. Attenuated total reflection micro FTIR characterisation of pigment-binder interaction in reconstructed paint films. *Anal. Bioanal. Chem.* **2008**, *392*, 65–76. [CrossRef]
57. Hermans, J.J.; Keune, K.; van Loon, A.; Corkery, R.W.; Iedema, P.D. Ionomer-like structure in mature oil paint binding media. *RSC Adv.* **2016**, *6*, 93363. [CrossRef]
58. Iorio, M.; Sodo, A.; Graziani, V.; Branchini, P.; Casanova Municchia, A.; Ricci, M.A.; Salvadori, O.; Fiorin, E.; Tortora, L. Mapping at the nanometer scale the effects of sea-salt derived chlorine on cinnabar and lead white by using delayed image extraction in ToF-SIMS. *Analyst* **2021**, *146*, 2392. [CrossRef] [PubMed]
59. Noun, M.; Van Elslande, E.; Touboul, D.; Glanville, H.; Bucklow, S.; Walter, P.; Brunelle, A. High mass and spatial resolution mass spectrometry imaging of Nicolaus Poussin painting cross section by cluster TOF-SIMS. *J. Mass Spectrom.* **2016**, *51*, 1196–1210. [CrossRef] [PubMed]

60. Bellei, S.; Nevin, A.; Cesaratto, A.; Capogrosso, V.; Vezin, H.; Tokarski, C.; Valentini, G.; Comelli, D. Multianalytical Study of Historical Luminescent Lithopone for the Detection of Impurities and Trace Metal Ions. *Anal. Chem.* **2015**, *87*, 6049–6056. [CrossRef] [PubMed]
61. Camacho Becerra, J.A. Génesis de un Estilo. Altares de la Catedral de Guadalajara, 1757–1919. In *La Catedral de Guadalajara. Su Historia y Significados*; El Colegio de Jalisco: Zapopan, Mexico, 2012; pp. 11–88.

 heritage

MDPI

Article

On the Hierarchical Use of Colourants in a 15th Century *Book of Hours*

Angelo Agostino [1], Eleonora Pellizzi [2], Maurizio Aceto [3,4,*], Simonetta Castronovo [5], Giovanna Saroni [6] and Monica Gulmini [1]

[1] Dipartimento di Chimica, Università degli Studi di Torino, via P. Giuria, 7-10125 Torino, Italy; angelo.agostino@unito.it (A.A.); monica.gulmini@unito.it (M.G.)
[2] Laboratoire Scientifique et Technique du Département de la Conservation, Bibliothèque Nationale de France, 14 Avenue Gutenberg, 77607 Bussy Saint-Georges, France; eleonora.pellizzi@bnf.fr
[3] Dipartimento di Scienze e Innovazione Tecnologica (DiSIT), Università degli Studi del Piemonte Orientale, viale Teresa Michel, 11-15121 Alessandria, Italy
[4] Centro Interdisciplinare per lo Studio e la Conservazione dei Beni Culturali (CenISCo), Università degli Studi del Piemonte Orientale, 11-15121 Alessandria, Italy
[5] Museo Civico di Arte Antica, p.zza Castello, 10123 Torino, Italy; simonetta.castronovo@fondazionetorinomusei.it
[6] Dipartimento di Studi Storici, Università degli Studi di Torino, via S. Ottavio, 20-10124 Torino, Italy; giovanna.saroni@unito.it
* Correspondence: maurizio.aceto@uniupo.it; Tel.: +39-011-360265

Abstract: An illuminated *Book of Hours* (in use in Chalon-sur-Saône) currently owned by the Museo Civico di Arte Antica and displayed in the prestigious Palazzo Madama in Torino (Italy) was investigated by means of optical microscopy, fibre optic reflectance spectroscopy, fibre optic molecular fluorimetry, X-ray fluorescence spectrometry and Raman spectroscopy. The aim of the scientific survey was to expand the knowledge of the manuscript itself and on the materials and techniques employed by Antoine the Lonhy, the versatile itinerant artist who decorated the book in the 15th century. The focus was to reveal the original colourants and to investigate the pigments used in rough retouches which were visible in some of the miniatures. The investigation was carried out in situ by portable instruments according to a non-invasive analytical sequence previously developed. It was evident that the use of different pigments by the master was ruled, at least partially, by a hierarchical scheme in which more precious materials were linked to the most important characters or details in the painted scene.

Keywords: non-invasive techniques; FORS; XRF; illuminated manuscripts; brazilwood; colourants; Antoine de Lonhy; Torino

Citation: Agostino, A.; Pellizzi, E.; Aceto, M.; Castronovo, S.; Saroni, G.; Gulmini, M. On the Hierarchical Use of Colourants in a 15th Century *Book of Hours*. *Heritage* **2021**, *4*, 1786–1806. https://doi.org/10.3390/heritage4030100

Academic Editor: Lucia Burgio

Received: 16 July 2021
Accepted: 11 August 2021
Published: 13 August 2021

Publisher's Note: MDPI stays neutral with regard to jurisdictional claims in published maps and institutional affiliations.

1. Introduction

The identification of the palette is among the various tasks of the scientific examination of an illuminated manuscript. Techniques and materials for illumination are reported, in principle, in some ancient manuals [1,2] which sometimes use colloquial names which may not be reliable descriptors of the substances actually employed. The combined contribution of researchers from the fields of natural sciences and humanities has shed light on this topic and now the set of the most relevant inorganic colourants employed in illumination is known [1–5]. Nevertheless, the materials used to decorate a specific codex may not be accurately recognised even by an expert eye, since different colourants may yield similar hues, they can be mixed to obtain a particular colour or applied in multilayers and the original colour can be modified and obscured by weathering [6,7]. Therefore, instrumental investigation still plays a crucial role in the recognition of the materials that were actually employed.

The characterisation of the palette gives useful information for conservation and also aids the selection of the proper conditions for preserving or exhibiting these fragile artworks. The determination of the original colourants within a manuscript may confirm or reject its authenticity [8] or may suggest a date for its production, since the period in which different colourants were in use is known. Moreover, a particular set of colourants, along with the compositions of the material employed, may lead to the identification of scriptoria and workshops or to the recognition of the intervention of different hands upon a series of codices or within a single manuscript. More generally, the knowledge of the materials employed expands the information available on a specific book, on the artists who decorated it and on the institutional or private clients for which the book was produced [9–12].

The choice of colourants used to decorate the various features is of particular interest to art historians: precious pigments may be chosen for the most important details, while more conventional materials might be used for less important details [1]. A hierarchy of colourants, following the market values of the materials, can reveal the schemes pursued by the artist in the representation of the various features. A commodity-related evaluation of the colourants used in a manuscript can be highly informative about its intrinsic value. These aspects usually receive very little attention in art historical studies.

With the aim of highlighting this specific topic, in this work we identified the colourants and evaluated the overall palette employed on a 15th century *Book of Hours* (in use in Chalon-sur-Saône) finely illuminated by the Burgundian painter Antoine de Lonhy.

Lonhy was a fascinating figure who was an itinerant artist active from circa 1446 to 1490 in Burgundy, Languedoc, Catalonia and in the Duchy of Savoy. His activity is presently documented by a large variety of artworks, including paintings on wood, frescoes, miniatures, cartoons for stained glass and embroidery [13–15]. A *Book of Hours* was a Christian devotional book with prayers to be said at certain times throughout the day. The small (17.2 × 11.7 cm) book investigated in this work consists of eighty-one pages of vellum with ten large arch-topped miniatures (Table 1) and decorated initials.

Table 1. The subjects of the miniatures.

Folio	Subject
1r	*Annunciation* (Figure 1a)
15r	*Visitation* (Figure 1b)
24r	*Nativity* (Figure 1c)
28v	*Angel announcing the Nativity to the shepherds* (Figure 1d)
31v	*Adoration of the Magi* (Figure 1e)
36v	*Holy Family in the run to Egypt* (Figure 1f)
42r	*God the Father and Jesus crowning the Virgin* (Figure 1g)
46r	*Crucifixion*, with the Virgin and St. John the Evangelist (Figure 1h)
49v	*Pentecost* (Figure 1i)
71v	*Burial service* (Figure 1j)

A rich naturalistic decoration, with acanthus, fruits and flowers, borders each page (Figure 1a).

The manuscript was made approximately between 1446 and 1449 in Burgundy, during the initial period of Lonhy's activity, but besides this, very little information was available about the manuscript until 1966, when it appeared in the catalogue of an auction house. A description of what is known about the book and its history, as well as an exhaustive review of the publications dealing with the versatile activity of Antoine de Lonhy has been produced by Saroni [16]. The manuscript has been owned by the Museo Civico di Arte Antica—MCAA—in Torino (Italy) since 2002 and it is currently displayed in the prestigious Palazzo Madama, home of MCAA, with the signature Inv. n. 399. The museum's collection includes some other artworks of the artist, who is strongly linked to Torino itself and to the surrounding region (Piedmont). The book is in fact an important element for documenting the figure of Antoine de Lonhy in his early years of activity.

Figure 1. (**a**) f. 1r, *Annunciation*. The main characteristics of the book (illuminated capital letters, rich four-side naturalistic decorations, arch topped miniatures) are shown. (**b**) f. 15r, *Visitation*. (**c**) f. 24r, *Nativity*. (**d**) f. 28v, *Angel announcing the Nativity to the shepherds*. (**e**) f. 31v, *Adoration of the Magi*. (**f**) f. 36v, *Holy Family in the run to Egypt*. (**g**) f. 42r, *God the Father and Jesus crowning the Virgin*. (**h**) f. 46r, *Crucifixion*, with the Virgin and St. John the Evangelist. (**i**) f. 49v, *Pentecost*. (**j**) f. 71v, *Burial service*.

Evidence of some conservation treatments and inpainting are present, although the exact extension and the date of these treatments are unknown: the original bookbinding was removed, the cover was substituted and some parts are missing whereas some pages have been added. Additionally, the sequence of the devotional texts has probably been modified. Rough retouches are evident on three miniatures and the presence of other, less evident, interventions in the upper part of most of the other miniatures was suggested by experts who examined the manuscript before its acquisition by the Museum.

As requested by the curator of the Museum's collection, non-invasive techniques were selected for investigating the colourants: optical microscopy (OM), fibre optic reflectance spectroscopy (FORS), fibre optic molecular fluorimetry (FOMF), X-ray fluorescence spectrometry (XRF) and micro-Raman spectroscopy (micro-Raman). Portable instrumentations were used in order to keep the precious manuscript in the Museum during the scientific investigation. Apart from identifying the original colourants, the investigation was extended to some evident retouches in order to suggest a possible date for the interventions on the miniatures.

The results from these analyses show that the palette used by the artist was rich and variegated and it is therefore a perfect candidate to testify how painters of manuscripts exploited different colourants to express their art in the 15th century, with intrinsic meaning beyond the mere colour.

2. Materials and Methods

2.1. Digital Optical Microscopy

A Dino-Lite (Naarden, The Netherlands) AM413TL-FVW model optical microscope was employed to record digital images from the parchment. The microscope allows magnification in the range 20–90×.

2.2. XRF

XRF measurements were performed with an EDXRF Thermo (Waltham, MA, USA) NITON spectrometer XL3T-900 GOLDD model, equipped with a Ag tube (max. 50 kV, 100 μA, 2 W), a large area SDD detector and an energy resolution of about 136 eV at 5.9 keV. The analysed spot had an average diameter of 3 or 8 mm and was focused by a CCD camera, with a working distance of 2 mm. The total time of analysis was 240 s. The instrument was held in position with a moving stage allowing micrometric shifts, in order to reach the desired probe-to-sample distance; the stage was laid on a tripod. The obtained spectra were processed with the commercial software bAxil, derived by the academic software QXAS from IAEA.

2.3. FORS

The FORS analysis was performed with an Avantes (Apeldoorn, The Netherlands) AvaSpec-ULS2048XL-USB2 model spectrophotometer and an AvaLight-HAL-S-IND tungsten halogen light source with a wavelength range of 360–2500 nm; the detector and light source were connected with fibre optic cables to an Avantes FCR-7UV200-2-1,5 × 100 reflection probe. In this configuration, light is sent and retrieved with a unique fibre bundle positioned at 45° from the surface normal, in order not to include specular reflectance. The spectral range of the detector was 200–1160 nm; depending on the features of the monochromator (slit width 50 μm, grating of UA type with 300 lines/mm) and of the detector (2048 pixels), the best spectra resolution was 2.4 nm calculated as FWHM. The diffuse reflectance spectra of the samples were referenced against the WS-2 reference tile provided by Avantes and guaranteed to be reflective at 98% or more in the spectral range investigated. The investigated area on the sample had a 1 mm diameter. In all measurements the distance between the probe and the sample was kept constant at 2 mm, corresponding to the focal length; the probe was inserted into a small aluminium block. To visualise the investigated area on the sample, the block contained a USB endoscope. The instrumental parameters were as follows: 10 ms integration time, 100 scans for a total acquisition time of 1.0 s for each spectrum. The system was managed with AvaSoft v. 8 dedicated software running under Windows 7.

2.4. FOMF

An Ocean Optics (Dunedin, FL, USA) Jaz model spectrophotometer was employed to measure the molecular fluorescence spectra. The instrument was equipped with a 365 nm Jaz-LED internal light source and an Avantes FCR-7UV200-2-1,5 × 100 reflection probe used to drive excitation light on the sample and to recover emitted light. The spectrophotometer was working in the range 191–886 nm; according to the features of the monochromator (200 μm slit width) and the detector (2048 elements), the spectral resolution available was 7.6 nm calculated as FWHM. The investigated area on the sample was 1 mm in diameter. In all measurements the distance between the probe and the sample was kept constant at 1.2 mm, corresponding to the focal length; the probe was inserted into a small aluminium block in order to exclude contributions from external light. To visualise the investigated area on the sample, the block contained a USB endoscope. Instrumental parameters were as follows: 3 s integration time, 3 scans for a total acquisition time of 9 s for every spectrum. The system was managed with SpectraSuite software running under Windows 7.

2.5. Micro-Raman

Raman analysis was performed with a portable system assembled by Horiba Jobin-Yvon (Villeneuve d'Ascq, France). The system is composed of a microHR spectrometer with a spectral resolution of 5 cm^{-1}; a CCD Synapse detector; a ModularHead analytical probe containing a video camera for visualisation of samples, a notch filter for elimination of Rayleigh radiation and a microscope objective (20×, 50× or 80×); a tripod to support the analytical probe, equipped with an XYZ stage with micrometric movements to allow accurate focusing; an Ar 532 nm and a He-Ne 632 nm lasers; optic fibres to carry laser radiation on the sample and Raman radiation to the spectrometer. The system is managed with LabSpec 5 dedicated software running under Windows XP.

2.6. Preparation of Standard Paints

In order to identify the colourants present in the manuscript under study, a spectral database was built by measuring FORS and micro-Raman spectra of reference paint samples applied on parchment. These were prepared in our laboratory by employing pictorial materials according to the indications contained in medieval treatises [1,2,4]. In consideration of the painting techniques mostly used by ancient illuminators, two binders were used to disperse pigments and lakes: egg white tempera and gum Arabic.

2.7. Analytical Workflow

In order to optimise the workflow, we employed an analytical protocol which had been developed previously [17]. A preliminary inspection of the painted areas under high magnification optical microscopy (up to 80×) was undertaken in order to highlight their micro-texture and to select the most significant spots for instrumental analysis. The chosen analytical techniques were employed sequentially, following the investigation protocol. Firstly, a large number of spots were investigated by FORS. This allowed areas to be identified with similar reflectance spectra, which can be attributed to the presence of the same materials. In several cases, the comparison of the obtained spectra with those collected in the spectral database (see Section 2.6), along with indications from literature [18,19], allowed for the colourants employed to be determined directly.

A second survey was performed with FOMF, which enabled some of the identifications obtained by means of FORS to be confirmed.

XRF spectrometry was then employed to obtain in-depth information on the underlying layers of paint and the selective identification of the colourants through the detection of key elements that can be linked to one specific pigment. XRF is of particular interest in identifying metal pigments and white/black colourants; in addition, this technique can yield information concerning the provenance of raw materials [20]. With only a few exceptions, organic dyes cannot be identified with this technique.

Lastly, Raman spectroscopy was employed to reveal the identity of those colourants that were still unknown and to confirm the presence of the pigments that had been identified by the other—less selective—techniques.

3. Results

The palette used by the miniaturist is described in the following paragraphs. In addition to the identification of colourants, the discussion will focus on the choice of the colourants in the pictorial scene.

3.1. Black Pigments

Most of the black areas in the illuminations consist of small details such as floor-tiles (ff. 1r and 49v, Figure 1a,i) or shoes (ff. 28v and 36v, Figure 1d,f), which were painted next to green areas. The spectral features shown by FORS analysis revealed the presence of a carbon-based pigment, but these areas also yielded intense signals from copper and zinc when analysed by XRF. This suggested the presence of a copper green pigment spread under the black one (see later).

3.2. Blue Pigments

Two main blue pigments were identified in the miniatures, i.e., azurite and ultramarine blue, according to their characteristic absorption bands at 640 and 600 nm, respectively. Two different cases were recognised in the spectra from the blue areas of the *Book of Hours*, each associated to a different blue hue. An exemplary case is reported in Figure 2a: the Virgin's mantle on f. 1r shows bright and dark blue hues employed to paint the drapery.

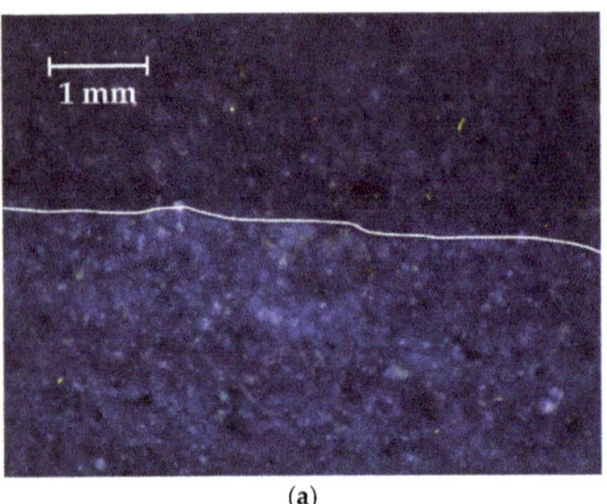

(**a**)

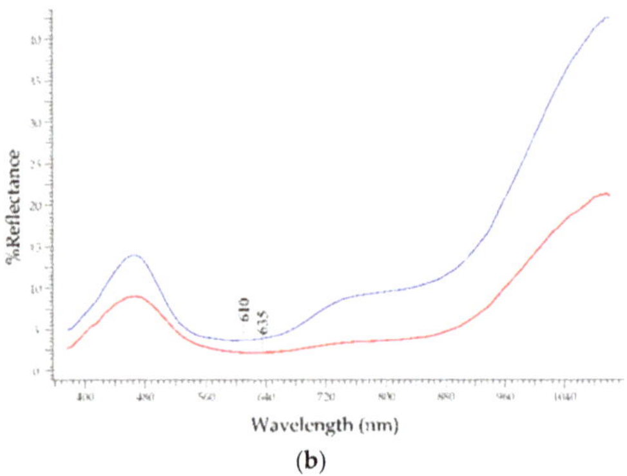

(**b**)

Figure 2. (**a**) Micro-photography (80×) of the mantle of the Virgin on folio 1r. The presence of different blue shades is made more evident by the white border line. (**b**) Selected FORS spectra representative of the different blue shades. The blue line represents spectra collected from darker blue areas within the miniatures and points to the presence of ultramarine, the red line represents spectra collected from lighter blue areas within the miniatures and of the blue areas in the decorations on page borders and capital letters and points to the presence of azurite.

The two different hues revealed by the FORS spectra can be related to the use of different pigments: the spectra from the brighter areas show a broad absorption band centred at about 635–640 nm, thus pointing to the presence of azurite, whereas the band centred at about 610–615 nm in the darker areas suggests the use of ultramarine blue (Figure 2b).

The presence of both pigments, and specifically of ultramarine blue applied over the azurite layer for darkening the colour in the miniatures, was confirmed by micro-Raman by focusing the analysis on grains of the two pigments. This was also was the case for the blue skies, for which the FORS spectra suggested the presence of azurite in the brighter areas and the use of ultramarine blue to darken the colour. The bathochromic shift of the absorption band for ultramarine blue (610–615 nm in place of the expected 600 nm) can be justified by considering the application of this pigment as a thin layer on top of the underlying azurite layer and therefore the final spectrum contained contributions from both pigments. This feature was verified with mock-ups prepared in the laboratory.

The blue decorations on the page borders instead gave FORS spectra that could be attributed to the presence of azurite alone, both in the lighter and darker areas: here the brightness of the colour was adjusted by mixing azurite with a white pigment.

All the XRF spectra collected from the dark and bright areas in the miniatures and from the blue decorations on the page borders showed intense copper signals, as well as signals of barium and zinc; these latter elements were indicated in the literature as possible impurities of natural azurite [11,21], since their presence can be related to the geo-chemical processes of formation of the deposits. A fairly linear relationship between the counts of copper and barium and between those of copper and zinc emerged (Figure 3) and seems to confirm the use of natural azurite throughout the manuscript. More evidence of the natural origin of azurite, as suggested by Aru et al. [22], could be the presence of iron oxides: we did not identify such accessory phases, but the counts of iron in the XRF spectra of the blue areas were on average 4 times higher than in the spectra of parchment.

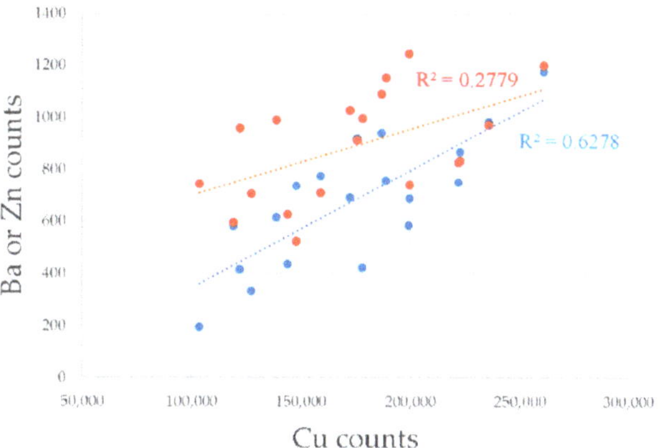

Figure 3. Bivariate plot for XRF signals of copper (K_α line, 8.05 keV) vs. barium (red dots, L_α line, 4.47 keV) and zinc (blue dots, K_α line, 8.64 keV) recorded from blue areas in the miniatures. Regression lines with coefficients R are indicated.

The results obtained by XRF on those areas where ultramarine blue was present deserve some discussion. Portable XRF instruments, that do not operate under vacuum, are generally not able to detect weak XRF signals from light elements such as sodium, aluminium and silicon, which are the main constituents of lazurite—$Na_3Ca(Al_3Si_3O_{12})S$—i.e., the blue mineral that gives the pigment its rich colour. The intensity of the calcium signals from the dark blue areas did not significantly differ from those obtained from the parchment, where this element is present due to the alkaline treatments with calcium hydroxide that occurred in the production procedures. Signals from sulphur, which is present in the structure of ultramarine blue, were detected from the dark blue areas, but this element cannot be considered as a selective marker for this pigment because it can be present in parchment itself; moreover, it is difficult to separate the S K_α line at 2.31 keV from the Pb M_α line at 2.34 keV and lead is usually present as lead white to tune the blue colour. Nevertheless, the XRF spectra recorded in dark blue areas, in which FORS and spectra Raman indicated the presence of ultramarine in the superficial layer of the miniature, differed from those recorded from the bright areas of pure azurite because of the systematic presence of high signals of potassium. No further indication could be derived from the other analytical techniques employed; therefore, possible explanations for the presence of this element in the ultramarine layers could be related to the phenomenon of vicariance between the sodium ions present in lazurite and the potassium ions present in the environment of formation of the rock (similar to that occurring in pyroxenes between Mg^{2+} and Mg-vicariant ions [23]), or to the processes adopted for the purification of the pigment (that employed potash according to Cennini [4]) or even to the possible addition of extenders to the precious pigment. Ultimately, the presence of potassium can be considered a potential elemental marker for ultramarine blue in cases where Al, Si and S cannot be clearly detected.

3.3. *Brown Pigments*

The brown colour was mainly used to depict architectural features, such as the inside of the huts on ff. 24r and 31v (Figure 1c,e). In all cases, the pigments used were iron oxides or iron-rich red earths, according to the features of the FORS spectra and to the dominant presence of iron evidenced by XRF. It was not possible to obtain a better identification by means of Raman spectroscopy. Some of these pigments, however, were probably related to later retouches (see Section 3.11 for further details).

3.4. Green Pigments

All the reflectance spectra of the green hues were bell-shaped in the visible region of the spectrum, with a reflectance maximum in the range 520–575 nm. The absorption band in these spectra systematically occurred at ca. 750 nm (Figure 4a), which is not fully compatible with the feature of verdigris (720 nm) nor with that of malachite (800 nm) [24].

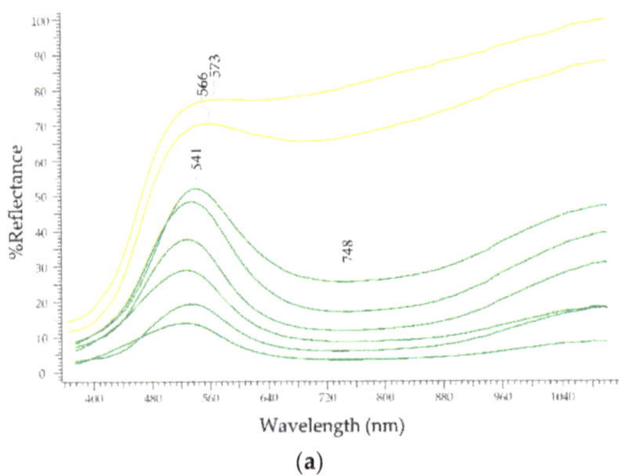

(a)

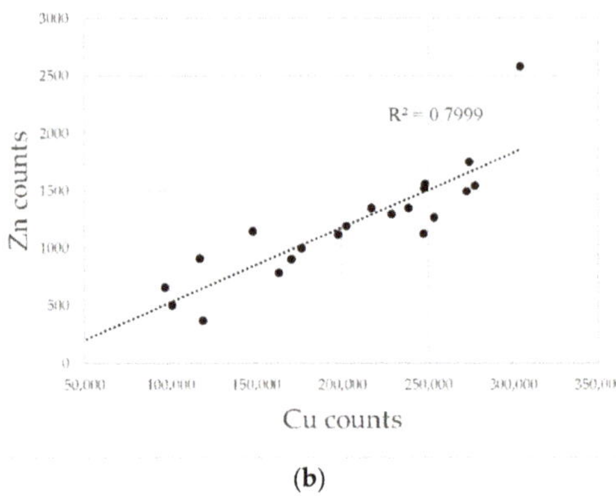

(b)

Figure 4. (**a**) selected FORS spectra representative of the different green shades. The green lines are spectra collected from green areas within the miniatures, the top yellow-green lines are spectra collected from yellow-green areas within the miniatures. (**b**) Bivariate plot for XRF signals of copper (K_α line, 8.05 keV) vs. zinc (K_α line, 8.64 keV) recorded from green areas in the miniatures; the regression line with coefficient R is indicated.

The XRF analysis detected strong copper signals, as well as weak zinc signals, that showed a linear trend when plotted on a graph (Figure 4b). Zinc ores are described as being contaminants of copper ores [21], therefore this relationship would suggest the use of a natural copper compound.

Unfortunately, with the configuration used, the micro-Raman equipment did not allow us to gain molecular information from the green pigment present and so the actual nature of the pigment remained unknown. A consideration of the features of the FORS spectra and a comparison with those obtained on other coeval *Book of Hours* manuscripts produced in the same area [25,26] suggest the use of a copper sulphate, such as brochantite or antlerite, as the green pigments. To support this tentative identification, the presence of S was suggested by the XRF spectrum, though again, as in the case of ultramarine blue, it could not be definitely confirmed due to the proximity of the S K_α line at 2.31 keV with the Pb M_α line at 2.34 keV.

The spectral range of the reflected light (see the two top spectra in Figure 4a with reflection maxima shifted towards NIR) determines the different shades (green to yellow-green) observed among the painted green areas. These shades were obtained by mixing a green and a yellow pigment, as was evident from observations of the green areas under the optical microscope (Figure 5).

In such yellow-green areas, significant tin and lead signals were also detected. Moreover, an evident correlation emerged among the XRF signals for tin and lead (not shown), and this strongly suggests the use of lead-tin yellow to brighten the colour. The presence of lead-tin yellow type I which was also used as pure yellow pigment (see Section 3.9), was also confirmed by micro-Raman analysis. The use of lead-tin yellow to modify the tonality of green pigments has already been noted in other works [25,26].

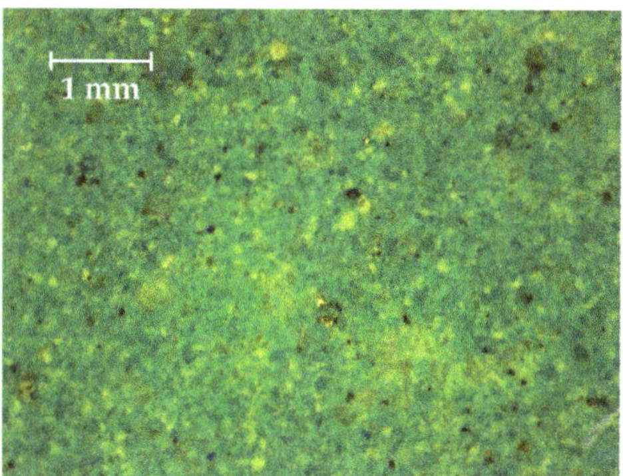

Figure 5. Micro-photography (80×) of yellow-green meadows in the background of Saint Elisabeth and the Virgin on f. 15r (*Visitation*, Figure 1b): green and yellow particles are clearly visible.

3.5. Grey Pigments

Grey horizontal bands are present at ff. 31v (Figure 1f) and 46r (Figure 1h) in the top part. At f. 71v (Figure 1j) the vaults are rendered in grey too. In all cases, silver was used as evidenced by the XRF analysis. As will be discussed in paragraph 3.11, it seems that this was a later intervention and was therefore not due to Antoine de Lonhy.

3.6. Pink/Purple/Violet Colourants

Hues ranging from pink to violet are used extensively throughout the miniatures. They were employed for painting the garments worn by anonymous shepherds as well as the rich mantles of God the Father and Christ. The observation under high magnification of areas painted in pink allowed the use of a mixture of blue and red pigments to be excluded as a potential explanation for obtaining this hue (Figure 6a). Instead, the presence of a pink colourant was also observed. Sporadically, the presence of blue and/or white particles was also evident, thus indicating the use of a blue pigment to tune the final hue towards violet and of a white pigment to brighten the colour.

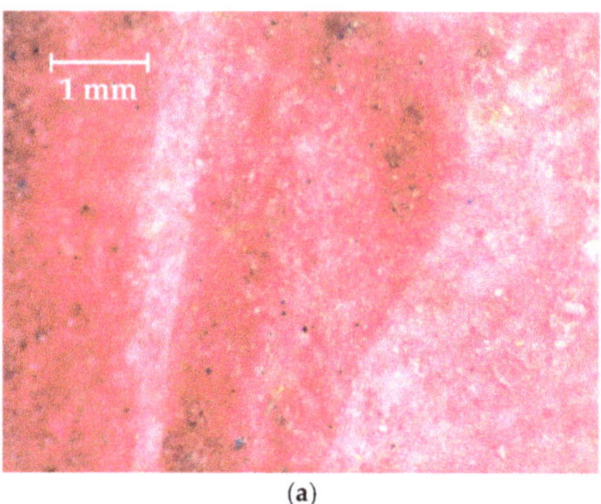

(a)

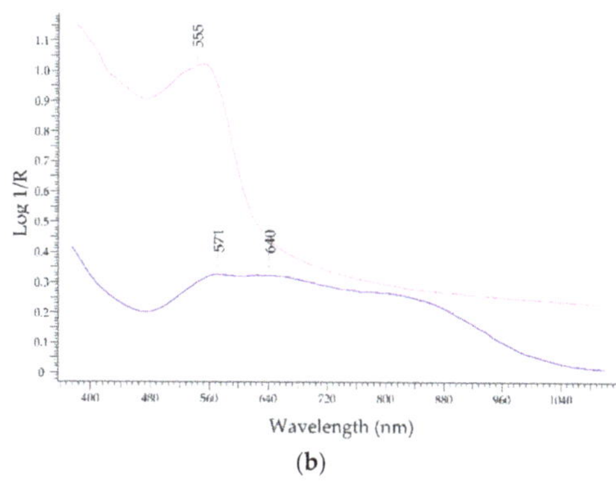

(b)

Figure 6. (**a**) Image under 80× magnification of a pink area. (**b**) FORS spectrum in Log(1/R) coordinates of a pink area (pink line, namely the coat of the Angel on the right side at f. 1r, Figure 1a) and of a violet area (violet line, the sky on top at f. 42r, Figure 1g).

The numerous FORS spectra collected from the pink areas showed rather homogeneous spectral features. In particular, they were characterised by an inflexion point in the range 595–600 nm, an absorption band centred at 555 nm and a reflection maximum in the range 475–500 nm (Figure 6b; spectra are shown in Log(1/R) coordinates to better appreciate the absorption features). These features, when compared with the spectra obtained from our reference palette and according to the indications given by Melo et al. [25] and by Roger et al. [27], point to the presence of brasilwood lake, a colourant obtained by adsorbing dyes extracted from the wood of various species of *Caesalpinia* trees on alum, chalk, lead white or gypsum. The identification was confirmed by the FOMF analysis that yielded emission bands at 604 and 625 (data not reported). Brasilwood was in use throughout the Middle Ages. It was mentioned by Cennini [4] with regards to improving the colour of ultramarine blue. It was also used by 15th century Parisian miniature painters [27] who preferred it to insect dyes and madder. In the present case, brasilwood was also used on a layer of azurite to obtain a violet-purple tone, such as in the wall behind the angel at f. 1r (Figure 1a) or in the wall and in the background at f. 42r (Figure 1g), or to highlight red areas, such as in the orange at f. 24r (Figure 1c).

XRF detected intense signals from lead in purple areas. This could be related both to the presence of lead white used to brighten the colour and to its use as the inorganic substrate for the organic dye to prepare the lake pigment. Weak signals of copper (and of barium), detected in a few of the analysed spots imply that the blue grains that were sporadically observed under 80× magnification are those of azurite, which is coherent with their absorption features detected at 571 and 640 nm in the violet spectrum shown in Figure 6b. However, it was not possible to obtain micro-Raman signals from the purple pigment due to the strong fluorescence that obscured the weak Raman signals (if any).

3.7. Red Pigments

Spectra recorded by FORS on red areas showed the typical sigmoid shape that characterises the reflectance spectra of warm hues within the visible region. For these hues, the wavelength associated to the point of inflection in the reflectance spectrum may be indicative of a specific pigment [24], though this spectral feature could shift depending on the concentration of the pigment or if an achromatic colourant is added. With regards to the FORS spectra recorded on red areas of the *Book of Hours* (both miniatures and floral decorations on page borders), two main groups of spectra were observed. These two groups can be clearly distinguished by the different position of the inflection point of the sigmoid-shaped portion of the spectra (Figure 7). In one group, this feature was observed at 600 nm, whereas in the other group it occurred at a significantly lower wavelength (565 nm).

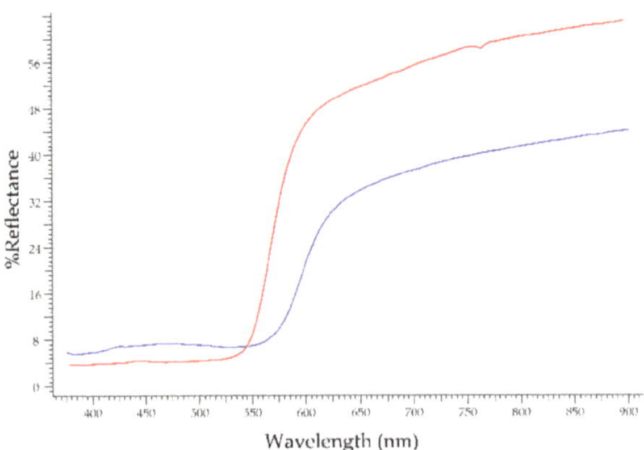

Figure 7. Selected FORS spectra representative of the two groups of spectra collected from red areas. Red line: spectrum obtained from the red vest of one of the Magi in the miniature on f. 31v (Figure 1e); blue line: spectrum obtained from the red acanthus on the upper part of f. 1v. The inflection points suggest the presence of red lead in the miniature and vermilion in the decoration on page borders.

These results suggested that two different pigments were used for reds: cinnabar or vermillion (HgS; it is not possible to distinguish between the natural and the synthetic version) and minium or red lead (Pb_3O_4), respectively. Both XRF and micro-Raman analysis confirmed the presence of these pigments. In fact, the XRF analysis revealed intense Pb signals in those areas in which the FORS analysis suggested the presence of red lead; moreover, the micro-Raman spectra of this pigment (with characteristic signals at 120, 150 and 550 cm^{-1}) were recorded from red particles in the same spots. The presence of cinnabar/vermillion was confirmed both by the occurrence of mercury, detected via XRF, and by the characteristic micro-Raman spectrum, i.e., a strong band at 254 cm^{-1} with a weak shoulder at 280 cm^{-1} and a less intense band at 340 cm^{-1}.

No red lead was detected in the decorations on the page borders and in the red letters throughout the written text, where only vermillion was used to paint red. This is consistent with a procedure involving another artisan with a different palette who decorated these parts of the manuscript. The hypothesis of a different artist working at the page borders, a less important part of the decoration, is more than reasonable if we consider that usually the decoration of a manuscript was more of a group work, in which the most important artist devoted himself to the most important aspects, namely the main miniatures.

3.8. White Pigments

The white pigment employed throughout the manuscript was identified as lead white. In fact, the micro-Raman spectra of the white particles showed the sharp band at 1049 cm^{-1} that characterises this pigment, whereas XRF data resulted in very high lead signals that emerged in most of the XRF spectra collected from the manuscript. This suggests that lead white was used, by itself or mixed with other pigments, as a primer under the illuminations. The most relevant exceptions are the areas painted with gold or mosaic gold, in which very weak signals of lead were detected.

3.9. Yellow and Yellow-Brown Pigments

The set of FORS spectra collected throughout the manuscript were mainly characterised by sigmoid-shaped spectra with points of inflection within 480 and 560 nm. According to this feature, four different yellow colourants could be hypothesised after the FORS survey and the use of the other techniques allowed us to confidently identify the following yellow pigments.

1. In one instance, e.g., in the spectrum recorded from the apostle's yellow tunic in the miniature representing the *Pentecost* on f. 49v (Figure 1i), the inflexion point was at ca. 480 nm. In this case, the comparison with the spectra collected from the reference palette indicates the presence of either lead-tin yellow type I or orpiment. The presence of lead-tin yellow was also suggested by the intense signals of these two elements in XRF spectra and finally confirmed by micro-Raman analysis, which also assessed the crystalline structure of the compound as type I [28].
2. As with other yellow-brown hues, the FORS spectra and XRF analysis suggested the presence of yellow ochre or other iron-based pigments, according to the minimum at ca. 890 nm, an inflection point at ca. 570 nm, the typical shoulder at 450 nm and the identification of iron by XRF. High fluorescence prevented obtaining Raman spectra. These features were shown, for example, on the inner part of the aedicule on f. 1r (Figure 1a), as well as the Cross and the ground of the Golgotha in the *Crucifixion* on f. 46r (Figure 1h).
3. Other yellow areas, appearing darker with respect to those mentioned above, showed glittering particles on a dark-yellow background when observed under 80× magnification. The texture of these particles showed a coarser grain size (Figure 8).

(a)

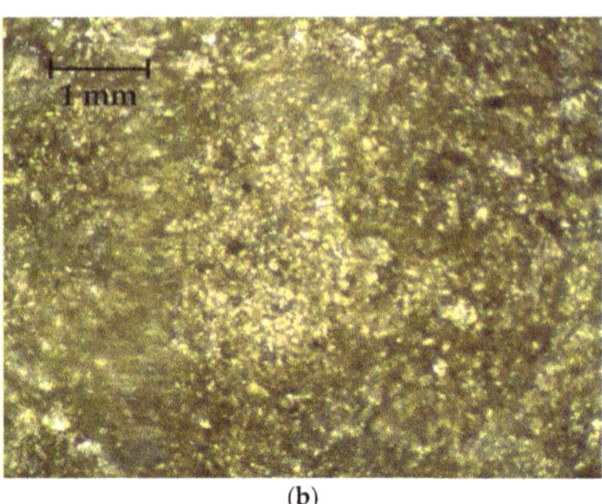

(b)

Figure 8. (**a**) Micro-photography (80×) of the golden architecture on f. 1r. (**b**) Micro-photography (80×) of the glittering details on the rock painted in the picture background on f. 15r (Figure 1b).

The FORS spectrum showed an inflexion point at ca. 530 nm, which is consistent with the feature of mosaic gold (also called purpurin–SnS_2) [24], a well-known substitute for gold which is typical of medieval miniature painting. XRF analysis confirmed the presence of tin and sulphur, as did the micro-Raman analysis that yielded spectra showing a very intense peak at 321 cm^{-1} (Figure 9).

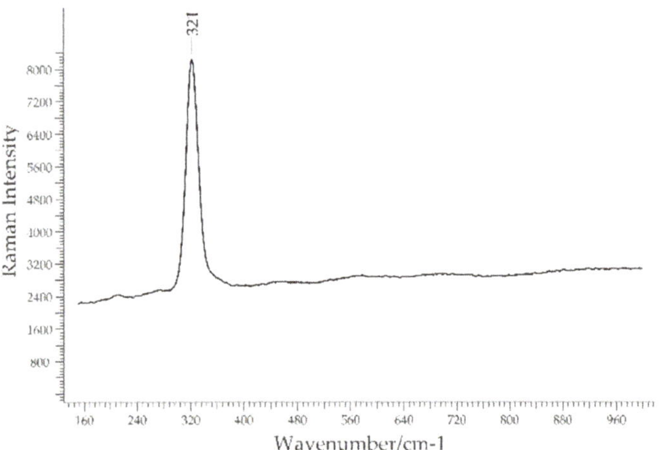

Figure 9. Raman spectrum obtained from brown-yellow areas revealing tin sulphide.

The Raman spectra of Sn(IV) sulphide reported in the literature [29] shows an intense signal at slightly lower frequency (313 cm^{-1}), therefore one can assume that different recipes could yield different molecular structures that may give slightly different Raman responses.

4. Finally, finer glittering particles were identified by XRF as gold; these areas also yielded a FORS spectrum with the typical inflexion point of gold at ca. 505 nm [24]. Powdered gold was used here, in the same way as other pigments, mixed with a medium and spread as a thin film with a brush; since the "ready to use" pigments were generally kept in shells, this technique is referred to as "shell gold", to distinguish it from the gold leaf technique, in which a very thin leaf of gold adheres on the vellum by means of an adhesive [30].

In the page borders decorations, the large yellowish leaves were obtained by spreading a thin layer of shell gold.

Gold was also employed for the haloes, executed using the gold leaf technique. In these areas XRF analysis also revealed calcium, sulphur and weak signals of copper and iron, in addition to gold. These traces of copper and iron are possibly related to impurities in the gold leaf while the high XRF signals from calcium and sulphur, significantly higher than those obtained from the unpainted parchment, indicate the use of gypsum in the typical *asiso* ground for gold, used in the Middle Ages starting from the 13th century [30].

3.10. Black Ink

The black ink used for writing all through the manuscript was composed of iron-gall ink, as revealed by the characteristic FORS spectrum with the reflectance level raising in the NIR [31]. The metal composition of ink is homogeneous in all the pages, being rich in iron and copper; at f. 9r a different ink, richer in copper and zinc, was highlighted (XRF spectra in Figure 10a), giving material evidence to the hypothesis of a later insertion for this folio, as suggested by the different style in the writing (Figure 10b,c).

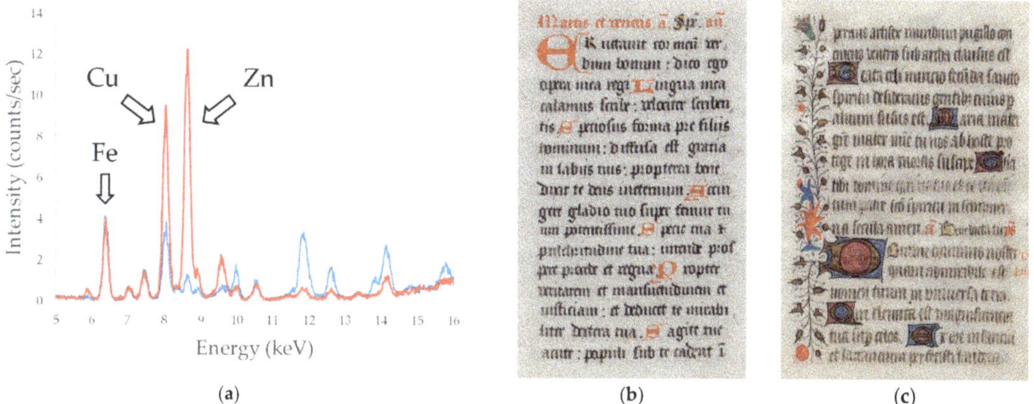

Figure 10. (**a**) XRF spectra of the black inks at f. 9r (red line) and at f. 42r. (blue line). (**b**) Black ink at f. 9r (later addition). (**c**) Black ink at f. 3r (original).

3.11. Retouches

Many of the (more or less evident) retouches are in the upper part of the miniatures, where the analysis detected, in some cases, the presence of pigments that differed from the original ones. The miniature on f. 31v (Figure 11a) appeared coarsely retouched: a large grey band was apparently added on the original picture, and it was apparent that the sky could have been completely repainted.

Inspection under high magnification supported these hypotheses, since the texture of the blue pigments was completely different from that observed in the original parts (Figure 11b), with blue particles of sub-micrometric size. Micro-Raman signals obtained in these areas are reported in Figure 12 and revealed the presence of phthalocyanine blue, a synthetic pigment available since 1935 (bands at 490, 596, 682, 747, 950, 1140, 1207, 1335, 1443, 1523 and 1569 cm^{-1}). Signals of lazurite (bands at 263, 550 and 1092 cm^{-1}), possibly pertinent to the original layers, were also detected.

(a)

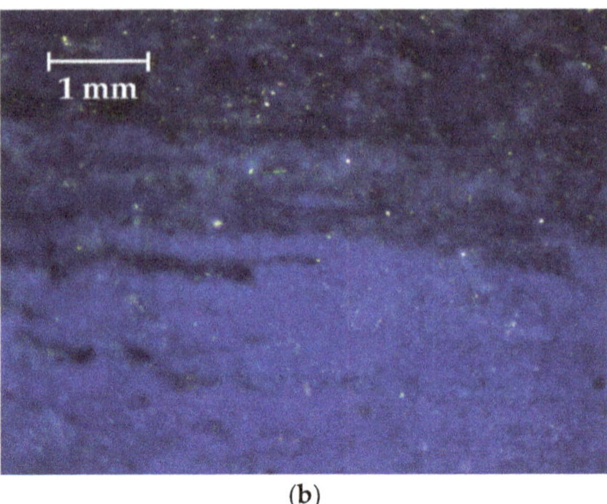

(b)

Figure 11. (**a**) Miniature on f. 31v. Retouches in the upper part of the picture and in the hut are visible. (**b**) Micro-photography (80×) of the blue sky close to the grey retouch.

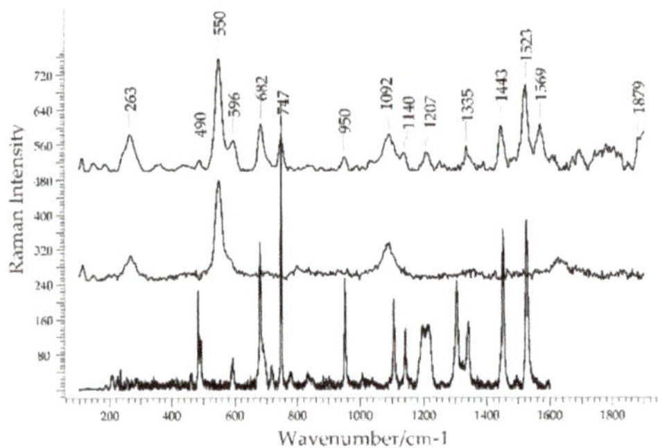

Figure 12. Raman signals recorded from the blue sky on folio 31v (**top**), from natural ultramarine spread on parchment with gum Arabic (**middle**) and from phthalocyanine blue spread on parchment with gum Arabic (**bottom**).

On examination, the grey band was attributed, via XRF, to a layer of powdered silver. Besides silver signals, XRF revealed here the presence of titanium, which points to the use of another modern pigment, i.e., titanium white (TiO_2). The lack of any micro-Raman signals from this compound, which is a strong Raman scatterer, suggests that titanium white was used to mask the underlying picture prior to the laying of silver powder which was presently darkened by weathering.

A similar rough intervention is visible on f. 46r (Figure 1h), where a grey band was added to the lower part of the sky; here again, a possible intervention on the sky itself was suggested. However, despite the various macroscopic similarities between the two restoration interventions on ff. 31v and 46r, instrumental investigation did not give us any incontrovertible evidence that they were carried out at the same time by the same restorer. In fact, glittering silver particles observed in the grey band under the microscope on f. 31v appeared significantly coarser than those observed on f. 46r (Figure 13).

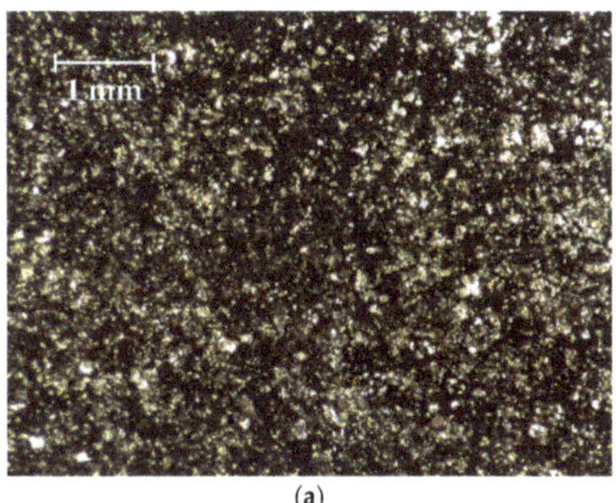

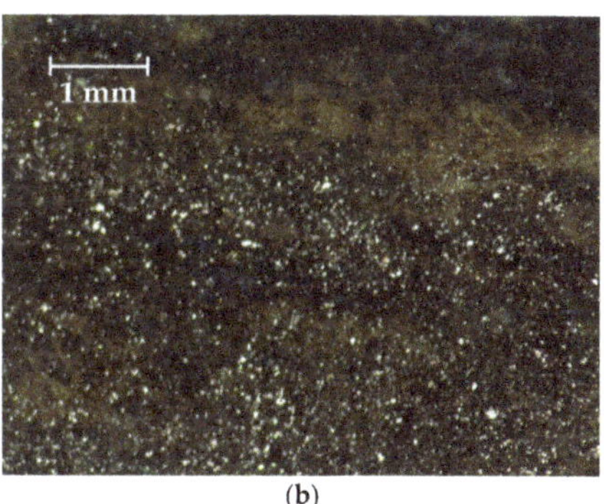

(a) (b)

Figure 13. (**a**) Micro-photography (40×) of the grey bands on f. 31v. (**b**) Micro-photography (40×) of the grey bands on f. 46r. Different dimensions of the silver particles are evident.

XRF confirmed that silver was employed in both interventions, but it did not show the presence of titanium on f. 46r, thus pointing to a different silvering technique. With regards to the blue pigments, the Raman investigation carried out on f. 46r indicated the presence of ultramarine blue, whereas the presence of phthalocyanine blue found on f. 31v was not detected; also, the FORS spectra pointed to the presence of azurite and ultramarine, i.e., the same pigments originally employed by Antoine de Lonhy. Moreover, under the microscope the sky of the miniature on f. 46r presents a micro-texture resembling that of other original parts painted with the same pigments. These results suggest that the restorer who painted the silvered band around the Cross on f. 46r spared the sky above it.

A peculiar case is the dark vaults above the main characters in the funerary scene at f. 71v (Figure 1j). Here, the morphology of the silver particles and the absence of titanium seems similar to those employed in retouches on f. 46r. However, the XRF analysis revealed the presence of gold, indicating the original presence of a golden architecture, possibly similar to that still visible in the miniature on f. 1r (Figure 1a) but covered at a later date with silver that now appears darkened.

The analysis of the sky also did not highlight any proof of retouched areas, suggesting that the sky here was spared by the restorer. The same holds for other miniatures: no clues of the use of modern materials emerged after the evaluation of micro-textural and compositional characteristics, revealing the sole presence of blue pigments that are fully comparable, from a compositional and morphological point of view, with the original ones.

Retouches on browns at ff. 24r and 31v (Figure 1c,e) are clearly visible to the naked eye. FORS spectra from the dark brown interior part of huts yielded evidence of iron oxide pigments, similar to those obtained from other brown untouched areas present in other folios. The XRF analysis of the repainted areas, however, showed lower iron signals with respect to the original iron oxide pigments; moreover, in all the retouched areas, signals of chromium and cobalt, very similar in intensity, were detected, whereas in the original brown pigments these elements were undetectable. These data strongly suggest that ff. 24r and 31v were modified by the same hand; unfortunately, a strong fluorescence band prevented the collection of micro-Raman signals which could have better elucidated the nature of the brown pigment.

4. Discussion

As highlighted by the results of the scientific investigation, the palette used by Antoine de Lonhy for this *Book of Hours* is rich and variegate (Table 2). The results also show how the different colourants were used in the decoration of the book, which may give further information and allow a comparison with the palette from the coeval manuscripts.

Table 2. The palette identified in the decoration of the *Book of Hours*.

Colour	Colourants
Black	carbon pigment
Blue	ultramarine blue azurite phthalocyanine blue [1]
Brown	iron oxides
	iron earths [1]
Green	brochantite
Grey	Silver [1]
Pink	brasilwood
Red	cinnabar
	red lead
Violet	brasilwood/azurite
White	lead white
Yellow	gold leaf shell gold mosaic gold lead-tin yellow type I yellow ochre

[1] possibly related to retouches.

4.1. *The Hierarchical Use of Colourants*

It emerges that—for blues, reds and yellows—different colourants were used for the same hue, and some kind of hierarchical scheme can be highlighted for the miniatures painted by Antoine de Lonhy. We are not talking here about the hierarchy of *colours*, an aspect on which there is a very wide coverage [32,33], but rather the hierarchy of the *colourants* inside the different colours. To investigate this topic, we can start from the commercial aspects discussed by S. Nash [34] in her study on the proofs of payment for artistic materials supplied to the court of the Dukes of Burgundy from the 14th to 15th centuries, which are now conserved in the Archives Départementales de la Côte-d'Or in Dijon (France).

4.1.1. The Use of Blues

In the Middle Ages the blue pigments were among the most expensive colourants. This is well known for ultramarine blue, due to the difficulty of supplying the stone lapis lazuli—from which ultramarine blue was prepared—from Afghanistan. Nevertheless, the best-quality azurite was also far more expensive than nearly all other colourants [34] (p. 130), including organic blue pigments such as indigo and woad. Therefore, it is interesting to understand how the artists managed the hierarchy of the blue pigments. There was a wide difference in price between ultramarine blue and azurite. Nash [34] (pp. 125–130) reported a cost of 128 FRF/lb for *fin azur d'Acre*, a term associated by the author to ultramarine blue, while *azur d'Alemaigne* (azurite from Germany) ranged between 1 and 20 FRF/lb. Delamare [35] (p. 132) estimated that in the Renaissance, the cost of ultramarine blue was up to 400–500 times that of azurite. The study by Kubersky-Piredda on the trade of colourants in Florence during Renaissance [36] reported a price ranging between 500 and 16.800 *soldi/libbra* for ultramarine blue and between 40 and 400 *soldi/libbra* for azurite.

However, in the miniatures of the *Book of Hours* it seems that Antoine de Lonhy used the two blue pigments in chromatic combination rather than as alternatives. In the mantle of the Virgin at f. 1r (Figure 1a), for example, he used azurite as the base and ultramarine blue for highlighting. The same holds true for the skies present in all the miniatures.

Therefore, we may suggest that the artist used the blue pigments according to their *tone* rather than their *price*.

4.1.2. The Use of Reds

Although the contextual presence of vermillion and red lead in a medieval manuscript is not surprising at all, the use of these two pigments within the *Book of Hours* deserves some discussion. The difference in price is well documented; Nash [34] (pp. 143–146) reports that vermillion was sold for about four times the price of red lead.

The hierarchy followed by the author for red pigments is somewhat different than that for the blues. Red lead was used throughout the miniatures, where it was the main red pigment employed, while vermillion was then applied over the red lead to darken the hue and to heighten the red mantles of some characters, in a way similar to the combination of azurite and ultramarine blue. However, a more hierarchical use is apparent in the depiction of the blood of Jesus trickling from his hands and chest in the *Crucifixion* (f. 46r, Figure 1h), for which Antoine de Lonhy reserved the more precious vermillion.

4.1.3. The Use of Yellows

The choice of yellow pigments is perhaps the most remarkable from the hierarchical point of view. Antoine de Lonhy used five different pigments for this colour: shell gold, gold leaf, mosaic gold, lead-tin yellow type I and yellow ochre. Examples are reported in Figure 14.

(a)

(b)

(c)

Figure 14. Examples of use of yellow pigments in the *Book of Hours*. □: mosaic gold; ■: lead-tin yellow type I; ○: gold leaf; •: shell gold; ◇: yellow ochre. (**a**) f. 15r, *Visitation*. (**b**) f. 46r, *Crucifixion*. (**c**) f. 49v, *Pentecost*.

Shell gold was of course the most expensive of the set. It was used by Antoine de Lonhy to paint various details in the miniatures, such as the bench that holds Christ and God the Father in the miniature depicting the *Coronation of the Virgin* on f. 42r (Figure 1g), as well as the garments of some characters: examples are Saint Elisabeth's tunic on f. 15r and the tunic of the apostle sitting on the right hand of the Virgin as well as the mantle of the apostle on the right part of the picture in the miniature depicting the *Pentecost* on f. 49r. Apparently, the painter chose the precious shell gold here in order to stress the relevance of the characters who deserve the most precious materials.

Lead-tin yellow type I or *giallolino* was considered quite valuable: its cost could be double that of orpiment [34] (pp. 151–152) and was much more than the various types of yellow ochres. Despite this, in the *Book of Hours* lead-tin yellow was used mostly in a mixture with a green copper pigment (see Section 3.4) to modify its tonality but rarely as pure yellow pigment, perhaps only for the coat of an apostle, probably St. Peter, praying in the left part of the miniature in the *Pentecost* on f. 49v. The fact that this pigment was reserved for this specific apostle would suggest an intentional use somehow linked to the identity of the character. In fact, St. Peter is an important apostle because it was to him that Christ entrusted the keys of heaven.

It is difficult to estimate the ranking of mosaic gold. Nash [34] does not report information for this synthetic pigment. A reference [37] can be found in a *Taxa*, i.e., a price list, issued in 1568 in the town of Liegnitz (Silesia, Poland) in which *Aurum musicum* (mosaic gold) is cited among the most expensive *Colores* or materials for painting, with a price 6 times that of *Aurum pigmentum* (orpiment), 12 times that of *Bley gelb* (lead-tin yellow or massicot) and 12 times that of *Ocker gelb* (yellow ochre). N. Turner, in her commentary study on Johannes Alcherius [38], suggested that mosaic gold, rather than being used by illuminators as an inexpensive substitute for gold leaf and shell gold, could have been used alongside them to expand the range of golden hues. In the *Book of Hours* considered here, mosaic gold was used to paint secondary subjects such as the rocks in the background on ff. 15r and 28v (Figure 1b,d), or the straw roof of the hut of the Nativity in the illuminations on ff. 24r and 31v (Figure 1c,e). Only one instance was found where shell gold had been used instead of the less precious mosaic gold, i.e., the rocks in the background of the illumination depicting the Virgin and the infant Jesus in the run to Egypt (f. 36v, Figure 1f), in order to highlight these details.

Last in ranking were the yellow ochres, which are by far the cheapest pigments in the lists reported by Nash [34] (pp. 155–157). Examples of their use are the inner part of the aedicule on f. 1r (Figure 1a) and the Cross and the ground of the Golgotha in the *Crucifixion* on f. 46r (Figure 1h).

4.2. Comparison with Other Works

There are few diagnostic works in which the hierarchy of colourants is discussed from the analytical point of view. Clark [9] cited the medieval practice of using the most valuable colourants, with particular reference to blue pigments, for the most important subjects inside a painting, but gave no specific information on the items analysed.

Bruni et al. [39] studied the decoration of a 15th century parchment folio kept at Archivio di Stato in Milan, originally commissioned by Francesco Sforza. The work is attributed to the Italian artist Michelino dei Molinari. The hierarchy of the colourants used by the artist was similar to that used by Antoine de Lonhy in the case of red pigments: the Child's vest, the tongue of a green dragon and the red initial capital letters in the text were painted with vermilion, whereas a red flower below the dragons was obtained by a mixture of red lead and vermilion. However, it was different in the case of blue pigments, since ultramarine blue and azurite were used separately for features with different symbolic value: the former was used for the mantle of the Virgin Mary and the Sforza's *Biscione* (the symbol of the Sforza dynasty), while the latter was used only for the initial capital letters in the text.

On the other hand, Bersani et al. [40], in a study on the ms. Pal. 212 (Biblioteca Palatina in Parma, Italy), a *Book of Hours* with 14th century miniatures produced in Bruxelles, found an unusual hierarchy of blue pigments, with azurite used in the central painting to realise the mantle of the Virgin Mary, one of the most important subjects of all illuminated books and lazurite used for the decoration of the frame, a feature undoubtedly less important. The authors attributed this unusual choice to the unavailability of lazurite for the artist who decorated the central painting. In the end, therefore, the choices of the artists follow their decorative schemes, but they are related to the availability of materials.

Of course, the hierarchical use of colourants can be found in other types of paintings. In the study by Edwards et al. [41] on the wall paintings of the Church of SS Cosmo and Damian in Basconcillos del Tozo (Castille y Léon, Spain), the authors highlighted the selective use of the precious cinnabar for the most important biblical figures.

4.3. Comparison with Coeval Manuscripts

It is interesting to note that a relatively similar palette was characterised by Melo et al. [25] in their analysis of three French 15th century *Book of Hours* presently in the Palacio Nacional de Mafra (Portugal). This information may support art historians in interpreting the context of production for these manuscripts and to (possibly) link other artworks to the production of Antoine de Lonhy.

5. Conclusions

The present work represents the first step towards the disclosure of materials and techniques employed by Antoine de Lonhy in the course of his artistic activity and it integrates the studies on style and iconography already carried out about the work of this versatile medieval artist. Further to this, the instrumental inspection of the untouched parts of the miniatures allowed us to recognise the colourants originally used by the master, and to highlight the criteria that he used for selecting a specific pigment. This choice appears not merely bound to technical or aesthetic reasons, since specific pigments were selected for specific parts in the miniatures. In particular, a hierarchical scheme is identifiable for yellows and partially for reds, while in the case of blues the colourant choice seems to be led by difference instances. Of course, the hierarchy of the colourants within this work offers only a partial view of what the *modus operandi* of the artist was.

The shine of gold is that of mosaic gold, while shell gold was used to paint objects and details that are somehow connected to characters whose sacredness required the use of the finest materials. As a whole, four different yellow colourants were hierarchically employed within the miniatures.

Vermilion was used to paint the stray of acanthus, strawberries and flowers in decorations on page borders (possibly by an associate of de Lonhy), while in the miniatures its use was reserved to the blood of Jesus Christ, in order to emphasise the sanctity of this detail, while less valuable red lead was used to obtain reds throughout the miniatures.

Another peculiarity concerns the use of lead-tin yellow; this pigment was largely used in all the miniatures to tune green colours. Only in one case was it used by itself, i.e., to paint the vest of the old apostle, probably St. Peter, praying in the left part of the miniature on f. 49v. It is possible that this specific use is linked to the identity of this important character.

The overall analytical results indicated that the palette used to paint the decorations bordering each page differs substantially from that used in the miniatures.

It is known that the painter who executed the miniatures generally did not work on other decorations on the page, but this investigation has also pointed out that different pigments could be used in the different stages of the decoration of the manuscript.

All in all, the palette chosen by the artist for the decoration of the ten miniatures, including the precious colourants such as gold, lapis lazuli and cinnabar, indicates the importance and credit of this work.

The investigation also gave information on the materials used for retouches. In particular, phthalocyanine blue, in use since 1935, and titanium white, also introduced during the 20th century, were found in the illumination on f. 31v, thus setting a *post quem* terminus for some of the coarse interventions clearly visible in the manuscript. Moreover, the presence of different materials would exclude, as suggested by autoptic inspection of the book, that the retouches could have been performed by the same hand.

Author Contributions: Conceptualisation, M.A. and M.G.; methodology, A.A.; software, E.P.; validation, A.A. and M.G.; formal analysis, A.A.; investigation, M.A., E.P. and A.A.; resources, S.C.; data curation, M.G.; writing—original draft preparation, M.A. and M.G.; writing—review and editing, S.C. and G.S.; visualisation, A.A.; supervision, M.G.; project administration, M.A. All authors have read and agreed to the published version of the manuscript.

Funding: This research received no external funding.

Acknowledgments: The authors would like to thank Enrica Pagella of the Museo Civico di Arte Antica in Torino, for allowing thorough inspection of the manuscript.

Conflicts of Interest: The authors declare no conflict of interest.

References

1. Brunello, F. *De Arte Illuminandi*; Neri Pozza: Vicenza, Italy, 1975.
2. Caffaro, A. *De Clarea, Manuale Medievale Di Tecnica Della Miniatura (Secolo XI)*; Arci Postiglione: Salerno, Italy, 2004.
3. Caffaro, A. *Scrivere in Oro. Ricettari Medievali d'arte e Artigianato (Secoli IX-XI). Codici Di Lucca e Ivrea*; Liguori: Napoli, Italy, 2003.
4. Frezzato, F. *Cennino Cennini: Il Libro Dell'arte*; Neri Pozza: Vicenza, Italy, 2009.
5. Clarke, M. *The Art of All Colours: Mediaeval Recipe Books for Painters and Illuminators*; Archetype Publications: London, UK, 2000.
6. Aceto, M.; Agostino, A.; Boccaleri, E.; Crivello, F.; Cerutti Garlanda, A. Identification of Copper Carboxylates as Degradation Residues on an Ancientmanuscript. *J. Raman Spectrosc.* **2010**, *41*, 1434–1440. [CrossRef]
7. Clark, R.J.H.; Gibbs, P.J. Raman Microscopy of a 13th Century Illuminated Text. *Anal. Chem.* **1998**, *70*, 99A–104A. [CrossRef]
8. Burgio, L.; Clark, R.J.H.; Hark, R.R. Spectroscopic Investigation of Modern Pigments on Purportedly Medieval Miniatures by the 'Spanish Forger'. *J. Raman Spectrosc.* **2009**, *40*, 2031–2036. [CrossRef]
9. Clark, R.J.H. Raman Microscopy: Application to the Identification of Pigments on Medieval Manuscripts. *Chem. Soc. Rev.* **1995**, *24*, 187. [CrossRef]
10. Chaplin, T.D.; Clark, R.J.H.; Jacobs, D.; Jensen, K.; Smith, G.D. The Gutenberg Bibles: Analysis of the Illuminations and Inks Using Raman Spectroscopy. *Anal. Chem.* **2005**, *77*, 3611–3622. [CrossRef] [PubMed]
11. Van Hooydonk, G.; De Reu, M.; Moens, L.; Van Aelst, J.; Milis, L. A TXRF and Micro-Raman Spectrometric Reconstruction of Palettes for Distinguishing Between Scriptoria of Related Medieval Manuscripts. *Eur. J. Inorg. Chem.* **1998**, *1998*, 639–644. [CrossRef]
12. Vandenabeele, P.; Wehling, B.; Moens, L.; Dekeyzer, B.; Cardon, B.; Von Bohlen, A.; Klockenkämper, R. Pigment Investigation of a Late-Medieval Manuscript with Total Reflection X-Ray Fluorescence and Micro-Raman Spectroscopy. *Analyst* **1999**, *124*, 169–172. [CrossRef]
13. Avril, F. Le Maître Des Heures de Saluces: Antoine de Lonhy. *Revue de l'Art* **1989**, *85*, 9–34. [CrossRef]
14. Elsig, F. *Antoine de Lonhy*; Silvana Editore: Cinisello Balsamo, Italy, 2018; ISBN 9788836638697.
15. AA.VV. *Il Rinascimento Europeo Di Antoine de Lonhy*; Baiocco, S., Castronovo, S., Eds.; SAGEP: Genova, Italy, 2021.
16. Saroni, G. Intorno a Un Libro d'Ore Di Antoine de Lonhy Giovane. *Palazzo Madama Stud. Not. Riv. Annu. Mus. Civ. d'Arte Antica di Torino* **2010**, *0*, 7–23.
17. Aceto, M.; Agostino, A.; Fenoglio, G.; Gulmini, M.; Bianco, V.; Pellizzi, E. Non Invasive Analysis of Miniature Paintings: Proposal for an Analytical Protocol. *Spectrochim. Acta Part A Mol. Biomol. Spectrosc.* **2012**, *91*, 352–359. [CrossRef]
18. Fiber Optics Reflectance Spectra (FORS) of Pictorial Materials in the 270–1700 Nm Range. Available online: http://fors.ifac.cnr.it (accessed on 3 July 2021).
19. Bacci, M. UV-VIS-NIR, FT-IR and FORS spectroscopy. In *Modern Analytical Methods in Art and Archaeology*; Ciliberto, E., Spoto, G., Eds.; John Wiley & Sons Ltd.: New York, NY, USA, 2000; pp. 321–361.
20. Smieska, L.M.; Mullett, R.; Ferri, L.; Woll, A.R. Trace Elements in Natural Azurite Pigments Found in Illuminated Manuscript Leaves Investigated by Synchrotron X-Ray Fluorescence and Diffraction Mapping. *Appl. Phys. A* **2017**, *123*, 484. [CrossRef]
21. Seccaroni, C.; Moioli, P. *Fluorescenza X, Prontuario per l'analisi XRF Portatile Applicata a Superfici Policrome*; Nardini: Firenze, Italy, 2002.
22. Aru, M.; Burgio, L.; Rumsey, M.S. Mineral Impurities in Azurite Pigments: Artistic or Natural Selection? *J. Raman Spectrosc.* **2014**, *45*, 1013–1018. [CrossRef]
23. Angelici, D.; Borghi, A.; Chiarelli, F.; Cossio, R.; Gariani, G.; Lo Giudice, A.; Re, A.; Pratesi, G.; Vaggelli, G. M -XRF Analysis of Trace Elements in Lapis Lazuli-Forming Minerals for a Provenance Study. *Microsc. Microanal.* **2015**, *21*, 526–533. [CrossRef]
24. Aceto, M.; Agostino, A.; Fenoglio, G.; Idone, A.; Gulmini, M.; Picollo, M.; Ricciardi, P.; Delaney, J.K. Characterisation of Colourants on Illuminated Manuscripts by Portable Fibre Optic UV-Visible-NIR Reflectance Spectrophotometry. *Anal. Methods* **2014**, *6*, 1488. [CrossRef]
25. Melo, M.J.; Otero, V.; Vitorino, T.; Araújo, R.; Muralha, V.S.F.; Lemos, A.; Picollo, M. A Spectroscopic Study of Brazilwood Paints in Medieval Books of Hours. *Appl. Spectrosc.* **2014**, *68*, 434–444. [CrossRef]
26. Trentelman, K.; Turner, N. Investigation of the Painting Materials and Techniques of the Late-15th Century Manuscript Illuminator Jean Bourdichon. *J. Raman Spectrosc.* **2009**, *40*, 577–584. [CrossRef]
27. Roger, P.; Villela-Petit, I.; Vandroy, S. The Brazil Colorant Lakes in Medieval Illuminations: Reconstitution Starting with Ancient Recipes. *Stud. Conserv.* **2003**, *48*, 155–170. [CrossRef]
28. Bell, I.M.; Clark, R.J.H.; Gibbs, P.J. Raman Spectroscopic Library of Natural and Synthetic Pigments (Pre-$\sim$ 1850 AD). *Spectrochim. Acta Part A Mol. Biomol. Spectrosc.* **1997**, *53*, 2159–2179. [CrossRef]
29. Edwards, H.G.M.; Farwell, D.W.; Newton, E.M.; Rull Perez, F.; Jorge Villar, S. Raman Spectroscopic Studies of a 13th Century Polychrome Statue: Identification of a 'Forgotten' Pigment. *J. Raman Spectrosc.* **2000**, *31*, 407–413. [CrossRef]
30. Whitley, K.P. *The Gilded Page: The History & Technique of Manuscript Gilding*; Oak Knoll Press: New Castle, DE, USA, 2000.
31. Aceto, M.; Calà, E. Analytical Evidences of the Use of Iron-Gall Ink as a Pigment on Miniature Paintings. *Spectrochim. Acta Part A Mol. Biomol. Spectrosc.* **2017**, *187*, 1–8. [CrossRef]
32. Gage, J. *Colour and Meaning: Art, Science and Symbolism*; Thames and Hudson: London, UK, 2000.
33. Pulliam, H. Color. *Stud. Iconogr.* **2012**, *33*, 3–14.

34. Nash, S. 'Pour couleurs et autres choses prise de lui . . . ': The Supply, Acquisition, Cost and Employment of Painters' Materials at the Burgundian Court, c. 1375–1419. In *Trade in Artists' Materials: Markets and Commerce in Europe to 1700*; Kirby, J., Nash, S., Cannon, J., Eds.; Archetype Publications Ltd.: London, UK, 2010; pp. 97–182.
35. Delamare, F. *Blue Pigments: 5000 Years of Art and Industry*; Archetype Publications: London, UK, 2013.
36. Kubersky-Piredda, S. The Market for Painters' Materials in Renaissance Florence. In *Trade in Artists' Materials: Markets and Commerce in Europe to 1700*; Kirby, J., Nash, S., Cannon, J., Eds.; Archetype Publications: London, UK, 2010; pp. 223–243.
37. Burmester, A.; Haller, U.; Krekel, C. Pigmenta et Colores: The Artist's Palette in Pharmacy Price Lists from Liegnitz (Silesia). In *Trade in Artists' Materials: Markets and Commerce in Europe to 1700*; Kirby, J., Nash, S., Cannon, J., Eds.; Archetype Publications Ltd.: London, UK, 2010; pp. 314–324.
38. Turner, N. The Recipe Collection of Johannes Alcherius and the Painting Materials Used in Manuscript Illumination in France and Northern Italy, c. 1380–1420. *Stud. Conserv.* **1998**, *43*, 45–50. [CrossRef]
39. Bruni, S.; Cariati, F.; Casadio, F.; Toniolo, L. Identification of Pigments on a XV Century Illuminated Parchment by Raman and FTIR Microspectroscopies. *Spectrochim. Acta Part A Mol. Biomol. Spectrosc.* **1999**, *55*, 1371–1377. [CrossRef]
40. Bersani, D.; Lottici, P.P.; Vignali, F.; Zanichelli, G. A Study of Medieval Illuminated Manuscripts by Means of Portable Raman Equipments. *J. Raman Spectrosc.* **2006**, *37*, 1012–1018. [CrossRef]
41. Edwards, H.G.M.; Farwell, D.W.; Rull Perez, F.; Jorge Villar, S. Spanish Mediaeval Frescoes at Basconcillos Del Tozo: A Fourier Transform Raman Spectroscopic Study. *J. Raman Spectrosc.* **1999**, *30*, 307–311. [CrossRef]

 heritage

MDPI

Article

Multi-Modal, Non-Invasive Investigation of Modern Colorants on Three Early Modern Prints by Maria Sibylla Merian

Olivia Dill, Marc Vermeulen, Alicia McGeachy and Marc Walton *

The Center for Scientific Studies in the Arts (NU-ACCESS), Northwestern University, Evanston, IL 60208, USA; oliviadill2024@u.northwestern.edu (O.D.); marc.vermeulen@northwestern.edu (M.V.); alicia-mcgeachy@northwestern.edu (A.M.)
* Correspondence: marc.walton@northwestern.edu

Abstract: Northwestern University's Charles Deering McCormick Library of Special Collections owns three hand-colored copperplate engravings that once belonged to an edition of *Matamorphosis Insectorum Surinamensium* by artist-naturalist Maria Sibylla Merian (1647–1717). Because early modern prints are often colored by early modern readers, or modern collectors, it was initially unclear whether the coloring on these prints should be attributed to the print maker, to subsequent owners or collectors, or to an art dealer. Such ambiguities posed challenges for the interpretation of these prints by art historians. Therefore, the prints underwent multi-modal, non-invasive technical analysis to assess the date and material composition of the prints' coloring. The work combined several different non-invasive analytical techniques: hyperspectral imaging (HSI), macro X-ray fluorescence (MA-XRF) mapping, surface normal mapping with photometric stereo, visible light photography, and visual comparative art historical analysis. As a result, the prints and paper were attributed to a late eighteenth-century posthumous edition of Merian's work while the colorants were dated to the early twentieth century. This information enables more thorough contextualization of these prints in their use as teaching and research tools in the University collection.

Keywords: Maria Sibylla Merian; colored prints; hyperspectral imaging; X-ray fluorescence spectroscopy; photometric stereo; Prussian blue; non-invasive pigment characterization

Citation: Dill, O.; Vermeulen, M.; McGeachy, A.; Walton, M. Multi-Modal, Non-Invasive Investigation of Modern Colorants on Three Early Modern Prints by Maria Sibylla Merian. *Heritage* **2021**, *4*, 1590–1604. https://doi.org/10.3390/heritage4030088

Academic Editor: Francesco Soldovieri

Received: 26 June 2021
Accepted: 30 July 2021
Published: 4 August 2021

1. Introduction

Metamorphosis Insectorum Surinamensium, first published by Maria Sibylla Merian (1647–1717) in Amsterdam in 1705, is among the most famous examples of early modern insect imagery. The volume consists of 60 copperplate engravings, designed by Merian, engraved by professional engravers, Pieter Sluyter, Joseph Mulder, and Daniel Stopendael, and printed in Merian's Amsterdam workshop. The original volume was published in Dutch and Latin editions and was sold either containing unadorned engravings or, for an added fee, hand-colored by Merian or her daughters, Dorothea Maria and Johanna Helena [1]. The product of two years of research conducted on site in Suriname, the images in *Metamorphosis* constitute a landmark in the history of scientific imagery. They contain detailed portrayals of the different stages of metamorphosis of insects native to Suriname and they set these depictions of transformation in the insects' botanical habitats. Although *Metamorphosis* contains inaccuracies, embellishments, and mistranslations, the images in the volume as a whole are unique for the depth of information that they provide about the flora and fauna of Suriname, for their vivid yet concise compositions, and, notably for this study, for the fact that they number among few early modern natural history prints to survive with original coloring.

Whether color could be used as reliable and reproducible descriptive feature of natural history specimens was hotly contested amongst early modern botanists, naturalists, and artists [2]. Reproducing the color of natural phenomena, especially in the context of expeditions like Merian's, challenged naturalists to develop means to systematize their terms and observations [3] and demanded expertise in art, craft and science, making the valence and materiality of early modern color a rich and active area of research for historians of art and science [4]. Merian herself chose to include color as a significant feature of insects' descriptions. In the texts accompanying her images she was careful to include lively textual descriptions of insects' color as identifying features [5]. In the images themselves she relied on a system of modelli, painted in watercolor on paper and vellum, in order to standardize the colors used to represent insects in colored editions of the prints [1]. *Metamorphosis* constitutes an important case study in the history of early modern colored prints.

The study of color in early modern European woodcuts and engravings poses challenges to print scholars, notably that it is often difficult to assess when and by whom coloring is applied [6]. It was common practice among early modern European print owners or book readers to interact with printed works [7] and to add their own annotations or coloring using ink, watercolor, or gouache [8]. It was also common for dealers and collectors especially in the nineteenth century to add color to prints either as enthusiasts or as forgers [9]. In the case of *Metamorphosis*, these challenges associated with authenticating colored prints more generally are compounded by the fact that Merian's original copper plates survived after her death. They were embellished and reprinted in several posthumous editions in the The Netherlands and France in 1726, 1730 and 1771, thus often making it challenging to assess the proximity of loose-leaf prints to Merian's workshop [10].

Analytical methodologies have been employed in the study of color in early modern works on paper; in particular, X-ray fluorescence (XRF) and multispectral photography have been shown to be effective methods of characterizing colorants used to color prints [9,11,12] and illuminated manuscripts [13]. Some have used art historical and technical analyses (XRF and Raman spectroscopy) to study the relationship between printing and color workshops and practices in the sixteenth and seventeenth centuries [6]. These studies concern themselves with works whose coloring can be verified to be early modern. Some have applied XRF to identify modern coloring on early modern prints [8] but there remains further work to be done in specifying and characterizing the practice and materials of modern coloring of earlier prints. Merian's work in particular has been the subject of some technical analyses, which used microscopic investigation to distinguish between prints and counterproofs but not in order to classify or propose dates for the pigments used [14].

Three painted prints from *Metamorphosis* (Call no. Folio 581.9 S357.m; Figure 1) reside in Northwestern University's Charles Deering McCormick Library of Special Collections. These prints are referenced in Merian's text as plate 9, plate 54 and plate 55 and in the following we refer to individual prints by these in-text plate numbers. The coloring on these prints is carefully done and thickly applied. With the exception of two peppers in plate 55, the color scheme of the Northwestern Special Collection prints follows that of counterproof and watercolor versions of the same images that are securely attributed to the Merian workshop, such as those in the British Museum [15–17]. The care of coloring, as well as the aforementioned factors surrounding the history of colored prints by Merian and others, made it initially unclear when, why, and by whom the Northwestern University Special Collections prints and their coloring were produced. Still, given the significance of *Metamoprhosis* in the histories of botany and entomology and with respect to the eighteenth century use of color in illustrated natural history, these prints represent an invaluable archival resource.

Figure 1. Prints from Maria Sibylla Merian's *Metamorphosis* in Northwestern University Chalres Deering McCormick Special Collections, call no: Folio 581.9 S357.m: plate 55 (**left**), plate 9 (**middle**), plate 54 (**right**).

In order to assess how, why, and by whom the prints were made, and in particular whether they were made by the same maker, the prints in Northwestern Special Collections underwent technical analysis with macro X-ray fluorescence (MA-XRF), hyperspectral imaging, transmitted light photography, and photometric stereo imaging. These data were analyzed in dialogue with archival art historical research to characterize and date three separate components of the prints: pigments, printed lines, and paper support. MA-XRF and, in two cases, hyperspectral data were used to investigate the prints' pigments, photometric stereo imaging was used to investigate how the printed marks were made, and transmitted light photography was used to visualize paper structure. Exhaustive characterization of pigments, including modern organic pigments, may require the use of molecular techniques such as Raman spectroscopy [18–21] or FTIR [22]. While the use of these two techniques would have been ideal in this study, no sampling was possible on these prints, and we did not have access to the necessary portable instrumentation for in situ analyses. Therefore, the investigation of the coloring materials could only be performed using MA-XRF and hyperspectral imaging, the combination of which may give good insights into inorganic colorants but limited insights into organic ones. The analysis of the prints presented here can better contextualize these specific prints, and shed light on whether and how the coloring on these prints may contribute to the study of the eighteenth century use of color in natural history. More broadly, this analysis provides valuable information for characterizing modern coloring of early modern prints and demonstrates the use of a multimodal workflow that combines XRF with other modes of analysis.

2. Materials and Methods

2.1. X-ray Fluorescence Spectroscopy

The prints were fully scanned using the NU-ACCESS MA-XRF system described in [23], for which the X-ray tube and polycapillary optic were replaced by a 50 W transmission Rh anode X-ray tube (Varex Imaging, Salt Lake City, UT, USA) and 500 μm collimator, respectively. MA-XRF scans were collected with a dwell time of 0.2 s, at 40 kV voltage and 1.250 A current. An XGLabs Elio X-ray fluorescence spectrometer operating at 40 kV and 600 μA was used to collect spot samples for comparison and confirmation. All XRF data processing to create elemental distribution maps was done in Python with the PyMCA framework [24].

2.2. Hyperspectral Imaging (HSI)

Hyperspectral images of the prints were acquired using a Resonon Pika II hyperspectral camera measuring between 400 and 900 nm with 2 nm spectral resolution. Only data between 410 and 900 nm were used for identification due to noise affecting lower wavelength channels. Hyperspectral data analysis was performed using the method and code provided in [25]. The method uses the Uniform Manifold Approximation and Projection (UMAP) algorithm to reduce the dimension of spectral data and then clusters hyperspectral data to produce characteristic spectra, called endmembers, for each color in each print.

2.3. Photometric Stereo Imaging

All three prints were imaged with a photometric stereo dome that is constructed of 81 LEDs, mounted at regular intervals on the concave interior of a hemispherical dome [26]. The prints were placed under the dome and illuminated with each LED in sequence while one photograph was captured for each incident lighting angle. Captures were made with a Canon EOS 50 Mark III DSLR fitted with a 20 mm prime lens. Data were processed using the near lighting model described in [26] and using open source software provided therein to produce three-dimensional false color maps indicating the orientation of the prints' surface (i.e., normal maps).

2.4. Transmitted Light Imaging

Watermarks are thinner regions of antique laid paper that when viewed in transmitted light appear lighter than the surrounding fibers. Early modern paper manufacturers placed watermarks into their papers by weaving of patterns into the wires of the screen-like molds used to shape paper sheets. Here, transmitted light images of the prints were captured in order to better visualize any watermarks within the sheets. After slipping a light sheet between each print and its mat as described in [27], one photograph was taken in ambient light and another with the print illuminated from below with the light sheet. A weighted subtraction of the two images was performed using the algorithm outlined in [28] in order to increase the visibility of watermark data within the sheets. Captures were made with a Canon EOS 50 Mark III DSLR, fitted with a Canon Zoom Lens EF 24–105 mm and with a Daylight Wafer 2 LED light table (model number D/E/U/A 35030).

3. Results and Discussion

3.1. Pigments

We propose the use of certain pigments on the basis of qualitative observation of MA-XRF maps, hyperspectral distribution maps and associated characteristic reflectance curves. In the case of Prussian blue and cadmium red the extraction of representative spectra from hypserspectral maps supported the identification of pigments. The earliest known manufacturing dates for potential pigments (Table 1) were used to propose *terminus post quem* for the prints. XRF data can be used to calculate elemental concentrations and therefore quantitatively identify the presence of certain elements; however, in our case approximate identifications were sufficient to address research questions concerning the prints' provenance. Summaries of all proposed identifications are given in Appendix A and data supporting this study are available in the associated supplementary information file.

Table 1. Earliest known date of commercial availability for pigments used to date the prints in NU Special Collections.

Pigment	Earliest Known Date
Prussian blue	1724 [29]
Cadmium red	1919 [30]
Cadmium barium red	1926 [31]
Cadmium yellow	1818 [30]
Lithopone	1878 [32]
Titanium white	1920 [33]

3.1.1. Brown

High pixel intensities in iron and manganese XRF maps in brown regions suggest the use of brown umber in both prints [34–36]. The co-ocurrence of iron and manganese is observed most clearly in the pomegranate husk in plate 9 and chrysalis on plate 55 (Figure 2). Umber was available to artists long before the eighteenth century and its use does not rule out an early modern date for the coloring [35]. However, high pixel intensities in XRF maps for cadmium and sulphur in brown and dark yellow regions of plate 9 suggest the presence of cadmium yellow, either mixed with brown umber or applied in a separate layer. Cadmium yellow's earliest use as an artist's pigment is 1818 [30].

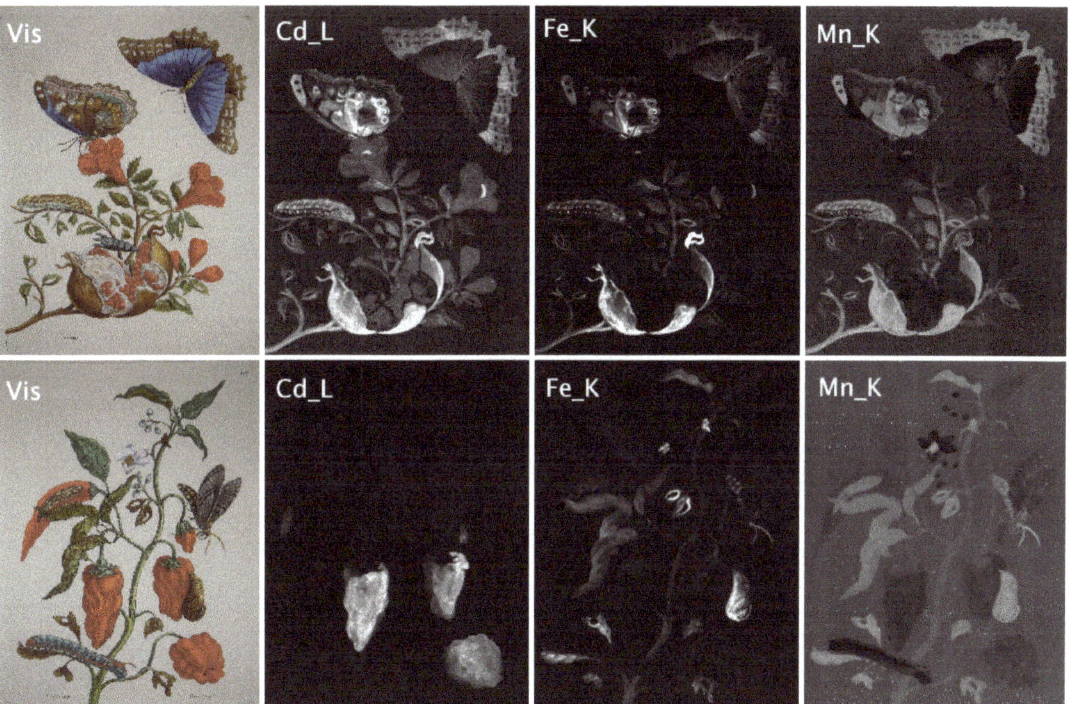

Figure 2. MA-XRF maps for cadmium (**left**), iron (**center**), and manganese (**right**) in plate 9 (**top**) and plate 55 (**bottom**).

3.1.2. White

High pixel intensities in the titanium MA-XRF map suggest the use of titanium white in white regions of plate 55, such as the white flower (Figure 3) [33,37]. Titanium white was not commonly available as an artist's pigment before around 1920, thus moving the *terminus post quem* for the coloring on plate 55 to the first quarter of the twentieth century [33]. Characteristic features in reflectance data for titanium white fall between 393 and 406 nm, which is outside the range of the hyperspectral camera used for this study [38,39]. Therefore in our case HSI is not useful in confirming this identification.

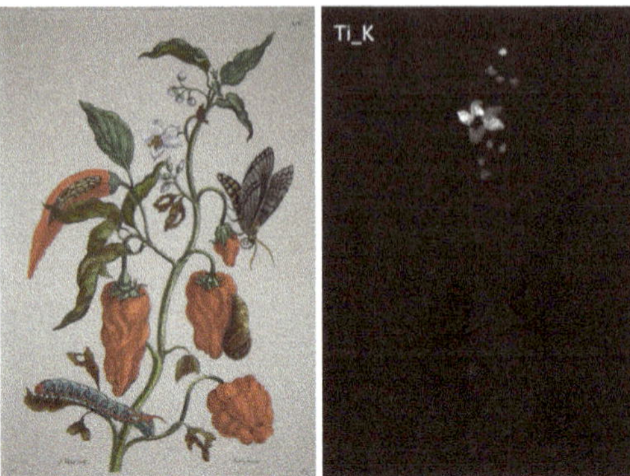

Figure 3. Visible light photograph and (**left**) and MA-XRF map for titanium in Plate 55 (**right**).

High pixel brightness in XRF maps for zinc, barium and sulphur occurred in several regions of all three prints (Figure 4). This combination is consistent with the use of lithopone, which consists of zinc sulfide and barium sulphate [32,40]. Lithopone is a white pigment, but is observed here in regions that visually appear green and blue, suggesting its use as a filler. Lithopone was first patented and commercially produced in 1878; thus, its presence on these prints further confirms that the coloring on the prints is modern [32]. The use of two whites may suggest the use of two campaigns of coloring, or the choice to differentiate areas of the print.

Figure 4. False color images indicating brightness in the XRF maps for Ba_L, S_K, and Zn_k in plates 9 (**left**) 54 (**center**) and plate 55 (**right**).

3.1.3. Red

High pixel brightness in cadmium, sulphur, and selenium XRF maps in several red regions of plates 54 and 55 (Figure 5) suggests the use of cadmium red to color red regions in both prints [30,41]. Characteristic spectra in hyperspectral data Figure 6) are consistent with those for cadmium red [30,41] namely the inflection point around 590 nm and absorption maximum around 500 nm. Cadmium red became commercially available in 1919, confirming the early twentieth century *terminus post quem* for the coloring [30].

Figure 5. False color images indicating brightness in the XRF maps for Cd_L, S_K, and Se_K, in plates 54 (**left**) and plate 55 (**right**).

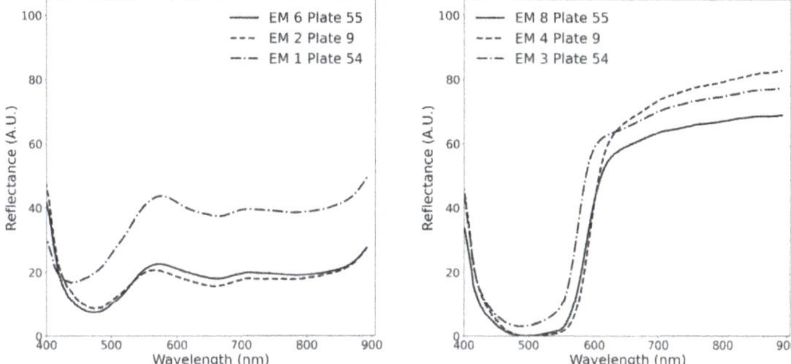

Figure 6. Characteristic spectra for unidentified green (**left**) and cadmium and barium-rich reds (**right**).

Additionally, we observe that in the upper left-most pepper of plate 55 and in the uppermost spike of the flower on plate 54, circled in red in Figure 7, pixel intensities in cadmium XRF maps are a factor of 4 lower than the pixel intensities in other red regions of the cadmium map of the same print. The pixel brightness for selenium is also low in these areas and barium remains constant throughout. Because visibly the red remains consistent, it is possible that in these regions a second red, possibly an organic red with a barium-rich substrate, such as barium oxide, was applied in addition to cadmium red [31].

Pixel brightness is low in all regions of the Selenium XRF map for Plate 9 (Figure 8), indicating that cadmium red was not used in Plate 9. Cadmium and barium are found together in red regions of plate 9 (Figure 7), suggesting the use of cadmium barium red [30,42] in plate 9 or possibly, as in plates 54 and 55, a barium-rich organic red. The presence of multiple reds could indicate multiple distinct campaigns of coloring or the choice to distinguish or retouch parts of the prints. The shared use of cadmium red between plates 54 and 55 suggests that both were produced using a shared palette.

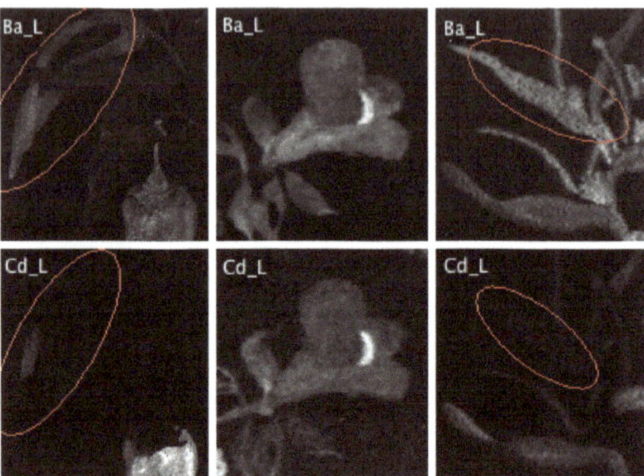

Figure 7. Details from plate 55 (**left**), plate 9 (**center**), and plate 54 (**right**) showing XRF maps for barium and cadium.

Figure 8. Visible light photograph and (**left**) and MA-XRF map for selenium in plate 9 (**right**).

3.1.4. Blue

The analysis of hyperspectral data suggests the use of Prussian blue in plate 9. The characteristic spectral curve in blue regions (Figure 9), in particular the local minimum around 675 nm and the inflection point around 554 nm, are consistent with characteristic spectra for Prussian blue published in other identifications of Prussian blue [43–45]. Prussian blue did not become commercially available until 1724, seven years after Merian's death, confirming that the coloring could not have been applied during Merian's lifetime [29].

Although Prussian blue is rich in iron, an element that can be detected with XRF, it has been observed that in practice Prussian blue requires relatively little pigment to produce strong coloration and thus it can be challenging to detect this pigment with XRF [25,46]. In our case we observe high pixel intensities in MA-XRF maps for iron in regions corresponding to darker shades of blue in the butterfly at the upper right of plate 9 (Figure 9). All blue regions of the butterfly's blue wings in hyperspectral images of plate 9 demonstrate the characteristic reflectance spectrum of Prussian blue, which together with the iron map allows us to confirm the identification of Prussian blue.

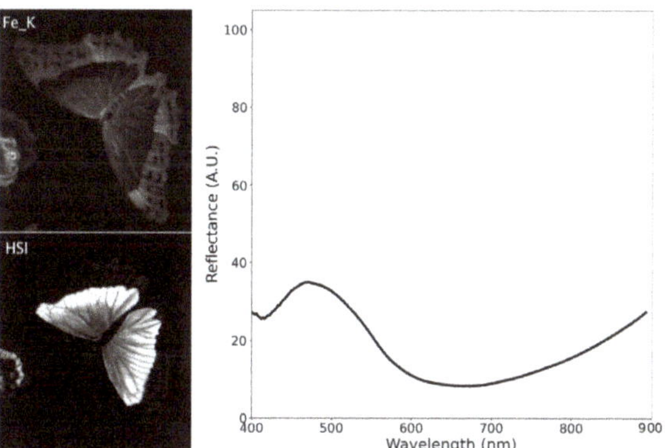

Figure 9. Details from MA-XRF map for iron (**top left**), blue endmember map (**bottom left**) and characteristic spectra (**right**) used to identify Prussian blue in plate 9.

3.1.5. Green and Yellow

We were unable to propose the presence of specific colorants in the case of a number of likely-organic green and yellow pigments. The hyperspectral data treatment was able to extract characteristic spectra for green regions, but these curves did not correspond to the curves for inorganic pigments [47]. XRF data in these regions did not register elements typically associated with green pigments (such as copper) or inorganic yellows (such as iron). This lack of association along with the identification of barium, sulphur, zinc, and, in some green regions, iron with varying pixel brightness throughout suggests the use of unidentified organic pigments or mixtures. Sampling and invasive analysis would be necessary to further specify these pigments.

3.1.6. Comparisons between Prints

The recurrence of several pigments (lithopone, barium-rich and cadmium reds, and brown umber) identified on the basis of XRF data and, in the case of cadmium red and Prussian blue, hyperspectral imaging suggest that there is a common palette between the three prints. This is supported by the close correspondence in hyperspectral data and characteristic spectra between prints. Characteristic spectra for red regions of plate 54 and plate 55, which conform to published spectra for cadmium red (Figure 6), support the identification of the same pigment in both prints [30,32]. Although we are unable to precisely identify which pigment they correspond to, characteristic spectra for the green regions in each print are nearly identical between the three plates (Figure 6), suggesting the repeated use of one green pigment between all three. The use of the same pigment between prints suggests that they were colored together by the same hand or with a systematic application of color between the three.

3.2. Printed Lines

Normal maps and comparison with other printed editions of *Metamorphosis* were used to determine whether lines that appeared to be printed were in fact the product of an eighteenth century intaglio printing process such as copperplate engraving, or if they were outright forgeries or reproductions.

3.2.1. Photometric Stereo Normal Maps

In normal maps for all three prints produced through photometric stereo data, a large-scale deformation of the sheet within the region of the plate mark can be observed (Figure 10). Because this deformation only occurs within the plate mark indicating the boundary of the copperplate we propose that these deformations resulted from the application of high pressure to a wet sheet during printing. Smaller-scale deformations in the surface were more difficult to register due to the thickness with which the paint is applied to the surface, but in plate 54 the printed lines are raised from the surface of the sheet (Figure 10), thus supporting the proposition that these were printed via an intaglio process.

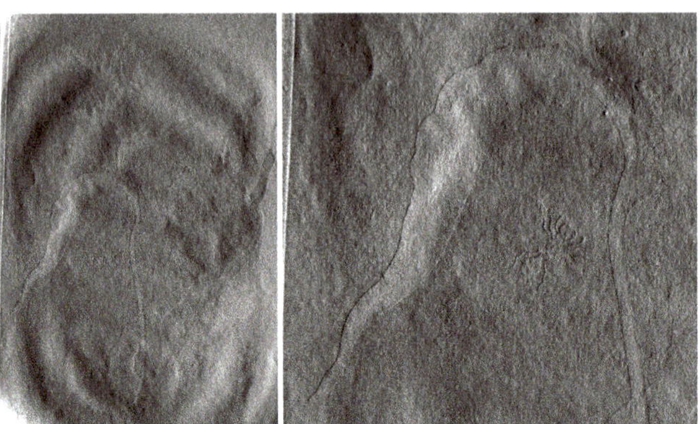

Figure 10. Magnitude of the normal direction along the horizontal axis for plate 54 (**left**) and detail (**right**).

3.2.2. Comparison with Other Editions

Prints in Northwestern University were compared to other editions of Merian's *Metamorphosis*: an edition published in Amsterdam by Merian in 1705 in the University of Chicago Special Collections (call no.: ff QL466.M57 1705), a 1726 edition published in Amsterdam by P. Gosse in Northwestern Medical School's Galter Health Sciences Library (call no.: XO 595.7 M54d 1726) and a 1771 edition translated into French published in Paris by LC Desnos (call no.: XO 595.7 M54h3 1771), also in Galter Health Sciences Library. All three editions are uncolored.

After visual inspection of plates from different editions it was observed that Northwestern Special Collections' prints have a Roman numeral plate number in the upper right of each print and a French name for the plant species at the bottom of each print. These two features are unique to the 1771 edition, *Histoire Generalé des Insectes de Surinam...*, which is the last known edition of Merian's work, published in Paris in 1771 by L.C. Desnos (Figure 11) [10]. The presence of printed features that are only present in the 1771 edition confirms that the NU Special Collections prints may be attributed to a 1771 edition.

Figure 11. Details from plate 54 from Galter Library's 1726 edition (**left**), NU Special Collections (**center**) and Galter Library's 1771 edition (**right**). A Roman numeral plate number in the upper right and a label in French were added to the plates between 1726 and 1771.

3.3. Paper

If documented in watermark databases, watermarks can provide information about the date and location of the paper production. Computed watermark images show that Plates 9 and 55 share a watermark in the shape of a bishop or abbot, wearing a mitre and a cross, bearing a staff and banner (Figure 12). The dark shadow around vertical chain lines indicates that the paper is antique laid paper, rather than laid or wove, suggesting that it was produced before the end of the eighteenth century [48].

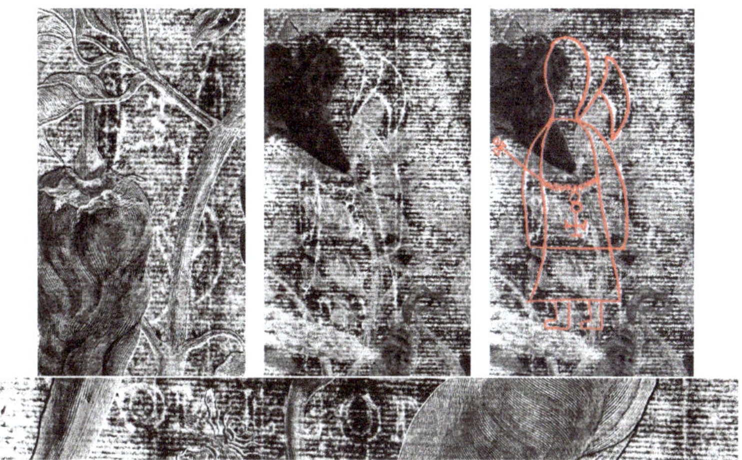

Figure 12. Computed images of watermarks found in plate 55 (**top left**), plate 9 (**top middle**) and plate 54 (**bottom**). The outline of a bishop watermark can be observed when manually reinforced in plate 9 (**top right**).

Dating of the watermarks themselves was inconclusive. This particular bishop watermark has not been documented in any of the common databases of eighteenth century watermarks, and does not match with watermarks in other editions of *Metamorphosis*, given in Figure 13.

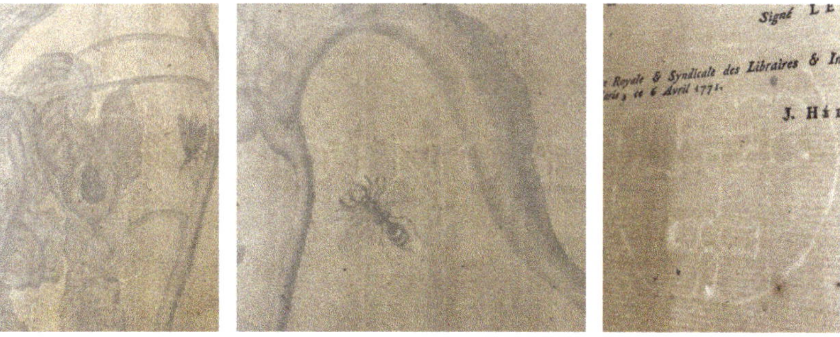

Figure 13. Transmitted light images showing watermarks in different editions of *Metamorphosis*. 1705 (not shown) and 1726 (**left**) editions both share the Strasbourg Lily watermark. The 1705 countermark is "PVL" (not shown) the 1726 countermark is "Honig" (**middle**). The 1771 edition has an unidentified watermark (**right**) and unintelligible countermark with text.

Watermarks are helpful, however, in addressing whether sheets were produced and procured from the same maker or edition. Plates 9 and 55 have the same watermark; however, the prints were impressed on the opposite side of the sheet with the watermarks vertically inverted with respect to one another. This suggests that these two sheets come from the same batch of paper and are likely mold-mates. This supports the hypothesis that plates 9 and 55 came from the same original source.

4. Conclusions

This paper has presented the technical analysis of three prints from *Metamorphosis Insectorum Surinamensium* by Maria Sibylla Merian in Northwestern University Library's Special Collections. The analysis was undertaken in order to determine a date, printer, and colorist for the prints, which were difficult to ascertain due to the complexities of the timeline of publication of Merian's works and the common practice of later coloring of early modern prints. This study employed a holistic analysis that would facilitate the non-invasive characterization of the artworks in three layers: pigment, print, and paper. Because of similarities in pigments—brown umber, barium-rich red, cadmium red, and lithopone—used in multiple plates, similar characteristic spectra for green pigments in all three plates, and the matching watermarks in plates 9 and 55 we conclude that the prints belonged to the same edition and were colored together. The use of two different red pigments, one barium-rich and the other cadmium red, and two different white pigments, titanium white and lithopone, alludes to the possibility of multiple campaigns of coloring, but this remains an open question. We date the coloring to post-1920, based on the use of cadmium red and titanium white, which were not commercially available to artists before 1919 and 1920, respectively. Based on the presence of features that are only present in posthumous editions of Merian's plates as they were reprinted in later editions, namely a Roman numeral plate number and French caption, we date the prints themselves to a French translation published in Paris in 1771. A more specific dating of the printed elements or the painted components would require further analysis of the organic components of the drawing and further research into the provenance of the watermarks.

The relative chronology of the different components of the prints in Northwestern Special Collections confirms that the prints are the product of a twentieth century art dealer or collector adding coloration to earlier prints. The dating of the coloring as well as the care with which it was undertaken suggest that this may have been an attempt to add value to relatively low-value later editions of Merian's works. As a result of this study the prints will be re-cataloged in the Northwestern Library catalog and the use of these prints by students and researchers will therefore be better contextualized in terms of this more complicated provenance.

Supplementary Materials: The following are available online at https://www.mdpi.com/article/10.3390/heritage4030088/s1, Figure S1: Elemental maps resulting from MA-XRF scan of plate 9, Table S2: Results of XRF point analysis for plate 9, Figure S2: Sample locations for XRF point analysis of plate 9, Figure S3: Endmember maps resulting from hyperspsectral imaging for plate 9 and associated reflectance spectra, Figure S4: Elemental maps resulting from MA-XRF scan of plate 54, Table S3: Results of XRF point analysis for plate 54, Figure S5: Sample locations for XRF point analysis of plate 54, Figure S6: Endmember maps resulting from hyperspectral imaging for plate 54 and associated reflectance spectra, Figure S7: Elemental maps resulting from MA-XRF scan of plate 55, Table S4: Results of XRF point analysis for plate 55, Figure S8: Sample locations for XRF point analysis of plate 55, Figure S9: Endmember maps resulting from hyperspsectral imaging for plate 55 and associated reflectance spectra.

Author Contributions: Conceptualization, O.D., M.V., A.M., M.W.; methodology, O.D., M.V., A.M., M.W.; investigation, O.D., M.V., A.M.; data curation, O.D., M.V., A.M.; writing—original draft preparation, O.D.; writing—review and editing, O.D., M.V., AM., M.W.; visualization, O.D., M.V.; supervision, M.W. All authors have read and agreed to the published version of the manuscript.

Funding: This collaborative initiative is part of NU-ACCESS's broad portfolio of activities, made possible by generous support of the Andrew W. Mellon Foundation.

Institutional Review Board Statement: Not Applicable.

Informed Consent Statement: Not Applicable.

Data Availability Statement: Data supporting this study are available in the associated supplementary information file.

Acknowledgments: The authors gratefully acknowledge support from and collaboration with Northwestern University Libraries Charles Deering McCormick Library of Special Collections Curator Scott Krafft and Northwestern University Libraries Book and Paper Conservator Roger Shaw. The authors wish to thank Katie Lattal, Special Collections Librarian at Galter Health Sciences Library and Learning Center, for facilitating access to Galter Library holdings.

Conflicts of Interest: The authors declare no conflict of interest. The funders had no role in the design of the study; in the collection, analyses, or interpretation of data; in the writing of the manuscript, or in the decision to publish the results.

Appendix A. Proposed Pigment Identifications and Corresponding Elements

Table A1. Summary of elements present in XRF analysis for plate 9.

Color	Elements Present	Possible Pigment
Red	Ba, Cd, S, Si, Zn	Organic red (inferred) with barium substrate, or cadmium-barium red
Yellow	Ba, Cd, Fe, Mn, S, Si, Zn	Brown umber and/or cadmium yellow, lithopone
Green	Ba, Fe, Mn, S, Si, Zn	Organic green (inferred) or Prussian blue mixture, lithopone
Green	Ba, Fe, S, Si, Ti, Zn	Organic green (inferred), or Prussian blue or iron oxide mixture, titanium white, lithopone
Mint Green	Ba, S, Ti, Zn	Organic green (inferred), titanium white, lithopone
Blue	Ba, Fe, Mn, S, Si, Zn	Prussian blue possibly with iron oxide, lithopone
Brown	Ba, Ca, Cd, Fe, Mn, S, Si, Zn	Brown umber, cadmium yellow, lithopone
Paper	As, Ba, Co, Cu, K, Mn, Ni, S, Si	Paper
Mat	Ca, Fe, Ti	Mat

Table A2. Summary of elements present in XRF analysis for plate 54.

Color	Elements Present	Possible Pigment
Red 1	Ba, Cd, S, Se, Si,Ti, Zn	Cadmium red, lithopone
Red 2	Ba, Cd, S, Si, Zn	Cadmium red, and/or organic red (inferred) with barium substrate
Pink	Ba, S, Zn	Organic red (inferred), zinc white or lithopone
Yellow 1	Ba, S, Zn	organic yellow (inferred), lithopone
Yellow 2	Ba, Cd, Fe, S, Ti, Zn	titanium white, cadmium-based yellow, lithopone
Light Green	Ba, Fe, S, Ti, Zn	organic green (inferred), titanium white, possibly Prussian blue mixture, lithopone
Green	Ba, Ca, Fe, S, Si, Ti, Zn	Titanium white, organic green(inferred) or possibly Prussian blue mixture, lithopone
Brown	Ca, Fe, Mn, S, Zn	Brown umber
Paper	As, Co, Cu, Fe, K, Mn, Ni, S, Si	Paper
Mat	Ca, Fe, Ti	Mat

Table A3. Summary of elements present in XRF analysis for plate 55.

Color	Elements Present	Possible Pigment
Red 1	Ba,Cd, S, Se, Si, Zn	Cadmium red, lithopone
Red 2	Ba, Cd, K P, Si,	Organic red (inferred) with barium substrate and/or cadmium red
Yellow	Fe	Iron Oxide
Dark Green	Ba, Fe, Mn, S, Si, Zn	Organic green (inferred), or mixture with Prussian blue, lithopone
Light Blue	S, Si, Ti, Zn,	Organic Blue (inferred) mixed with titanium White
Brown	Ba, Fe, Mn, S, Si, Zn	Brown umber, lithopone
White	Si, Ti	Titanium white
Paper	As, Cu, K, Mn, S	Paper
Mat	Ca, Fe, P, Ti	Mat

References

1. Reitsma, E. *Maria Sibylla Merian & Daughters: Women of Art and Science*; Rembrandt House Museum: Amsterdam, The Netherlands; JPaul Getty Museum: Los Angeles, CA, USA; ZwolleWaanders: Zwolle, The Netherlands, 2008; pp. 184–204.
2. Freedberg, D. The Failure of Colour. In *Sight & Insight: Essays on Art and Culture in Honour of E.H. Gombrich at 85*; Onians, J., Ed.; Phaidon Press: London, UK, 1994; pp. 245–263.
3. Lack, H.W.; Ibáñez, V. Recording colour in late eighteenth century botanical drawings: Sydney Parkinson, Ferdinand Bauer and Thaddäus Haenke. *Curtis's Bot. Mag.* **1997**, *14*. [CrossRef]
4. Baker, T.; Dupré, S.; Kusukawa, S.; Leonhard, K. (Eds.) *Early Modern Color Worlds*; Brill: Leiden, The Netherlands, 2016.
5. Merian, M.S. *Metamorphosis insectorum Surinamensium, ofte Verandering der Surinaamsche Insecten, Waar in de Surinaamsche Rupsen en Wormen Met Alle Des zelfs Veranderingen na Het Leven Afgebeeld en Beschreven Worden...*; Gerard Valck: Amsterdam, The Netherlands, 1705; pp. 9, 49, 53.
6. Dackerman, S. *Painted Prints: The Revelation of Color in Northern Renaissance & Baroque Engravings, Etchings, & Woodcuts*; Baltimore Museum of Art: Baltimore, MD, USA; Pennsylvania State University Press: University Park, PA, USA, 2002; pp. 1–9, 49–80.
7. Schmidt, S.K.K. *Altered and Adorned: Using Renaissance Prints in Daily Life*, 1st ed.; Art Institute of Chicago: Chicago, IL, USA, 2011.
8. Stillo, S.E. Putting the World in Its "Proper Colour": Exploring Hand-Coloring in Early Modern Maps. *J. Map Geogr. Libr.* **2016**, *12*, 158–186. [CrossRef]
9. Dackerman, S.; Primeau, T. The History and Technology of Renaissance and Baroque Hand-couloured prints. In *The Broad Spectrum: Studies in the Materials, Techniques, and Conservations of Color on Paper*; Salvesen, B., Stratis, H.K., Webber, P., Wheeler, M., Eds.; Archetype: London, UK, 2002; p. 78.
10. Schmidt-Loske, K. *Die Tierwelt der Maria Sibylla Merian (1647–1717): Arten, Beschreibungen und Illustrationen*; Number 10 in Acta Biohistorica; Basilisken-Presse: Marburg, Germany, 2007; pp. 128–130.
11. Hahn, O.; Oltrogge, D.; Bevers, H. Coloured Prints of the 16th Century: Non-Destructive Analyses on Coloured Engravings from Albrecht DÜrer and Contemporary Artists. *Archaeometry* **2004**, *46*. [CrossRef]
12. Zannini, P.; Baraldi, P.; Aceto, M.; Agostino, A.; Fenoglio, G.; Bersani, D.; Canobbio, E.; Schiavon, E.; Zanichelli, G.; De Pasquale, A. Identification of colorants on XVIII century scientific hand-coloured print volumes. *J. Raman Spectrosc.* **2012**, *43*. [CrossRef]
13. Ricciardi, P.; Legrand, S.; Bertolotti, G.; Janssens, K. Macro X-ray fluorescence (MA-XRF) scanning of illuminated manuscript fragments: Potentialities and challenges. *Microchem. J.* **2016**, *124*, 785–791. [CrossRef]
14. Schrader, S.; Turner, N.; Yocco, N. Naturalism under the Microscope: A Technical Study of Maria Sibylla Merian's "Metamorphosis of the Insects of Surinam". *Getty Res. J.* **2012**, *4*. [CrossRef]
15. Merian, M.S. *Red Maize-Like Flower, Possibly a Banana Plant, from an Album of 91 Drawings Entitled 'Merian's Drawings of Surinam Insects'*; Reference No. SL,5275.54; British Museum: London, UK, 1701.
16. Merian, M.S. *Sweet Pepper with Examples of a Butterfly, Chryslais and Caterpillar, from an Album of 91 Drawings Entitled 'Merian's Drawings of Surinam Insects'*; Reference No. SL,5275.55; British Museum: London, UK, 1701.
17. Merian, M.S. *Two Blue Butterflies and a Caterpillar, from an Album of 91 Drawings Entitled 'Merian's Drawings of Surinam Insects'*; Reference No. SL,5275.9; British Museum: London, UK, 1701.
18. Casadio, F.; Daher, C.; Bellot-Gurlet, L. Raman Spectroscopy of cultural heritage Materials: Overview of Applications and New Frontiers in Instrumentation, Sampling Modalities, and Data Processing. In *Analytical Chemistry for Cultural Heritage*; Mazzeo, R., Ed.; Topics in Current Chemistry Collections; Springer International Publishing: Cham, Switzerland, 2017; pp. 161–211. [CrossRef]
19. Bersani, D.; Madariaga, J.M. Applications of Raman spectroscopy in art and archaeology. *J. Raman Spectrosc.* **2012**, *43*. [CrossRef]
20. Saverwyns, S. Russian avant-garde... or not? A micro-Raman spectroscopy study of six paintings attributed to Liubov Popova. *J. Raman Spectrosc.* **2010**, *41*. [CrossRef]
21. Fremout, W.; Saverwyns, S. Identification of synthetic organic pigments: The role of a comprehensive digital Raman spectral library. *J. Raman Spectrosc.* **2012**, *43*. [CrossRef]
22. Prati, S.; Sciutto, G.; Bonacini, I.; Mazzeo, R. New Frontiers in Application of FTIR Microscopy for Characterization of Cultural Heritage Materials. *Top. Curr. Chem.* **2016**, *374*, 26. [CrossRef]
23. Pouyet, E.; Barbi, N.; Chopp, H.; Healy, O.; Katsaggelos, A.; Moak, S.; Mott, R.; Vermeulen, M.; Walton, M. Development of a highly mobile and versatile large MA-XRF scanner for in situ analyses of painted work of arts. *X-ray Spectrom.* **2020**. [CrossRef]
24. Solé, V.A.; Papillon, E.; Cotte, M.; Walter, P.; Susini, J. A multiplatform code for the analysis of energy-dispersive X-ray fluorescence spectra. *Spectrochim. Acta Part At. Spectrosc.* **2007**, *62*, 63–68. [CrossRef]
25. Vermeulen, M.; Smith, K.; Eremin, K.; Rayner, G.; Walton, M. Application of Uniform Manifold Approximation and Projection (UMAP) in Spectral Imaging of Artworks. *Spectrochim. Acta Part Mol. Biomol. Spectrosc.* **2021**, *252*. [CrossRef]
26. Huang, X.; Walton, M.; Bearman, G.; Cossairt, O. Near light correction for image relighting and 3D shape recovery. In Proceedings of the 2015 Digital Heritage, Granada, Spain, 28 September–2 October 2015; pp. 215–222. [CrossRef]
27. Ruiz, P.; Raju, G.; Dill, O.; Cossairt, O.; Walton, M.; Katsaggelos, A. Visible Transmission Imaging of Watermarks by Suppression of Occluding Text or Drawings. *Digit. Appl. Archaeol. Cult. Herit.* **2017**, *15*, e00121. [CrossRef]
28. Boyle, R.D.; Hiary, H. Watermark location via back-lighting and recto removal. *Int. J. Doc. Anal. Recognit.* **2009**, *12*. [CrossRef]
29. Berie, B.H. Prussian Blue. In *Artists' Pigments: A Handbook of Their History and Characteristics*; Fitzhugh, E.W., Ed.; National Gallery of Art: Washington, DC, USA, 1986; Volume 3, pp. 191–219.

30. Fiedler, I.; Bayard, M. Cadmium Yellows, Oranges, and Reds. In *Artists' Pigments: A Handbook of Their History and Characteristics*; Feller, R.L., Ed.; National Gallery of Art: Washington, DC, USA, 1986; Volume 1, pp. 65–109.
31. Barium Oxide—CAMEO. Available online: http://cameo.mfa.org/wiki/Barium_oxide (accessed on 15 May 2021).
32. Capua, R. The Obscure History of a Ubiquitous Pigment: Phosphorescent Lithopone and Its Appearance on Drawings by John La Farge. *J. Am. Inst. Conserv.* **2014**, *53*. [CrossRef]
33. Laver, M. Titanium Dioxide Whites. In *Artists' Pigments: A Handbook of Their History and Characteristics*; Fitzhugh, E.W., Ed.; National Gallery of Art: Washington, DC, USA, 1986; Volume 3, pp. 295–357.
34. D'Elia, E.; Buscaglia, P.; Piccirillo, A.; Picollo, M.; Casini, A.; Cucci, C.; Stefani, L.; Romano, F.P.; Caliri, C.; Gulmini, M. Macro X-ray fluorescence and VNIR hyperspectral imaging in the investigation of two panels by Marco d'Oggiono. *Microchem. J.* **2020**, *154*, 104541. [CrossRef]
35. Helwig, K. Iron Oxide Pigments; natural and synthetic. In *Artists' Pigments: A Handbook of Their History and Characteristics*; Anderson, B.H., Ed.; National Gallery of Art: Washington, DC, USA, 1986; Volume 4.
36. Laclavetine, K.; Boust, C.; Clivet, L.; Le Hô, A.S.; Laval, E.; Mathis, R.; Menu, M.; Pagliano, E.; Salmon, X.; Selbach, V.; et al. Non-invasive study of 16th century Northern European chiaroscuro woodcuts: First insights. *Microchem. J.* **2019**, *144*, 419–430. [CrossRef]
37. Duran, A.; Franquelo, M.L.; Centeno, M.A.; Espejo, T.; Perez-Rodriguez, J.L. Forgery detection on an Arabic illuminated manuscript by micro-Raman and X-ray fluorescence spectroscopy. *J. Raman Spectrosc.* **2011**, *42*. [CrossRef]
38. Pronti, L.; Felici, A.C.; Ménager, M.; Vieillescazes, C.; Piacentini, M. Spectral Behavior of White Pigment Mixtures Using Reflectance, Ultraviolet—Fluorescence Spectroscopy, and Multispectral Imaging. *Appl. Spectrosc.* **2017**, *71*. [CrossRef]
39. Picollo, M.; Bacci, M.; Magrini, D.; Trumpy, G.; Tsukada, M.; Kunzelman, D. Modern White Pigments: Their Identification by Means of non Invasive Ultraviolet, Visible and Infrared Fiber Optic Reflectance Spectroscopy. In *Modern Paints Uncovered, Proceedings from the Modern Paints Uncovered Symposium, Tate Modern, London, UK, 16–19 May 2006*; Tate Modern: London, UK, 2007; pp.129–139.
40. Burnstock, A.; Reissner, E.; Richardson, C. Analysis of Inorganic Materials from Paintings and Watercolours by Paul Cézanne from the Courtauld Gallery Using Two Methods of Non-Invasive Portable XRF with Light Microscopy and SEM/EDX Spectroscopy. In Proceedings of the 9th International Conference on NDT of Art, Jerusalem, Israel, 25–30 May 2008; p. 10.
41. Paulus, J.; Knuutinen, U. Cadmium colours: Composition and properties. *Appl. Phys. A* **2004**, *79*, 397–400. [CrossRef]
42. Rogge, C.E.; Epley, B.A. Behind the Bocour Label: Identification of Pigments and Binders in Historic Bocour Oil and Acrylic Paints. *J. Am. Inst. Conserv.* **2017**, *56*, 15–42. [CrossRef]
43. Biron, C.; Mounier, A.; Bourdon, G.L.; Servant, L.; Chapoulie, R.; Daniel, F. A blue can conceal another! Noninvasive multispectroscopic analyses of mixtures of indigo and Prussian blue. *Color Res. Appl.* **2020**, *45*. [CrossRef]
44. Vermeulen, M.; Leona, M. Evidence of early amorphous arsenic sulfide production and use in Edo period Japanese woodblock prints by Hokusai and Kunisada. *Herit. Sci.* **2019**, *7*, 73. [CrossRef]
45. Vermeulen, M.; Muller, E.; Leona, M. Non-Invasive Study of the Evolution of Pigments and Colourants Use in 19th-Century Ukiyo-e. *Arts Asia* **2020**, *50*, 103.
46. Glinsman, L.D. The practical application of air-path X-ray fluorescence spectrometry in the analysis of museum objects. *Stud. Conserv.* **2005**, *50*. [CrossRef]
47. Aceto, M.; Agostino, A.; Fenoglio, G.; Idone, A.; Gulmini, M.; Picollo, M.; Ricciardi, P.; Delaney, J.K. Characterisation of colourants on illuminated manuscripts by portable fibre optic UV-visible-NIR reflectance spectrophotometry. *Anal. Methods* **2014**, *6*. [CrossRef]
48. Hunter, D. *Papermaking: The History and Technique of an Ancient Craft*; Dover: New York, NY, USA, 1978.

 heritage

MDPI

Article

Identification of Pre-1950 Synthetic Organic Pigments in Artists' Paints. A Non-Invasive Approach Using Handheld Raman Spectroscopy

Rika Pause *, Inez Dorothé van der Werf and Klaas Jan van den Berg

Cultural Heritage Agency of the Netherlands, Hobbemastraat 22, 1071ZC Amsterdam, The Netherlands; I.van.der.Werf@cultureelerfgoed.nl (I.D.v.d.W.); K.van.den.Berg@cultureelerfgoed.nl (K.J.v.d.B.)
* Correspondence: r.pause@cultureelerfgoed.nl

Abstract: There is little information on the actual use of early synthetic organic pigments (SOPs) in art objects, especially those from before 1950. Their presence can, however, pose a challenge to conservation because their chemical composition, as well as their lightfastness and sensitivity to solvents, are often unknown. Here, a study on the non-invasive identification of SOPs in historic pre-1950 varnished paint-outs from artists' materials manufacturer Royal Talens is presented. The paints were analysed using a handheld Raman device. Spectra were evaluated by recording the spectra of the same samples with a benchtop instrument. This study demonstrated that the identification of SOPs in varnished oil paints with a non-invasive approach is possible and rather straightforward. The handheld Raman device allowed us to identify fourteen SOPs from eight pigment classes. Besides the occurrence of expected and the known SOPs of this time period, there were also some surprising results, like the detection of the triarylcarbonium pigments PG2 and PB8, and the monoazo Mordant Yellow 1.

Keywords: synthetic organic pigments; royal talens; handheld raman spectroscopy; microraman spectroscopy; modern artist oil paint

Citation: Pause, R.; van der Werf, I.D.; van den Berg, K.J. Identification of Pre-1950 Synthetic Organic Pigments in Artists' Paints. A Non-Invasive Approach Using Handheld Raman Spectroscopy. *Heritage* **2021**, *4*, 1348–1365. https://doi.org/10.3390/heritage4030073

Academic Editor: Diego Tamburini

Received: 30 June 2021
Accepted: 13 July 2021
Published: 17 July 2021

1. Introduction

After the accidental discovery of mauveine in 1856, a new class of dyestuffs and pigments was developed and introduced to the market. These were formerly defined as coal-tar dyes and are nowadays called synthetic organic pigments (SOPs). SOPs are "manufactured colourants that have a carbocyclic ring skeleton" [1]. In contrast to natural organic pigments, they are synthesized in a laboratory environment and produced on an industrial scale, so that natural organic pigments like indigo and alizarin can be considered as SOPs when they are produced industrially [1].

After their introduction, paint manufacturers started to use SOPs for their tube paints, which led to new shades in artists' colours and to the imitation of traditional pigments [2–4]. As it was initially not mandatory for manufacturers to indicate the use of SOPs in their paint formulations, artists often lacked knowledge of the exact content of their supplies. Considering the minor fastness characteristics of the early SOPs compared to traditional pigments, their use also led to material problems previously unknown to the artists, who tended to take a critical view on these new pigments [2,5]. In contrast to the traditional artists' pigments, the availability of certain SOPs is linked to an exact date of introduction, which can be traced by means of patents. Therefore, the identification of SOPs can not only provide information about the dating of an artwork, but can also expose forgeries [6]. Furthermore, knowledge of the occurrence of certain SOPs can provide useful information for restoration and conservation approaches or art technological studies.

SOPs can be difficult to detect in paint samples, because, due to high tinting strength, relatively small amounts are generally present [7]. Furthermore, these pigments have a very small particle size and are usually part of a complex system of ingredients in artists' tube paints. Therefore, traditional methods for pigment identification, such as polarised light microscopy, can hardly be used. Identification with Fourier transform infrared (FTIR) spectroscopy may be equally complicated, because the strong and broad peaks of paint binders, inorganic substrates, or extenders can interfere with the SOPs' information [8]. Invasive and destructive methods like pyrolysis gas-chromatography/mass-spectrometry (Py-GC/MS) can sometimes provide useful information on the presence of certain SOPs, such as azo-pigments. Others may not be suitable for this technique due to their thermal and chemical instability [9,10]. High performance liquid chromatography-mass spectrometry (HPLC-MS) can be challenging too, as getting the pigment to dissolve can be difficult and the effects of dissolution on the pigment structure may be problematic for interpretation [11]. MicroRaman spectroscopy (μ-RS) has been demonstrated to be a suitable analytical tool for the identification of SOPs. It is relatively sensitive towards these pigments, requires little sample material, and allows samples to be used for further, complementary analysis [12–14]. Approaches for μ-RS SOPs identification in samples from artworks are under constant development [9,10,15,16]. The main limitation in using μ-RS to identify SOPs in artists' paints, however, is related to the fluorescence associated with organic binders, varnish layers and additives, or sometimes even with the target pigments themselves. In order to obtain an interpretable spectrum, it is important to perform measurements in equilibrium between fluorescence and the intensity of the Raman signal, which makes this approach time-consuming [8,17].

All these analytical techniques, however, require the sampling of small paint fragments, which, besides being destructive, can be challenging, as the probability that the samples are not representative for the materials studied might be quite high [18]. An analytical approach using non-invasive devices that allow for unlimited measurements therefore seems desirable [19]. The use of visible reflectance spectroscopy and visible-excited spectrofluorimetry was successful for the non-invasive identification of contemporary SOPs [20]. Yet, for the non-invasive study of artworks, Longoni et al. [20] recommend the use of further complementary non-invasive analysis. The usage of portable Raman systems seems to be promising [21,22]. A portable device that minimizes fluorescence and can operate at low energy is needed. Damage by strong laser absorption should be avoided to prevent local burns, which may occur especially on opaque or dark materials [8].

Different handheld Raman systems have been reviewed and tested for their application in the field of cultural heritage [19,23–25]. Technological developments have countered the most difficult problems, such as low Raman signal, high fluorescence, and environmental interactions, as well as the positioning and focusing of the laser [26,27]. To date, the Bravo handheld Raman instrument developed by Bruker seems to be the most promising device [19] and its applicability to the study of modern art was evaluated [28]. This device employs a DUO Laser system (785 nm and 853 nm), recording spectra in two separate spectral ranges of 300–2200 and 1200–3200 cm^{-1}, respectively. The patented sequential shifted excitation (SSETM) works by temperature-shifting the lasers over a small wavelength range and recording three spectra per laser, finally merging them into a seventh averaged spectrum. In this way, and through the use of near infrared lasers, the system is able to reduce fluorescence and excite colour with low efficiency, also enabling the measurement of, for example, varnished paint samples [19,23,27]. However, Pozzi et al. [19] observed that if the peaks and curves of the individual spectra differ too much from each other, there is a risk of creating artefacts within the averaged spectrum. Another disadvantage of the device is related to the fact that the only two adjustable settings are exposure time and acquisition number, but not the energy of the laser, which can be a risk for certain art-technological studies [19]. These aspects, regarding the spectral quality and applicability, are discussed within this study.

The aim of this research is to study the potential of a handheld Raman device in identifying SOPs in historical varnished paint-outs (1932–1950) with consideration to future applications in the non-invasive study of early SOP-containing paintings.

Ongoing work on the collection of SOPs Raman reference spectra (https://soprano.kikirpa.be, accessed on 30 June 2021 [14,29]) enables and simplifies pigments' identification. These reference spectra were often obtained by studying modern artists' supplies and can give an idea of the actual SOPs that can be expected.

Before 1950, the following SOP groups were already developed and established: β-Naphthol pigment lakes; BON pigment lakes (pigments produced with 2-hydroxy-3-naphthoic acid as coupling component); β-Naphthol; Pyrazolones; Hansa Yellows; Diarylides; Naphthol AS; Iron complex Pigment Green B; Phthalocyanines; Nickel Azo Yellow [30]. However, as already mentioned, the actual distribution and use of this "new" group of pigments within artists' materials is still unclear today. Attempts to provide an overview of the SOPs that were known before 1950 and of their occurrence in artists' paints demonstrated the need for more research [7,30]. A starting point could be the study of the production archives of artists' material suppliers.

Our research in the archives of Royal Talens and its main pigment suppliers before 1950, BASF and Bayer, shows that there are considerably more SOPs used in artists' paints than hitherto known [4,11,31]. Notably, most of them have not been traced in paintings.

This study focuses on SOPs in Royal Talens' artists' oil paints from before 1950. Soon after its foundation in 1899, Talens developed into a major international player on the market, already selling more than 1,500,000 oil paint tubes in 1918 [32]. European artists such as van Dongen, Kirchner, Appel, and Munch used Talens' Rembrandt oil paints [33]. In this position, Talens declared the use of SOPs in fine artists' oil paint publicly in the beginning of the 1930s.

Based on pre-1950 oil paint recipes [4,16], a selected set of paint-outs (1932–1950) from the Royal Talens' archive was investigated. For the assessment and evaluation of the handheld Raman spectra, micro samples of the historic Talens paint-outs were also analysed with a benchtop microRaman spectrometer.

Handheld Raman spectra were evaluated, taking into account their major points of criticism, like background fluorescence, the intensity of the Raman signal, spectral resolution, and the risks of laser energy induced damages.

2. Materials and Methods

2.1. Samples

The sample selection consisted of paint-outs from 1932 to 1950. These were included within colour charts specifically dedicated to coal-tar pigments, labelled as unalterable Teerkleurstoffen [coal-tar dyes] (Figure 1), which were published in six editions of the book Kunstschildersmaterialen en Schildertechniek [Painting materials and painting technique] [34–40]. The books deal with Talens' fine Rembrandt artists' oil paints, which were on the market since 1904 [4,33]. In addition, some paint-outs that were listed under inorganic names within the inorganic paint colour charts, but suspected to contain SOPs according to recipe interpretation, were examined as well. To obtain results on a larger data set, it was decided to measure (if available) the same set of 42 potentially SOP-containing paint-outs in each Kerdijk edition, even if, according to the recipes, the same SOPs were to be expected. In total, 193 samples were analysed with handheld Raman spectroscopy and for comparison with benchtop microRaman spectroscopy. Samples were labelled with a "K" (Kerdijk) and the year of publication. All paint-outs were varnished.

2.1.1. Instrumentation

The handheld Raman measurements were conducted with a Bravo Spectrometer (Bruker). The device works with two excitation wavelengths, recording spectra in two separate spectral ranges of 300–2200 and of 1200–3200 cm^{-1} with a DUO Laser system (785 nm and 853 nm). The near infrared lasers (NIR) aid in reducing fluorescence, as the photon does not possess enough energy to induce high molecular fluorescence [22,41]. Furthermore, using two lasers enables researchers to obtain better quality data in the overlap region of 1200–2200 cm^{-1} and lowers fluorescence interference [19]. The energy reaching the surface during the measurement is officially stated to be less than 100 mW (Table 1). Since the energy cannot be set manually, this indication is too unspecific to be taken as a guide value when analysing delicate cultural heritage objects. In order to estimate the energy that actually reaches the surface of the sample, measurements were carried out with a "Power and Energy Meter" from Thorlabs, using the same settings of the historic Talens paint-outs analyses. The measured energy value was about 45 mW, with measurements conducted at a distance of about half a millimetre, with a spot size of 1 mm. This value also agrees with that determined by Pozzi et al. of about 50 mW [19].

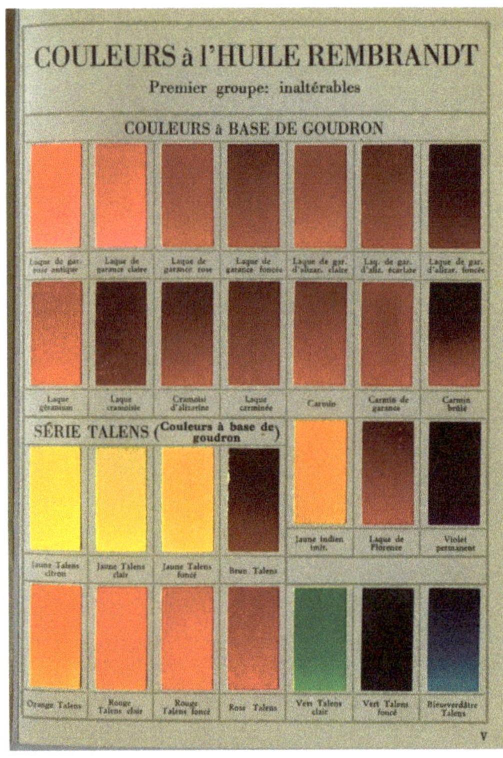

Figure 1. Kerdijk colour chart from 1932, presenting a selection of Talens' Rembrandt artists' oil paints 'à base de goudron' (based on coal tar). Please note that the colours are defined as 'inaltérables' (unalterable).

Table 1. Characteristics of the microRaman and handheld Raman devices.

Name/Producer	MicroRaman [a]	Handheld Raman [b]
Laser system	350 mW near infrared laser (class3B laser device)	Dual laser excitation
Wavelength	785 nm	785 nm and 853 nm
Max. laser power of device	100 mW	<100 mW (both lasers)
Max laser output on sample	21.7 mW (50× magnification)	45 mW (working distance)
Spectral range	95–3500 cm^{-1}	300–3200 cm^{-1}
Spectral resolution	4 cm^{-1}	10–12 cm^{-1}
Spot size	10–20 µm	10–15 µm [c]
Working distance	(depending on magnification)	ca. 0.5 mm

[a] Perkin Elmer, RamanMicro 300, coupled to Raman Station 400F; Available online: https://www.perkinelmer.com/CMSResources/Images/46-74804SPC_RamanStation400RRamanStation400F.pdf (accessed of 14 July 2020). [b] Bravo Bruker [13,23,32]. [c] Due to the lack of a camera and the size of the laser pinhole of 3 mm, it was not possible to precisely focus on this spot size.

To evaluate the handheld device's performance for the identification of SOPs, the colour charts were also analysed with a benchtop microRaman instrument. These reference spectra were recorded with a RamanMicro 300 in combination with a RamanStation 400F (Perkin Elmer, Waltham, MA, USA) and an Olympus BX51M microscope (Table 1). Micro-Raman spectra were acquired from very small samples of the same paint-outs. After the mechanical removal of the varnish from the sampling area, minute paint fragments were taken at the upper right edge where the paint film is thickest. The samples were placed on an aluminium tape, mounted on a glass slide, and positioned as close to perpendicular to the optical axis as possible to reduce fluorescence [42].

Spectra were recorded in the range of 300–2000 cm^{-1}, in order to perform the analysis within the same range as the handheld Raman. The microRaman reference spectra were only baseline corrected when necessary for their readability.

2.1.2. Acquisition of Spectra

As the Bruker Bravo does not include a camera, the measurement preparation demanded precise positioning. To locate and define the measuring spot on the paint-outs, a Melinex® template in the shape of the optical head was used. Measurements were taken at the closest possible distance without touching the surface, which was approximately half a millimetre. Since this study was undertaken to investigate the possibility of analysing SOPs in real artworks where focusing on well-defined paint spots is mandatory, it was decided not to use the de-focusing tip. Handheld Raman spectra were recorded with initial settings of 100 ms and 20 acquisitions within the spectral range of 300–2000 cm^{-1}. These settings proved to be efficient to verify the Raman signal and fluorescence. Depending on the signal to noise ratio of the spectrum, these settings were tuned to: acquisition times 100–300 ms; 50–200 acquisitions.

A common limitation of portable Raman devices is the reduced possibilities to control the measurement settings [19]. For the Bravo spectrometer, the lack of direct control of the laser energy was considered potentially problematic. As the maximum energy reaching the surface is quite high in comparison to the Perkin Elmer microRaman instrument (Table 1), the measurement spot was checked under the microscope before and after analysis in order to be able to track possible laser induced damage.

Measurements were recorded without removing the varnish layer from the paint-outs at the upper middle edge of the samples, in order to measure a spot with high pigment density.

2.1.3. Interpretation of Raman spectra

Handheld Raman spectra were interpreted using reference spectral databases. Raman reference spectra of SOPs have been published listing the peaks with their relative intensity or in digital databases [10,14,29]. Furthermore, the continuously updated in-house built Raman spectral library was used.

It should, however, be taken into account that reference spectra are mostly acquired on pure materials, like pigment powders. When compared with paint systems, differences in heat dispersion or fluorescence behaviour may lead to slight differences in the spectra [43,44]. The excitation wavelength and power of the laser can influence the reference spectra too. As KIK-IRPA documents the exact measurement settings for each reference spectrum (https://soprano.kikirpa.be, accessed on 21 August 2020), their database was used as the main reference for the interpretation of the spectra, in addition to the in-house database.

In some cases, it was not possible to interpret the Raman spectrum and assign it to a defined pigment. In these specific cases, additional methods, such as ultra-high performance liquid chromatography mass spectrometry (UHPLC-MS), were used to try to identify the unknown samples. These results were verified by analysing corresponding references from the in-house reference collection.

3. Results and Discussion

As a pre-interpretation step of the Raman spectra acquired from the Kerdijk paint-outs, production recipes collected at the Talens' archives were studied and evaluated and a list of expected SOPs was made [4]. Where possible, the commercial names used in the recipes were related to the Colour Index definitions, e.g., the pigment Hansa Geel 10 [Hansa Yellow 10] corresponds to PY3. Then, the acquired Raman spectra were compared with the reference spectra of the expected SOPs.

Table 2 gives a detailed overview of the analysed paint-outs and their interpretation. The expected SOPs according to the recipes are reported. The settings for acquisition refer to the averaged settings yielding the best result. It was found that the handheld Raman produces readable and interpretable spectra with defined peaks at almost constant settings of 150–200 ms exposure time and 100–200 acquisitions.

3.1. Identified SOPs

The Talens Rembrandt oil paints of the Kerdijk colour charts were revealed to contain the SOPs of eight different pigment groups. Furthermore, it was also possible to identify paints that contained a mixture of SOPs (Table 2).

Some of these SOPs belong to groups that are already well known to have been used in fine artists' materials and can be expected in this time period. This concerns SOPs belonging to the BONA, monoazo, and phthalocyanine group as well as the ß-Naphthol pigments PR3 and PO5, which have indeed been found in artworks from before 1950 [30]. In other cases, the Kraplak [madder], Karmijn [carmine], and Alizarine [alizarine] paint-outs provided well defined spectra, but the SOPs have not yet been identified and are labeled as "unknowns".

Among the identified SOPs, there are also some pigments that were not yet known to have been used for the production of fine artists' paints dated before 1950. The occurrence of the mordant dyes and triarylcarbonium pigments are among the most surprising results of this study and emphasizes the importance of searching unexpected possible pigment groups in pre-1950 artists' materials. Three examples are discussed in more detail.

The paint-out Gamboge from 1932 was revealed to contain Mordant Yellow 1. This SOP was discovered by Nietzki in 1887 and is also known as Alizaringelb GG [Alizarine Yellow GG] or Beizengelb TG [Mordant Yellow TG] [45]. Wagner (1928) reports that this pigment was not very popular because of the elaborate preparation needed to obtain a pure product. It is therefore surprising that, according to the recipe records, Talens has used this SOP for fine artists' material production for almost 20 years [34,46]. Figure 2 displays the handheld Raman spectrum of the paint-out Gamboge and the microRaman spectrum of Mordant Yellow 1 from the in-house reference collection. The handheld Raman spectrum displays the peaks that allow an interpretation, but also contains artifacts.

Within the paint-out Talens Groen Licht [Talens Green Light] the pigment PG9, a SOP belonging to the group of Indamine, was detected. This pigment, a barium salt of an iron complex, was discovered in 1909 and labelled as Hansagrün GS [45]. It was recommended by the Colour Index to be used "particularly in mixture with CI PY 1 or 3" [45–47]. Schultz (1931) describes the pigment as a "very lively green when mixed with Hansa Yellow for all purposes (almost as fiery as Schweinfurter Green)" [48]. This suggests its use in artists' colours in mixture with monoazo yellows as a substitute of the toxic inorganic pigment and fits therefore to Talens' application, documented in the recipe of Talens Green Light.

Figure 3 shows the handheld Raman spectrum of the paint-out Talens Green Light and, for comparison, the microRaman spectrum of a Hansagrün GS reference (paint-out from Wagner, 1928). The peaks of the PY3 component in the reference spectrum are very faint, which can be related to different PY3 and PG9 ratios within the samples.

Table 2. Summary of the results obtained with the handheld Raman device on Kerdijk colour charts with Talens' Rembrandt oil paint-outs dated 1932–1950.

Sample Name [English Translation]	Commercial Name of the SOPs from the Recipe [a]	CI	Kerdijkbook	Exposure Time (ms)/Accumulations (No.)	Identification
Kraplak rose antique *[madder lake pink antique]*	Wurzelkrapplack 30101 Huber	-	K32-K50	200/200	Unknown A
Kraplak licht *[madder lake light]*	Wurzelkrapplack B6	-	K32-K44	200/200	Unknown A
	Kraplak 31417	-	K50	200/200	PR83 + Unknown A
Kraplak rose *[madder lake pink]*	Kraplak B6/Kraplak donker 30029 Huber	-	K32-K37	250/100	Unknown A (?)
	Kraplak 30101/Kraplak 30992	-	K41-K50	250/200	Unknown A (?)
Kraplak donker *[madder lake dark]*	Kraplak donker Huber 30029	-	K32-K44	200/100	Unknown A + Unknown C
	Kraplak 30992	-	K50	200/200	PR83
Alizarine Kraplak Licht *[Alizarine madder lake light]*	Kraplak IWG	-	K32-K42	200/100	PR83 + Unknown A
	Krapplack 3077	-	K44-K50	150/200	PR83 + Unknown A
Alizarine Kraplak donker *[Alizarine madder lake dark]*	Alizarine Kraplak 72	-	K32-K37	200/200	PR83 + Unknown A
	Permanentviolet/Karmijnlak/Kraplak	-	K41-K50	200/200	PR83 + Unknown A
Alizarine Crimson *[Alizarine crimson]*	Kraplak HE	-	K32-K50	150/200	PR83 (+Unknown A?)
Geraniumlak *[Geranium lake]*	Hansarood B/Kraplak Z	PR3 +?	K32	200/200	PR3 + PO5 [d]
	Isolechtrood B/Kraplak Z	PR3 +?	K37	200/200	PR3 + PO5 [d]
	Karmijnlak fijn/Cadmium red mid	-	K41-K44	200/200	PR3 + PR83
	Talens donker fijn/Karmijnlak fijn	PR3 +?	K50	200/200	PR3 + PR83
Karmijnlak *[Carmine lake]*	Alizarine madder 34549	-	K32-K50	200/200	PR83 + Unknown A
Crimsonlak *[Crimson lake]*	Alizarine Kraplak 45	-	K32	200/200	PR83 + Unknown A
Karmijn *[Carmine]*	-no recipe-	-	K32-K50	200/200	PR83

Table 2. *Cont.*

Sample Name [English Translation]	Commercial Name of the SOPs from the Recipe [a]	CI	Kerdijkbook	Exposure Time (ms)/Accumulations (No.)	Identification
Krapkarmijn *[Madder Carmine]*	-no recipe-	-	K32-K50	150/300	Unknown A
Karmijn donker *[Carmine dark]*	Kraplak donker Huber 30029/Ultramarijn VU 861E/Alizarine oranje 6231	-	K32	200/200	Unknown B?
Florentijn lake *[Florentine lake]*	Alizarine lak 45/Ultramarijn VU861E	- + PB29	K32	200/200	PR83 + Unknown A
Talens geel citroen *[Talens yellow lemon]*	Hansageel 10G	PY3	K32-K44	200/200	PY3
	Fanchon Yellow YH4	PY3	K50	200/200	PY3
Talens geel licht *[Talens yellow light]*	Hansageel GGR/Hansageel 10G	PY1 + PY3	K32-K37	200/200	PY3 +?
Talens geel middle *[Talens yellow middle]*	Hansageel GR	PY2	K32-K37	200/200	PY2
Talens geel donker *[Talens yellow dark]*	Hansageel GR/ Permanentrood 2G extra	PY2 + PO5	K32-K37	200/200	PY2 + PO5
Talens oranje *[Talens orange]*	Hansageel GR/ Permanenentrood 2G extra	PY2 + PO5	K32-K37	200/200	PY2 + PO5
Talens rood licht *[Talens red light]*	Permanentrood 2G extra	PO5	K32-K44	200/200	PO5
	Siegle Oranje 3TR	PO5	K50	200/100	PO5
Talens rood donker *[Talens red dark]*	Hansarood B	PR3	K32-K37	200/200	PR3
	Isolechtrood B	PR3	K41-K50	200/100	PR3
Talens rose *[Talens pink]*	Permanentrood F6R extra	PR57	K32-K44	200/200	PR57:1
	Rhodamin Toner	PR57	K50	200/200	PR57:1
Talens bruin *[Talens brown]*	Fanalgelb G	PY18 +?	K32-K37	200/200	?
	Jaune electra 100/Ultramarijn dnk 100/Kraplak dnk. 100	-	K41-K50	150/100	?

Table 2. *Cont.*

Sample Name [English Translation]	Commercial Name of the SOPs from the Recipe [a]	CI	Kerdijkbook	Exposure Time (ms)/Accumulations (No.)	Identification
Talens groen licht *[Talens green light]*	Hansageel 10G/ Permanentgroen	PY3 + PG18	K32-K37	150/200	PY3 + PG9
	Talens geel 100/ Rembrandtgroen 100	PY3 + PG7	K41-K50	200/100	PY3 + PG7
Talens groen donker *[Talens green dark]*	Fanalgelbgruen G supra	PG2	K32-K37	200/200	PB8
	-no recipe-	-	K41-K50	150/100	PG2
Talensgroenblauw *[Talens green blue]*	Fanalbremerblau G neu	-	K32-K37	200/200	PB8
	-no recipe-	-	K41-K50	150/100	PB15 + PG7
Indischgeel imit *[Indian Yellow imitation]*	Indischgeel imit Siegle 41258	-	K32-K44	200/200	PY1
	Transparent yellow 6202/Cadm. Dnk 101/Cadm. Oranje 101	-	K50	200/200	PY1 + PY3
Permanentviolet *[Permanent violet]*	Violet A 2145 Remmert	Natural Red 9	K32-K44	200/100	Unknown B
	Maroon 2142/Monastraalblauw B	Natural Red 9	K50	150/300	Unknown B
Rembrandtblauw *[Rembrandt Blue]*	Monastraalblauw	PB15	K41-K50	100/50	PB15
Rembrandtgroen *[Rembrandt Green]*	Heliogengroen G	PG7	K41-K50	100/50	PG7
Paul Veronesegroen *[Paul Veronese green]*	-no recipe-	PG7	K41-K50	200/100	PG7
Paul Veronesegoen tint *[Paul Veronese green tone]*	-no recipe-	PG7	K41-K50	150/100	PG7
Dekgroen *[Deck green]*	Hansageel 10G/ Monastraalgroen G	PG7 + PY3	K41-K50	200/100	PG7 (+PY3)
Dekgroen tint *[Deck green tone]*	Hansageel 10G/ Monastraalgroen G	PG7 + PY3	K41-K50	200/100	PG7 (+PY3)

Table 2. *Cont.*

Sample Name [English Translation]	Commercial Name of the SOPs from the Recipe [a]	CI	Kerdijkbook	Exposure Time (ms)/Accumulations (No.)	Identification
Gamboge *[Gamboge]*	Indgeel 3030 Huber of Echtgelb S	-	K32	200/200	Mordant Yellow 1 [b]
Alizarine oranje *[Alizarine orange]*	Alizarine oranje 6213	-	K32-K50	200/200	Mordant Orange 1 [c]
Groene lak licht *[Green lake light]*	-no recipe-	-	K32-K50	200/200	PG12
Groene lak donker *[Green lake dark]*	-no recipe-	-	K32-K50	200/100	PG12
Stil de grain geel *[Stil de grain yellow]*	Echtgelb S	-	K32-K44	200/200	Mordant Yellow 1 [b]
	Cone Yellow Lake ER 387		K50	200/200	Mordant Yellow 1 [b]
Vert de olive *[Olive green]*	Groene lak 2365/Alizarine oranje/Echtgelb S	-	K32	100/200	PG12
Vert de vessie *[Sap green]*	Groene lak 2548	-	K32	100/200	PG12

[a] Information based on archival research [3,4]. Further information will be available in R. Pause, PhD thesis, University of Amsterdam, due in 2022. [b] Identified by A.N, Proaño Gaibor, UHPLC-MS analysis. [c] Identified by B. Sundberg, UHPLC-MS analysis (Sundberg et al., 2020). [d] Additional peaks at 1343 cm^{-1} and 1368 cm^{-1}, which are not attributable to PR3 or PO5, are not yet interpreted.

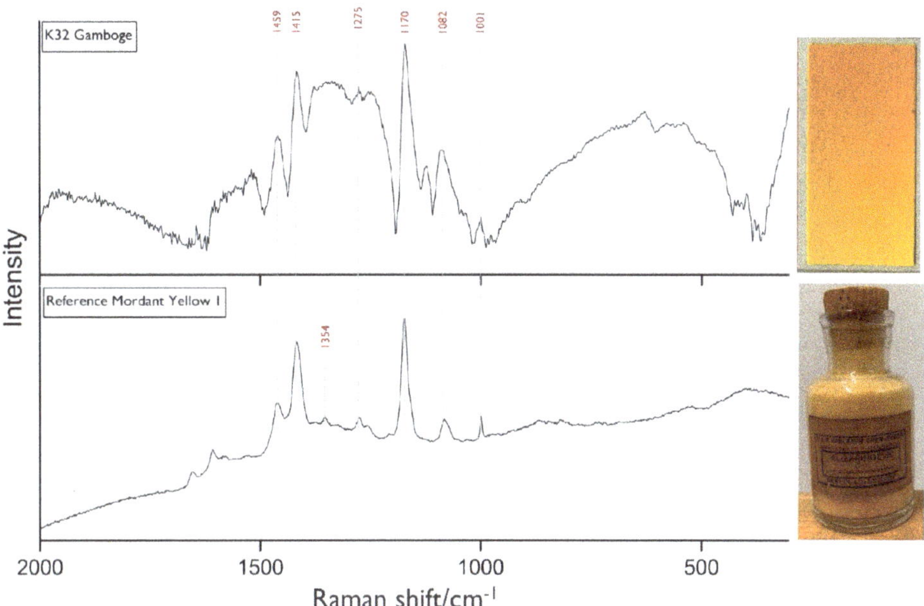

Figure 2. Handheld Raman spectrum of the paint-out Gamboge (K32) and the microRaman spectrum of Mordant Yellow 1, labelled as Alizarine Yellow GG (reference collection of the Cultural Heritage Agency of the Netherlands, Amsterdam, The Netherlands).

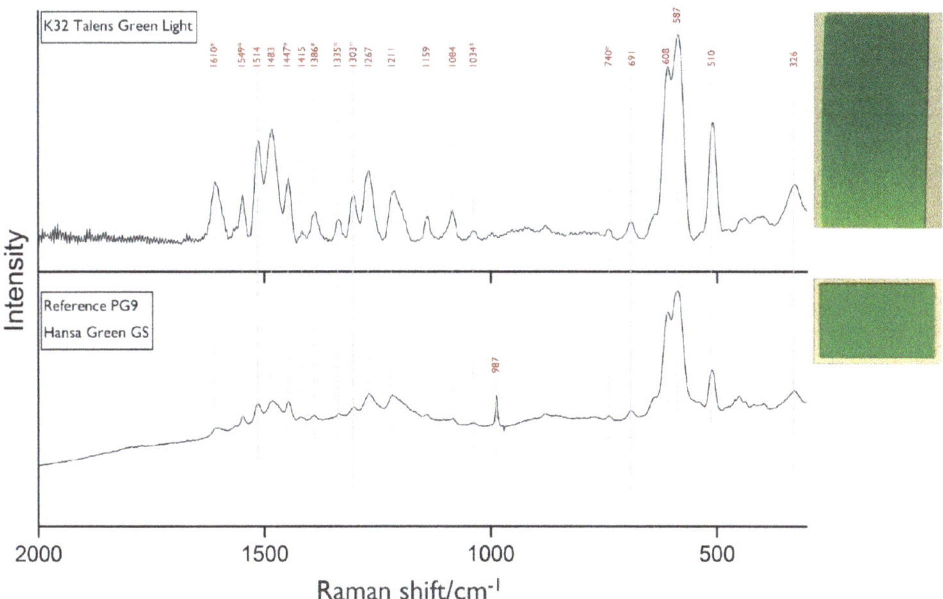

Figure 3. Handheld Raman spectrum of paint-out Talens Green Light (K32) and the microRaman spectrum of reference Hansagreen GS (Wagner, 1927), mixed with barium sulphate (987 cm^{-1}), both showing the characteristic peaks of PG9 in mixture with PY3. The peaks that clearly correspond to the KIK-IRPA reference for PY3 are highlighted with a star.

Two paints of the Talens in-house brand, labelled as "Talens paints", showed triarylcarbonium pigments. In the paint-out Talens Groen Donker [Talens Green Dark] of the 1941–1950 editions, the green pigment PG2 was identified. PG2, brought to the market under the product name Fanalgelbgrün [Fanal Yellow Green], is defined as either a mixture of Basic Green 1 and Basic Yellow 1, precipitated with phosphotungstomolybdic acid (PTMA), or Basic Green 1 with Auramin O, precipitated with PTMA [31,45,48].

The other triarylcarbonium pigment that was identified is PB8. It was detected in the paint-out Talens Green Dark (1932), contradicting the recipe (Table 2), and in the Talens Groen Blauw [Talens Green Blue] in the editions from 1932 and 1937. PB8, labelled in the Talens recipes as Fanalbremerblau G neu [Fanal Bremer Blue G new], is a mixture of Basic Blue 5 and Basic Green 1, precipitated with PTMA [17]. Figure 4 displays the handheld Raman spectrum of Talens Green Blue and the reference microRaman spectrum of a reference sample, labelled as a mixture of Basic Blue 5 and Basic Green 1 precipitated with PTMA, provided from the Historic dye collection of the technical university in Dresden.

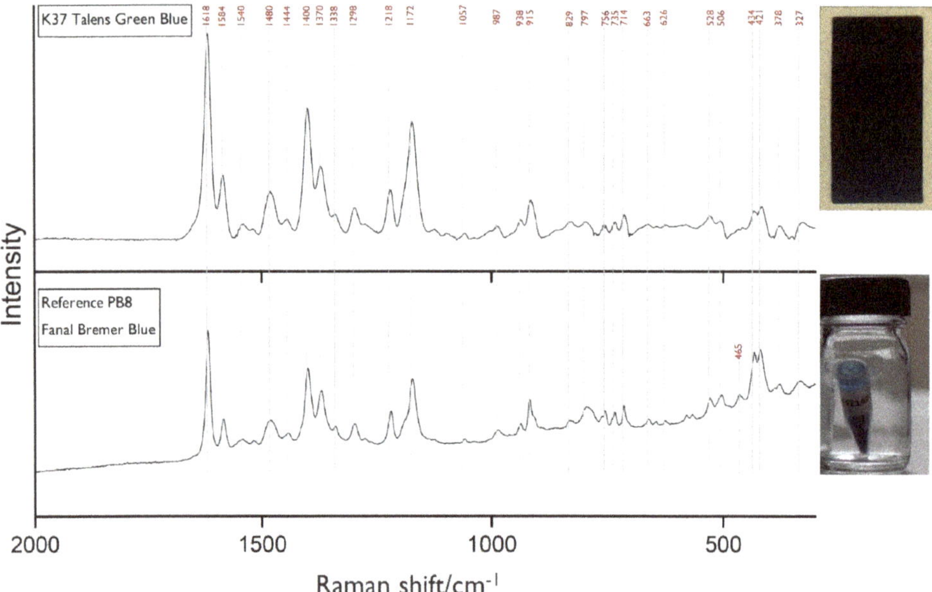

Figure 4. Handheld Raman Spectrum of paint-out Talens Green Blue (K32) and the microRaman spectrum of a reference sample containing Basic Blue 5 and Basic Green 1, precipitated with PTMA (Historic Dye Collection, Dresden, Germany).

Both PB8 and PG2 belong to the Fanal brand, which was introduced by I.G. Farben in 1924 [31]. Lomax et al. [1] point out that "the major use of these pigments [triarylcarbonium] is in the printing ink industry" and it seems surprising that Talens chose these SOPs for their fine artists' oil paint brand. However, Steger et al. detected two pigments (PV2 and PG1), belonging to this group, in a reverse glass painting from 1945 [49].

3.2. *Evaluation of the Handheld Raman's Applicability*

Raman spectra were not only interpreted for pigments' identification, but also evaluated for readability and quality. When dealing with the handheld Raman spectra that were acquired from our samples, we came across some of the issues that generally arise in the discussions about the applicability of handheld Raman spectroscopy [19].

A common problem which can lead to uninterpretable spectra is the occurrence of strong fluorescence that obscures the Raman scattering signal and affects the accuracy and sensitivity of the measurements [50]. As mentioned, the Bravo-Raman device uses sequentially shifted excitation, a patented fluorescence minimization method [19].

As an example, a paint-out containing a Hansa Yellow pigment was selected to demonstrate the result of the automatic fluorescence reduction. This important class of SOPs, which is commonly found in modern artworks [30], often exhibits strong fluorescence in Raman spectroscopy. Figure 5 shows the handheld Raman spectrum of the paint-out Talens geel citroen [Talens yellow lemon], where Hansa Yellow (PY3) was identified. The spectrum recorded with the microRaman was not further processed after acquisition and shows a distinct fluorescence, whereas the spectrum recorded with the handheld Raman appears to be corrected in this respect, thanks to the SSE approach and the two lasers. The spectrum presents intense peaks, although the spectral resolution is not very high, as demonstrated by the lack of a peak at ~1593 cm^{-1}, which was present in the microRaman spectrum.

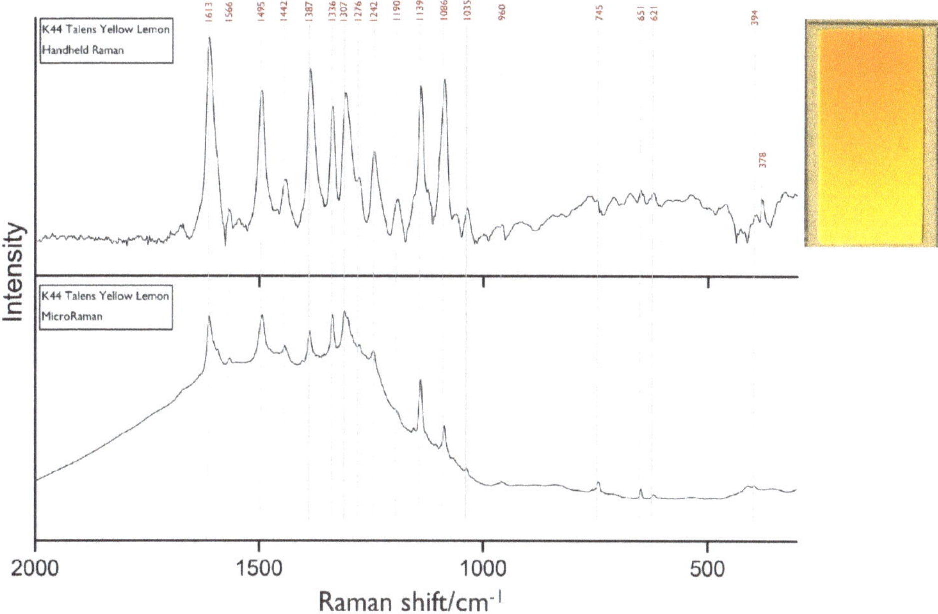

Figure 5. Handheld and microRaman spectra of paint-out K44 Talens Yellow Lemon.

Another issue we considered was the merging or shifting of peaks as a consequence of over excitation, when measuring with a high, non-adjustable energy. This was observed in the Handheld Raman analysis of the paint-out Talens Rood Donker (Talens Red Dark), containing PR3 (Figure 6). Peaks that are close to each other and similar in intensity seem to be smoothed. This could be due to the high laser power, which might be responsible for a shift in peaks to lower wavenumbers, peak broadening, or twin peaks merging (for example at 1621 cm^{-1} and 1605 cm^{-1}).

Another example of possible overexcitation is reported in Figure 7. Here the handheld Raman spectrum of paint-out K41 Rembrandtgroen [Rembrandtgreen] is shown where the green copper phthalocyanine pigment PG7 was identified. Analysing phthalocyanine pigments with Raman spectroscopy can lead to overexcitation and peak shifts [29,51]. Fremout et al. (2012) made this observation while analysing the related PB15 pigment and state that the exact reason for this phenomenon is unclear and that a reversible, heat induced deformation of the crystal structure is possible [29]. The handheld Raman and microRaman spectra seem indeed to be shifted when compared to the KIK-IRPA reference spectrum (https://soprano.kikirpa.be, accessed on 30 June 2021 [30]) (Figure 7). However, the PG7 spectrum published by Schulte et al. fits our Raman spectrum [10]. These discrepancies might be explained by the spectral differences observed for dry pigment powder and paint systems. Defeyt et al. noticed peak shifts when comparing the Raman spectra of blue copper phthalocyanine PB15 as a dry pigment powder and mixed with a binder [43].

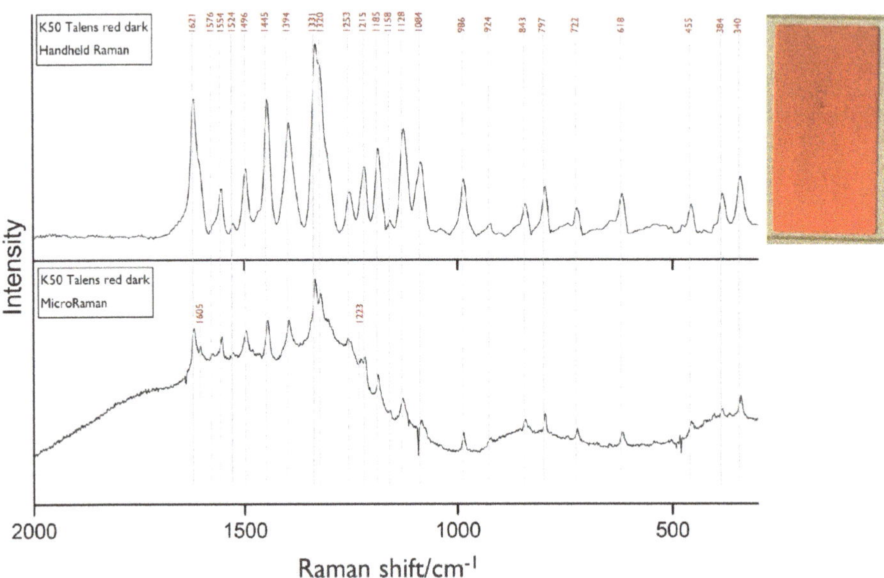

Figure 6. Handheld and microRaman spectra of paint-out K50 Talens red dark, mixed with barium sulphate (986 cm^{-1}).

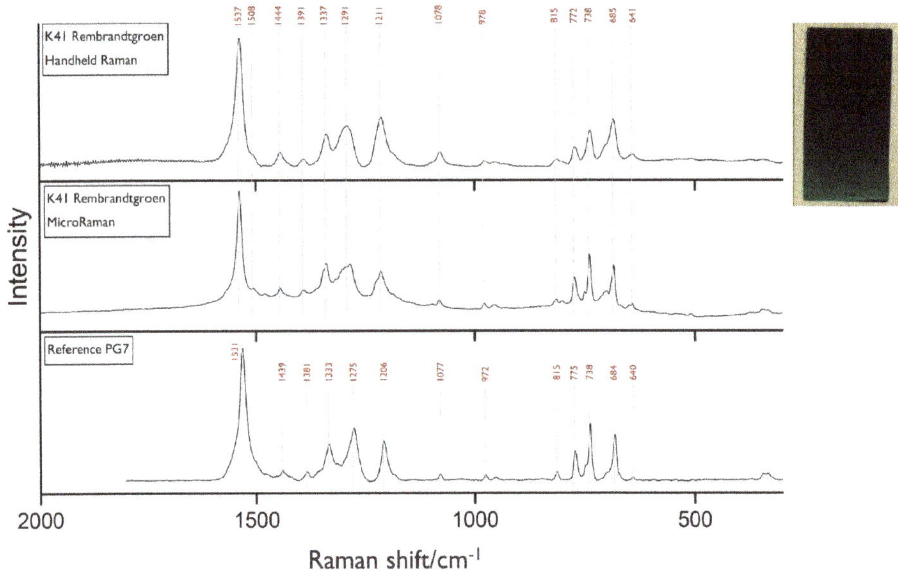

Figure 7. Handheld and microRaman spectra of paint-out K41 Rembrandtgroen and the KIK-IRPA reference spectrum for PG7.

Figure 8 reports the spectrum of Talens rose [Talens pink] in comparison with the reference spectra of the BON pigment lake Lithol Rubine, precipitated on two different substrates. This example was chosen to illustrate the possible effects of the minor spectral resolution of the handheld Raman device when examining closely related pigments. The KIK-IRPA references of Lithol Rubine PR57:1 (calcium carbonate salt) and PR57:2 (barium salt) show slight spectral differences, allowing a differentiation, given that the recorded spectrum is well defined. The spectral resolution of the Bravo device is 10–12 cm^{-1}, which turns out to be almost three times lower than that of our microRaman device (Table 1). Pozzi et al. also noted that "the inferior spectral resolution compared with benchtop spectrometers may cause difficulties in differentiating among closely related molecules with similar Raman fingerprints, especially if the analyses have poor Raman cross section or their spectra are characterized by low signal-to-noise ratio" [19]. It was, however, shown that the handheld Raman spectrum of Talens rose [Talens pink] contains the small spectral features that allow the assignment to PR57:1. In this case, the minor spectral resolution of the handheld Raman proved to be sufficient to allow a clear identification.

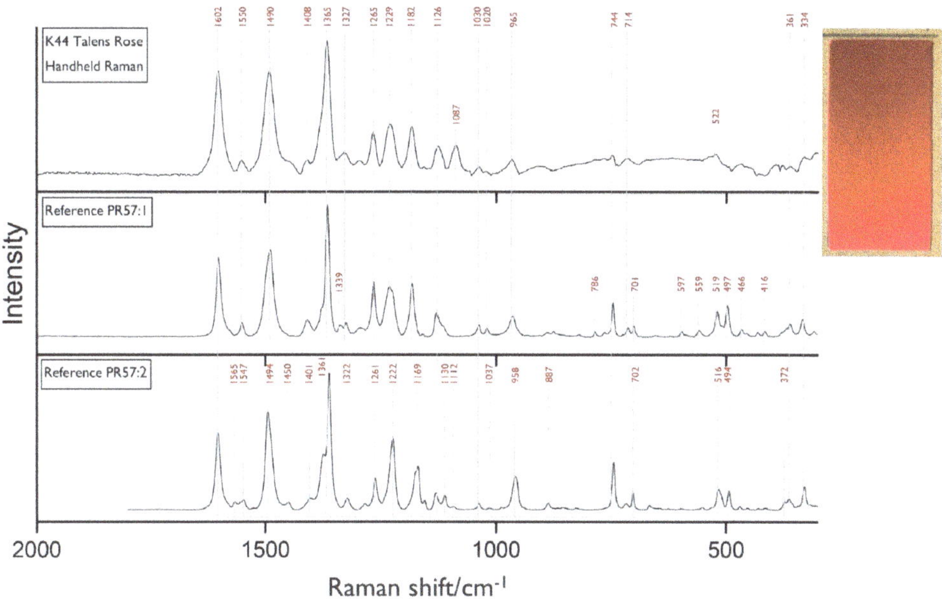

Figure 8. Handheld Raman spectrum of paint-out Talens rose and Raman spectra of the KIK-IRPA PR57:1 and PR57:2 references.

Two paint-outs (K37 Paul Veronesegroen [Paul Veronese Green] and K37 Dekgroen [Dekgreen]) were revealed to be rather sensitive to the handheld Raman lasers. The after-measurement check of the surfaces showed local burns, resulting in a small (~0.5 mm) black discoloration. According to Talens' recipes, these samples should contain a green SOP, but the X-ray fluorescence (XRF) measurements identified a copper arsenic pigment, which is rather sensitive to Raman laser irradiation. The analysis of these paint-outs highlights a strong disadvantage of the handheld Bravo, namely the non-adjustable laser energy, which is about 50 mW, more than two times higher compared to the microRaman instrument we used. The laser power can, however, be reduced by increasing the working distance to 1–2 mm. In this way, potential changes or thermal damage can be minimized or even prevented. Pozzi et al. [19] address this problem by proposing preventive measurement settings for sensitive samples. Other recommended options to avoid damage are the placement of a neutral density filter in front of the laser spot or the use of a defocusing tip.

It is worth mentioning that no SOP-containing paint-out showed any comparable damage. Since the sensitivity of emerald green is known, preliminary elemental analysis with XRF spectroscopy would have raised attention to this risk and demanded preventive action by adjusting the measurements settings.

4. Conclusions

This study demonstrated that identification of SOPs with a non-invasive approach in modern oil paint systems, also in the presence of a varnish layer, is possible. The handheld Raman was able to identify a large variety of SOPs—belonging to eight different pigment groups—within the historic varnished paint-outs of a leading artists' paint manufacturer and provided in the meantime a very useful insight into the use of early SOPs in modern artists' oil paints dated 1932–1950.

This research intends to provide a basis for future research into the non-invasive identification of SOPs in pre-1950 works of art. Detailed knowledge about the type and the characteristics of modern SOPs contributes to a better understanding of artists' materials from this period and helps to assess the restoration and conservation needs. Setting these pigments into their art technological context will contribute to studying the history of certain works of art and add to the knowledge available to detect forgeries.

Some of the SOPs identified within this study have not yet been analytically detected within works of art from before 1950. This outcome emphasizes the importance of further and intensive study of the use of SOPs in modern art. On the one hand, the knowledge of how to analyse SOPs without taking samples in a scan-like examination setup opens up new possibilities for their further investigation. On the other hand, this non-invasive approach can be used to clearly determine adequate spots for sampling, when necessary.

The advantages and drawbacks of the use of handheld Raman spectroscopy in the investigation of SOPs in artists' materials were considered. Thanks to the efficient suppression of fluorescence, varnished paint systems could be analysed and it was even possible to identify the mixtures of SOPs. The rather low resolution was a limiting factor, but turned out to be not too problematic for this set of samples. In some cases subtle spectral differences could still be detected, enabling a distinction between pigment substrates. The issue of non-adjustable energy was revealed to only be risky for one inorganic pigment, but this can be addressed with other non-invasive complementary techniques, such as handheld X-ray fluorescence spectroscopy. Another challenge is the lack of a camera for exact positioning. In this study a Melinex®template was successfully used for the analysis of the paint-outs, but the examination of specific and/or detailed paint areas in artworks could be hindered or difficult, depending on the surface and type of object.

Author Contributions: Conceptualization, R.P., I.D.v.d.W., and K.J.v.d.B.; methodology, R.P., I.D.v.d.W., and K.J.v.d.B.; investigation, R.P.; writing—original draft preparation, R.P., I.D.v.d.W., and K.J.v.d.B.; writing—review and editing, R.P., I.D.v.d.W., and K.J.v.d.B.; supervision, I.D.v.d.W. and K.J.v.d.B. All authors have read and agreed to the published version of the manuscript.

Funding: This research received no external funding.

Institutional Review Board Statement: Not applicable.

Informed Consent Statement: Not applicable.

Data Availability Statement: Data available on request. The data presented in this study are available on request from the corresponding author.

Acknowledgments: The authors would like to thank Royal Talens for allowing access to their production archive and the Historical collection of dyes of the TU Dresden for providing reference samples, as well as Art Ness Proaño Gaibor, Sanne Berbers (Cultural Heritage Laboratory, Cultural Heritage Agency of the Netherlands, Amsterdam, the Netherlands) and Brynn Sundberg (University of Amsterdam, Amsterdam, the Netherlands) for conducting conclusive UHPLC-MS analyses. Furthermore we would like to thank the SOPRANO network, especially Wim Fremout (KIK-IRPA), for providing reference spectra.

Conflicts of Interest: The authors declare no conflict of interest.

References

1. Lomax, S.Q.; Learner, T. A review of the classes, structures, and methods of analysis of synthetic organic pigments. *J. Am. Inst. Conserv.* **2006**, *45*, 107–125. [CrossRef]
2. Schäning, A. Synthetische Organische Farbmittel aus einer Technologischen Materialsammlung des 19./20. Jahrhunderts: Identifizierung, Klassifizierung und ihre Verwendung Sowie Akzeptanz in (Künstler) Farben Anfang des 20. Jahrhunderts. Ph.D. Thesis, Akademie der Bildenden Künste Wien, Wien, Austria, 2010.
3. Stege, H.; Krekel, C. *Keiner Hat Diese Farben Wie Ich—Kirchner Malt*; Schick, K., Skowranek, H., Eds.; Hatje Cantz Verlag Ostfildern: Berlin, Germany, 2011.
4. Pause, R.; Neevel, J.; van den Berg, K.J. Synthetic organic pigments in talens oil paint 1920–1950—The case of vermillion imit. In *Conservation of Modern Oil Paintings*; van den Berg, K.J., Bonaduce, I., Burnstock, A., Ormsby, B., Scharff, M., Carlyle, L., Heydenreich, G., Keune, K., Eds.; Springer International Publishing: Cham, Switzerland, 2019; pp. 109–118.
5. Doerner, M. Zur Abwehr! *Tech. Mitt. Malerei* **1931**, *47*, 13.
6. Saverwyns, S. Russian avant-garde ... or not? A micro-Raman spectroscopy study of six paintings attributed to Liubov Popova. *J. Raman Spectrosc.* **2011**, *41*, 1525–1532. [CrossRef]
7. Quillen Lomax, S.; Lomax, J.F.; De Luca-Westrate, A. The use of Raman microscopy and laser desorption ionization mass spectrometry in the examination of synthetic organic pigments in modern works of art. *J. Raman Spectrosc.* **2014**, *2*, 37–38.
8. Bouchard, M.; Rivenc, R.; Menke, C.; Learner, T. Micro-FTIR and micro-Raman study of paints used by Sam Francis. *E Preserv. Sci.* **2009**, *6*, 27–37.
9. Lutzenberger, K.; Stege, H. Künstlerfarben im Wandel—Synthetische organische pigmente des 20. Jahrhunderts und Möglichkeiten ihrer zerstörungsarmen, analytischen Identifizierung. *E Preserv. Sci.* **2009**, *6*, 89–100.
10. Schulte, F.; Brzezinka, K.W.; Lutzenberger, K.; Stege, H.; Panne, U. Raman spectroscopy of synthetic organic pigments in 20th century works of art. *J. Raman Spectrosc.* **2008**, *39*, 1455–1463. [CrossRef]
11. Sundberg, B.N.; Pause, R.; van der Werf, I.D.; Astefanei, A.; van den Berg, K.J.; van Bommel, M. Analytical approaches for the characterization of early synthetic organic pigments in artists' paints. *Microchem. J.* **2021**. under review.
12. Bell, I.M.; Clark, R.J.H.; Gibbs, P.J. Raman spectroscopic library of natural and synthetic pigments (pre- ≈ 1850 AD). *Spectrochim. Acta Part A Mol. Biomol. Spectrosc.* **1997**, *53*, 12. [CrossRef]
13. Saverwyns, S.; Fremout, W. Genuine or fake: A Micro-Raman spectroscopy study of an abstract painting attributed to vasily kandinsky. In Proceedings of the 10th International Conference on Non-Destructive Investigations and Microanalysis for the Diagnostics and Conservation of Cultural and Environmental Heritage, Florence, Italy, 13–15 April 2011.
14. Scherrer, N.C.; Zumbühl, S.; Delavy, F.; Fritsch, A.; Kühnen, R. Synthetic organic pigments of the 20th and 21st century relevant to artist's paints: Raman spectra reference collection. *Spectrochim. Acta Part A Mol. Biomol. Spectrosc.* **2009**, *73*, 505–524. [CrossRef]
15. Standeven, H.A.L. The history and manufacture of Lithol Red, a Pigment used by Mark Rothko in his Seagram and Harvard Murals of the 1950s and 1960s. *Tate Papers* **2008**, *10*, 1–8.
16. Vandenabeele, P.; Moens, L. *The Application of Raman Spectroscopy for the Nondestructive Analysis of Art Objects*; The International Society for Optical Engineering: Bellingham, WA, USA, 2000; Volume 4098.
17. Vandenabeele, P.; Edwards, H.G.M.; Moens, L. A decade of Raman spectroscopy in art and archaeology. *Chem. Rev.* **2007**, *107*, 675–686. [CrossRef] [PubMed]
18. Miliani, C.; Rosi, F.; Brunetti, B.G.; Sgamellotti, A. In situ noninvasive study of artworks: The MOLAB multitechnique approach. *Acc. Chem. Res.* **2010**, *43*, 728–738. [CrossRef]
19. Pozzi, F.; Basso, E.; Rizzo, A.; Cesaratto, A.; Tague, T.J., Jr. Evaluation and optimization of the potential of a handheld Raman spectrometer: In situ, noninvasive materials characterization in artworks. *J. Raman Spectrosc.* **2019**, *50*, 1–12. [CrossRef]
20. Longoni, M.; Freschi, A.; Cicala, N.; Bruni, S. Non-invasive identification of synthetic organic pigments in contemporary art paints by visible—Excited spectrofluorimetry and visible reflectance spectroscopy. *Spectrochim. Acta Part A Mol. Biomol. Spectroscopy.* **2020**, *229*, 117907. [CrossRef]
21. Casadio, F.; Daher, C.; Bellot-Gurlet, L. *Analytical Chemistry for Cultural Heritage, Topics in Current Chemistry Collections*; Mazzeo, R., Ed.; Springer: Cham, Switzerland, 2016; pp. 1–51.
22. Colomban, P. The on-site/remote Raman analysis with mobile instruments: A review of drawbacks and success in cultural heritage studies and other associated fields. *J. Raman Spectrosc.* **2012**, *43*, 1529–1535. [CrossRef]
23. Conti, C.; Botteon, A.; Bertasa, M.; Colombo, C.; Realini, M.; Sali, D. Portable sequentially shifted excitation Raman spectroscopy as an innovative tool for in situ chemical interrogation on painted surface. *Analyst* **2016**, *141*, 4599–4607. [CrossRef] [PubMed]
24. Jehlička, J.; Culka, A.; Bersani, D.; Vandenabeele, P. Comparison of seven portable Raman spectrometers: Beryl as a case study. *J. Raman Spectrosc.* **2017**, *48*, 1289–1299. [CrossRef]
25. Kantoğlu, Ö.; Ergun, E.; Kirmaz, R.; Kalayci, Y.; Zararsiz, A.; Bayir, Ö. Colour and ink characterization of ottoman diplomatic documents dating from the 13th to the 20th century. *Int. J. Preserv. Libr. Arch. Mater. Restaur.* **2018**, *39*, 4. [CrossRef]
26. Germinario, C.; Francesco, I.; Mercurio, M.; Langella, A.; Sali, D.; Kakoulli, I.; De Bonis, A.; Grifa, C. Multi-analytical and non-invasive characterization of the polychromy of wall paintings at the Domus of Octavius Quartio in Pompeii. *EPJP* **2018**, *133*, 359. [CrossRef]

27. Rousaki, A.; Costa, M.; Saelens, D.; Lycke, S.; Sanchez, A.; Tuñón, J.; Ceprián, B.; Amate, P.; Montejo, M.; Mirão, J.; et al. A comparative mobile Raman study for the on field analysis of the mosaico de los amores of the cástulo archaeological site (Linares, Spain). *J. Raman Spectrosc.* **2020**, *5*, 1913–1923. [CrossRef]
28. Vagnini, M.; Gabrieli, F.; Daveri, A.; Sali, D. Handheld new technology Raman and portable FT-IR spectrometers as complementary tools for the in situ identification of organic materials in modern art. *Spectrochim. Acta A Mol. Biomol. Spectrosc.* **2017**, *76*, 174–182. [CrossRef]
29. Fremout, W.; Saverwyns, S. Identification of synthetic organic pigments: The role of a comprehensive digital Raman spectral library. *J. Raman Spectrosc.* **2012**, *43*, 1536–1544. [CrossRef]
30. De Keijzer, M. The delight of modern pigment creations. In *Issues in Contemporary Oil Paint*; van den Berg, K.J., Burnstock, A., de Keijzer, M., Krueger, J., Learner, T., de Tagle, A., Heydenreich, G., Eds.; Springer: Cham, Switzerland, 2014; pp. 45–73.
31. Pause, R.; de Keijzer, M.; van den Berg, K.J. Phosphor, tungsten and molybdenum—From Brilliant to Fanal, unusual precipitation methods of triphenylmethane dyes in the early 20th century. *Stud. Conserv.* **2020**. currently under review.
32. Eskes, P. *Terugblik op de Historie van Talens*, 1st ed.; Koninklijke Talens, B.V., Ed.; Koninklijke Talens: Apeldoorn, The Netherlands, 1989.
33. Van den Berg, K.J.; van Gurp, F.; Bayliss, S.; Burnstock, A.; Ovink, B.K. Making paint in the 20th century: The Talens Archive. In *Sources on Art Technol. Back to Basic*; Archetyp Publications Ltd.: London, UK, 2016; pp. 43–50.
34. Pause, R. Synthetic Organic Pigments in Talens Oil Paint in the First Half of the 20th Century. Master's Thesis, Staatliche Akademie der Bildenden Künste Stuttgart, Stuttgart, Germany, 2018.
35. Kerdijk, F. *Kunstschildersmaterialen en Schildertechnik—Kleurstoffen, Bindmiddelen en Vernissen in de Olieverftechniek*, 1st ed.; Talens & Zoon: Apeldoorn, The Netherlands, 1932.
36. Kerdijk, F. *Kunstschildersmaterialen en Schildertechnik—Kleurstoffen, Bindmiddelen en Vernissen in de Olieverftechniek*, 2nd ed.; Talens & Zoon: Apeldoorn, The Netherlands, 1937.
37. Kerdijk, F. *Kunstschildersmaterialen en Schildertechnik—Kleurstoffen, Bindmiddelen en Vernissen in de Olieverftechniek*, 3rd ed.; Talens & Zoon: Apeldoorn, The Netherlands, 1941.
38. Kerdijk, F. *Kunstschildersmaterialen en Schildertechnik—Kleurstoffen, Bindmiddelen en Vernissen in de Olieverftechniek*, 4th ed.; Talens & Zoon: Apeldoorn, The Netherlands, 1942.
39. Kerdijk, F. *Kunstschildersmaterialen en Schildertechnik—Kleurstoffen, Bindmiddelen en Vernissen in de Olieverftechniek*, 5th ed.; Talens & Zoon: Apeldoorn, The Netherlands, 1944.
40. Kerdijk, F. *Kunstschildersmaterialen en Schildertechnik—Kleurstoffen, Bindmiddelen en Vernissen in de Olieverftechniek*, 6th ed.; Talens & Zoon: Apeldoorn, The Netherlands, 1950.
41. Wahadoszamen, M.; Rahaman, A.; Hoque, N.M.R.; Talukder, A.I.; Abedin, K.M.; Haider, A.F.M.Y. Laser Raman spectroscopy with different excitation sources and extension to surface enhanced raman spectroscopy. *J. Spectrosc.* **2015**, *2015*, 1–8. [CrossRef]
42. Colomban, P. On-site Raman study of artwork: Procedure and illustrative examples. *J. Raman Spectrosc.* **2018**, *49*, 921–934. [CrossRef]
43. Defeyt, C.; Vandenabeele, P.; Gilbert, B.; Van Pevenage, J.; Clootse, R.; Strivaya, D. Contribution to the identification of α-, β- and ε-copper phthalocyanine blue pigments in modern artists' paints by X-ray powder diffraction, attenuated total reflectance micro-fourier transform infrared spectroscopy and micro-Raman spectroscopy. *J. Raman Spectrosc.* **2012**, *43*, 1772–1780. [CrossRef]
44. González-Vidal, J.J.; Pérez-Pueyo, R.; Soneira, M.J.; Ruiz-Moreno, S. Automatic classification system of Raman spectra applied to pigments analysis. *J. Raman Spectrosc.* **2016**, *47*, 1408–1414. [CrossRef]
45. Colour Index (Ed.) *Society of Dyers and Colourists, and American Association of Textile Chemists and Colorists*, 3rd ed.; Society of Dyers and Colourists and Associates: Bradford, UK, 1971.
46. Colour Index (Ed.) *Society of Dyers and Colourists, and American Association of Textile Chemists and Colorists*, 2nd ed.; Society of Dyers and Colourists and Associates: Bradford, UK, 1956.
47. Wagner, H. *Chemie in Einzeldarstellungen*, 1st ed.; Wissenschaftliche Verlagsgesellschaft MBH: Stuttgart, Germany, 1927; Volume 13.
48. Schultz, G. *Farbstofftabellen*, 7th ed.; Akademische Verlagsgesellschaft MBH: Leipzig, Germany, 1931.
49. Steger, S.; Stege, H.; Bretz, S.; Hahn, O. A complementary spectroscopic approach for the non-invasive in-situ identification of synthetic organic pigments in modern reverse paintings on glass (1913–1946). *J. Cult. Herit.* **2019**, *38*, 20–28. [CrossRef]
50. Sichel, W. Properties and economics. In *Pigment Handbook*; Lewis, P.A., Ed.; Wiley: New York, NY, USA, 1988; Volume 1, pp. 613–615.
51. Scherrer, N.C. Laser dependent shifting of Raman bands with phthalocyanine pigments presented at RAA2011. In Proceedings of the 6th international Congress on the Application of Raman Spectroscopy in Art and Archaeology, Parma, Italy, 5–8 September 2011; pp. 203–204.

heritage

MDPI

Article

Exploring Liu Kang's Paris Practice (1929–1932): Insight into Painting Materials and Technique

Damian Lizun [1,*], Teresa Kurkiewicz [2] and Bogusław Szczupak [3]

1 Heritage Conservation Centre, National Heritage Board, 32 Jurong Port Rd, Singapore 619104, Singapore
2 Department of Painting Technology and Techniques, Institute for Conservation, Restoration and Study of Cultural Heritage, Nicolaus Copernicus University, ul. Sienkiewicza 30/32, 87-100 Toruń, Poland; teresak@umk.pl or teresa.kurkiewicz@googlemail.com
3 Department of Telecommunications and Teleinformatics, Wrocław University of Science and Technology, Wybrzeże Stanisława Wyspiańskiego 27, 50-370 Wrocław, Poland; boguslaw.szczupak@pwr.edu.pl
* Correspondence: damian_lizun@nhb.gov.sg or d.lizun@fineartconservation.ie

Abstract: This paper presents the results of an extensive study of 14 paintings by the pioneering Singapore artist Liu Kang (1911–2004). The paintings are from the National Gallery Singapore and Liu family collections. The aim of the study is to elucidate the painting technique and materials from the artist's early oeuvre, Paris, spanning the period from 1929 to 1932. The artworks were studied with a wide array of non- and micro-invasive analytical techniques, supplemented with the historical information derived from the Liu family archives and contemporary colourmen catalogues. The results showed that the artist was able to create compositions with a limited colour palette and had a preferential use of commercially available ultramarine, viridian, chrome yellow, iron oxides, organic reds, lead white, and bone black bound in oil that was highlighted. This study identified other minor pigments that appeared as hue modifications or were used sporadically, such as cobalt blue, Prussian blue, emerald green, cadmium yellow, cobalt yellow, and zinc white. With regard to the painting technique, the artist explored different styles and demonstrated a continuous development of his brushwork and was undoubtedly influenced by Modernists' artworks. This comprehensive technical study of Liu Kang's paintings from the Paris phase may assist art historians and conservators in the evaluation of the artist's early career and aid conservation diagnostics and treatment of his artworks. Furthermore, the identified painting materials can be compared with those used by other artists active in Paris during the same period.

Keywords: Liu Kang; SEM-EDS; MA-XRF; FTIR; IRFC; X-RAY; RTI; hidden paintings; pigments

Citation: Lizun, D.; Kurkiewicz, T.; Szczupak, B. Exploring Liu Kang's Paris Practice (1929–1932): Insight into Painting Materials and Technique. *Heritage* **2021**, *4*, 828–863. https://doi.org/10.3390/heritage4020046

Academic Editor: Diego Tamburini

Received: 26 April 2021
Accepted: 16 May 2021
Published: 19 May 2021

Publisher's Note: MDPI stays neutral with regard to jurisdictional claims in published maps and institutional affiliations.

1. Introduction

Liu Kang (1911–2004) was one of the most influential figures in the early development of modern art in Singapore. He was born in Yongchun, Fujian province, China. After graduating from Xinhua Arts Academy in Shanghai in 1928, he moved to Paris, where he stayed from February 1929 to April 1932. In the early decades of the 20th century, there was a growing enthusiasm in China towards the study of Western culture. Western art appealed to Chinese educational modernisers because of its realism and supposed association with science and progress [1]. Hence, the Chinese government encouraged graduates as well as established artists to further their art education in France and to promote modernisation ideas in a rapidly transforming China upon their return [2]. Liu Kang's stay in Paris had a significant influence on his career. He attended the Académie de la Grande Chaumière in Montparnasse and studied Impressionist, Post-Impressionist, and Fauvist styles [3]. In an essay from 1970, Liu Kang made a reference to his Paris phase: "We visited fine art museums and studied the masterpieces of past generations of artists, toured famous art galleries to admire recent works by contemporary artists, and gained much from this initiation" [4].

Although his Paris paintings may seem to show a chaotic mixture of different styles at first glance, a close examination reveals a variety of influences and the search for an individual technique. Liu Kang was attracted to Post-Impressionist and Fauvist styles for their expressive use of colour and form to depict a subjective view of the world [5,6]. It is worth noting that his artistic explorations of the Western painting styles achieved some success—his works were accepted by Salon d'Automne in 1930 and 1931. The variety of styles adopted by the artist in Paris invites the question of whether he actively explored different painting materials. Hence, this study aims to characterise Liu Kang's painting technique and the pigments he used during the period under review. This study expands the scope of the earlier preliminary investigation of two of Liu Kang's paintings [7,8] and the overview of his painting supports from Paris phase [9]. The collected data contribute to the knowledge of Liu Kang's painting materials. It may be useful to art historians exploring Liu Kang's workmanship and conservators planning conservation treatments of his paintings.

2. Materials and Methods

2.1. Materials

2.1.1. Investigated Paintings

The discussion focuses on seven paintings from the National Gallery Singapore (NGS) and six paintings from the Liu Kang Family Collection (Figures 1 and 2). The inclusion of the paintings from the Liu collection ensured that the overall research base represents a range of the artist's genres and painting technique during the period from 1929 to 1932. It needs to be noted that the Paris phase also includes artworks he painted during a trip to Saint Gingolph on the French–Swiss border between 8 August and 20 September 1929 [10]. The painting supports of the investigated artworks are not in the scope of this research as a comprehensive study of the supports was carried out earlier [9].

2.1.2. Samples

A total of 59 samples of the paint layer were collected only from the NGS paintings (Figure 1). There was no intention of carrying out any invasive sampling procedures on the paintings from the Liu collection.

2.2. Methods

The range of the applied analytical methods differed between the NGS and Liu collections, mainly due to the Liu family's wish to conduct only non-invasive in situ examination of their paintings. Thus, only visible light (VIS) photography of the characteristic features of the paint layers was conducted on paintings from the Liu collection to provide further insights into the artist's painting technique. The NGS paintings were studied first by means of non-invasive imaging techniques, comprising VIS, ultraviolet fluorescence (UVF), reflected ultraviolet (UVR), and near-infrared (NIR) photography. The aim of these non-invasive techniques was to conduct a preliminary characterisation of the pigments by recording their optical features under different wavelengths of the electromagnetic spectrum. An investigation of the paint textures was conducted with reflectance transformation imaging (RTI). To verify the presence of underlying paint layers, NIR and X-ray radiography (XRR) were carried out. The obtained results guided a sampling of the paint layers from the areas with prior loss, for characterising the paint mixtures in detail. Thus, the samples were analysed using optical microscopy (OM), polarised light microscopy (PLM), and field emission scanning electron microscope with energy dispersive spectroscopy (FE-SEM-EDS). A handheld portable X-ray fluorescence (XRF) was used to analyse some areas of the *Countryside in France* where sampling was not safe. A macro X-ray fluorescence (MA-XRF) scanner was engaged to perform the elemental mapping of *Landscape in Switzerland* and to support interpretation of data collected from the painting. The initial indication of the organic components of the paint samples was given by attenuated total reflectance-Fourier transform infrared spectroscopy (ATR-FTIR). The results are summarised in Appendix A, Table A1.

Figure 1. Paintings by Liu Kang from the collection of National Gallery Singapore: (**a**) *Autumn colours*, 1930, oil on canvas, 38.3 × 45.3 cm; (**b**) *Countryside in France*, 1930, oil on canvas, 46 × 54.7 cm, gift of the artist's family; (**c**) *Landscape in Switzerland*, 1930, oil on canvas, 45.6 × 55.7 cm, gift of C Y Hwang; (**d**) *Village scene*, 1931, oil on canvas, 46 × 55 cm, gift of the artist's family; (**e**) *French lady*, 1931, oil on canvas, 60.7 × 45.8 cm; (**f**) *Boat near the cliff*, 1931, oil on canvas, 53.7 × 72.4 cm, gift of the artist's family; and (**g**) *Breakfast*, 1932, oil on canvas, 46 × 54 cm. Images courtesy of National Heritage Board. White arrows indicate sampling areas.

Figure 2. Paintings by Liu Kang from the Liu Kang Family Collection: (**a**) *St Gingolph, Lac Leman, Switzerland*, 1929, oil on canvas, 45.5 × 37.5 cm; (**b**) *Landscape*, 1930, oil on canvas, 46 × 38 cm; (**c**) *Still life with books, Paris*, 1931, oil on canvas, 45 × 38 cm; (**d**) *Portrait of a man with his pipe, Paris*, 1931, oil on canvas, 45 × 38 cm; (**e**) *Self-portrait*, 1931, oil on canvas, 55 × 46 cm; and (**f**) *My landlady, Madame Normand*, 1932, oil on canvas, 54 × 45 cm. Images courtesy of Liu family.

2.2.1. Technical Photography

High-resolution technical images were acquired according to the workflow proposed by Cosentino [11–13]. A Nikon D90 DSLR modified camera with a sensitivity of between about 360 and 1100 nm was used. The camera's white balance preset for VIS photography was created with the X-Rite White Balance Target. The American Institute of Conservation Photo Documentation target was used to additionally finetune the white balance and determine optimum exposure for the VIS imaging. The photographs were shot RAW and further processed by Adobe Photoshop CC according to the standards described by the American Institute of Conservation [14].

VIS and UVF photography at 365 nm were taken with X-Nite CC1 and B + W 415 filters mounted together. NIR photography at 1000 nm, with an additional aim of false-colour infrared imaging (IRFC), was taken with Heliopan RG1000. UVR photography was achieved with Andrea "U" MK II filter.

A pair of 500 W halogen lamps were used for the VIS and NIR illumination system. The light source for UVF and UVR imaging consisted of two lamps equipped with eight 40 W 365 nm UV fluorescence tubes.

2.2.2. RTI

RTI was carried out according to the workflow proposed by the Cultural Heritage Imaging [15]. The images were processed using Adobe Photoshop CC, followed by RTIBuilder and RTIViewer software, as proposed by the Cultural Heritage Imaging [16,17].

2.2.3. XRR

The paintings from the NGS collection were digitally XRR using a Siemens Ysio Max digital XRR system with a detector size of 350 × 430 mm and a resolution of over 7 million pixels. The X-ray tube operated at 40 kV and 0.5–2 mAs. The images were processed with iQ-LITE imaging software and then exported to Adobe Photoshop CC for final alignment and merging.

2.2.4. XRF

XRF measurements were performed using Thermo Scientific™ Niton™ XL3t 970 spectrometer with a GOLDD+ detector and an Ag anode X-ray tube with a 6–50 kV voltage and up to 0.2 mA current. A mining mode with four elemental ranges and a measurement duration of 50 s each (total acquisition time of 200 s) was used to better differentiate the light elements from the heavy ones. The spectra were collected from a 3 mm diameter spot size. The acquired spectra were collected and processed using Thermo Scientific™ Niton Data Transfer (NDT™) 8.4.3 software.

2.2.5. MA-XRF

The elemental mapping of *Landscape in Switzerland* was conducted with the M6 Jetstream from Bruker Nano GmbH. The instrument's measuring head is equipped with a Rh-target microfocus X-ray tube (30 W, maximum voltage of 50 kV, maximum current of 0.6 mA), and a 30 mm^2 active area XFlash Silicon Drift Detector (energy resolution of <145 eV for Mn-Kα). The measuring head is mounted on an X-Y-Z motorised stage with a maximum travel range of 800 × 600 × 90 mm. The instrument offers an adjustable spot size from 100 µm to approximately 500 µm [18]. The paintings' elemental distribution maps were acquired with a dwell time of 10 ms/pixel, a pixel size of 300 µm, and an anode current of 599 µA. The data were collected and analysed with the Bruker's M6 software.

2.2.6. FE-SEM-EDS

The paint samples' cross-sections were mounted on carbon tapes and examined with a Hitachi SU5000 FE-SEM coupled with Bruker XFlash® 6/60 EDS. The SEM, backscattered electron mode (BSE) was operated with 60 Pa chamber pressure, accelerating voltage of

20 kV, 50–60 intensity spot and a working distance of 10 mm. The acquisition of data and processing were performed using the Bruker ESPIRIT 2.0 software.

2.2.7. ATR-FTIR

ATR-FTIR analyses were performed using a Bruker Hyperion 3000 FTIR microscope equipped with a mid-band MCT detector, coupled to a Vertex 80 FTIR spectrometer. For each sample, 64 scans were recorded in the spectral range of 4000–600 cm^{-1} and resolution of 4 cm^{-1}. The spectra were interpreted with Bruker Opus 7.5 software.

2.2.8. OM and PLM

The samples' structure was studied in reflected VIS and UV light on a Leica DMRX polarised microscope at magnifications of ×50, ×100, ×200, and ×400 coupled with a Leica DFC295 digital camera. PLM was carried out in transmitted VIS light at magnifications of ×100, ×200, and ×400 using the methodology developed by Peter and Ann Mactaggart [19].

2.2.9. Preparation of Samples

The samples selected for the cross-sections were embedded in a fast-curing acrylic resin, ClaroCit from Struers (Cleveland, OH, USA), and fine polished. The PLM pigment dispersions were prepared with a Meltmount nD = 1.662 mounting medium from Cargille (Cedar Grove, NJ, USA).

2.3. *Archival Sources*

The technical examination of paintings was supplemented with materials from the Liu family archives to elucidate the artist's painting practice. These materials include Liu Kang's photographs, watercolours artwork, and a Lefranc artists' colourmen price list of oil colours from October 1928. The latter may reflect the artist's interest in Lefranc colours as he had brought the price list home [8]. However, firm conclusions about any links between the Lefranc colours and materials used by Liu Kang should not be made due to weak evidence at the current stage of the research. From observing the advertisement sections of Le Salon's 1930 and 1932 exhibition catalogues [20,21] and Salon des indépendants 1930 [22], it is deducible that Liu Kang might have had access to a great range of painting materials from other local manufacturers and retailers [9]. He also might have had an opportunity to purchase the overseas brands from retailers, such as Lechertier Barbe LTD, Paris American Art Co., and S. C. & P. Harding (Paris) LTD (Figure 3). The advertisements of these retailers listed a selection of local brands such as Lefranc, Bourgeois Ainé, as well as imported materials from Rowney, Winsor & Newton (W&N), Talens, and Schmincke. While it remains uncertain what brand of colours the artist used, some references are made to the contemporary colourmen catalogues, such as the W&N catalogue from 1928, Lefranc catalogues from 1928 to 1931, and Bourgeois Ainé catalogue from 1929 to provide insights into the availability of certain pigments that are found in the paintings under investigation.

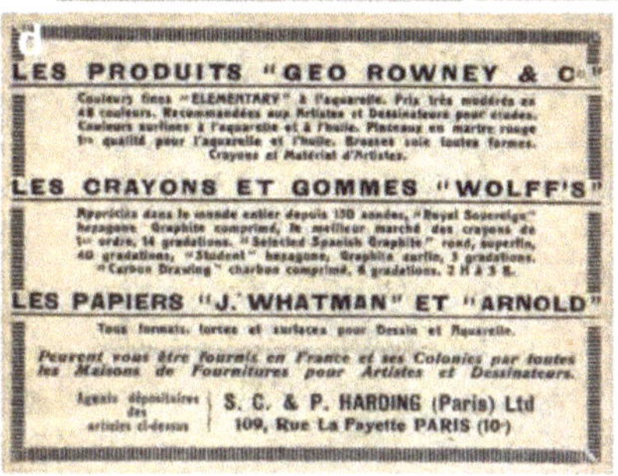

Figure 3. Advertisements by official retailers of local and overseas art materials, available in Paris in 1930 and 1932: (**a**) Lechertier Barbe LTD; (**b**) Paris American Art Co.; (**c**) Talens and Zoon, S.A. advertisement indicating Paris American Art Co. as its Paris representative; and (**d**) S. C. & P. Harding (Paris) LTD.

3. Results and Discussion

3.1. Pigments

3.1.1. Blue

Blue painted areas of the examined paintings are very often imaged purple in an IRFC, suggesting a use of ultramarine and/or co-containing blue. The analyses confirmed that the different blue hues primarily involve ultramarine (sulphur-containing sodium aluminium silicate), observed with PLM by particles with low refractive index that appear red with Chelsea filter and by SEM-EDS detection of Na, Al, Si, and S. Admixtures of viridian (hydrated chromium oxide) were confirmed by the detection of Cr and coincident PLM observation of large and rough particles with high refractive index. The presence of ultramarine and viridian is compliant with the purple appearance of the paint mixture in IRFC. Admixtures of cobalt blue (cobalt aluminium oxide) were confirmed with SEM-EDS by the concomitant presence of Co and Al as well with PLM (particles with a high refractive index that appear red with Chelsea filter) in *Boat near the cliff* (sample 26) and *French lady* (sample 20). Very dark blue in *Countryside in France* (sample 5) was probably made by mixing ultramarine with carbon black. It is worth noting that no Prussian blue (hydrated iron hexacyanoferrate) was found in the investigated blue mixtures, despite its discovery in the earlier study [7]. Prussian blue is known for its high tinting strength, which may be easy for artists to handle. It also produces dark greenish-blue shades that Liu Kang probably did not find suitable for painting sky or water. He might also be aware of the Prussian blue's tendency to fade or change in colour over time [23,24].

In almost all examined blue paint samples, except those from *Landscape in Switzerland* and *Boat near the cliff*, a high concentration of Pb seems to point to the consistent use of lead white (lead carbonate) for achieving lighter tints. Meanwhile, the presence of Ba, S, and Zn elements could be related to lithopone (a mixture of barium sulfate and zinc sulfide) and/or zinc white (zinc oxide) and barium white (barium sulfate), a common extender for lead white [25] as well as for viridian [26].

3.1.2. Green

The acquired PLM and SEM-EDS data revealed that viridian was Liu Kang's consistent choice for green. However, it is very often present in combination with other pigments. For instance, positive identification of emerald green was possible in sample 18 from *Autumn colours* with SEM-EDS detection of Cu and As elements and with FTIR by absorption bands at 1555, 1451, 816, 762, 690, and 635 cm^{-1} (Appendix A, Figure 1) [27]. However, based only on the concomitant presence of Cu and As, elements in other green paint mixtures, emerald green (copper acetoarsenite), and/or Scheele's green (copper arsenite) were considered.

According to Bourgeois Ainé and Lefranc's catalogues of oil paints, *Vert Veronese* (emerald green), *vert de Scheele* (Scheele's green), and its variant, *vert minérale,* were available during that time in Paris (Appendix A, Figures 3 and 4). Emerald green was also offered by W&N (Appendix A, Figure 2). Although it is known that some grades of emerald green were commercially adulterated with chromium pigments [28,29], MA-XRF distribution maps of Cr and Cu elements from *Landscape in Switzerland* showed only a partial co-location of Cu and Cr in the green areas (Figure 4). Hence, it could be said that the Cu–As-based green was not modified by the manufacturer but deliberately mixed with viridian where it suited the artist.

The analyses of the light and warm green hues of the investigated paintings confirmed a consistent use of chrome yellow (lead chromate) as an admixture of viridian. This can be exemplified by sample 3 from *Village scene* (Figure 5a). In this sample, besides viridian, chrome yellow was detected with PLM by anisotropic, rod-shaped particles with a high refractive index. The SEM-EDS analyses of the green and yellow clusters of not properly mixed paint indicated a varied intensity of Pb-, Ca-, and Cr-peaks contributing to viridian and chrome yellow, probably extended with chalk (calcium carbonate) or calcium chromate [30]. The presence of viridian in the examined paint is in agreement with the purple imaging of the sampling area (Figure 5b,c).

The co-location of Cr- and Cd-signals recorded with MA-XRF of *Landscape in Switzerland* (Figure 4) suggested an addition of cadmium yellow (cadmium sulfide) to viridian. This observation was corroborated with red UV fluorescence of the yellow particles [11], SEM-EDS detection of strong Cd- and S-signals, and PLM observation of sample 21 from *Landscape in Switzerland* (yellow, anisotropic particles with a high refractive index turn green in crossed polarised filters) (Figure 6). However, the presence of S, Ba, and Zn in the investigated mixture may suggest that, instead of pure cadmium yellow with lithopone and/or zinc and barium whites, zinc-modified light cadmium yellow or cadmopone (co-precipitated cadmium sulfide and barium sulfate) was used [31]. A high concentration of Pb in most of the examined light-green samples may suggest concurrent admixtures of chrome yellow and/or lead white to viridian. A combination of viridian and lead white to obtain a light-green tone was detected in *Breakfast* (sample 4). Meanwhile, strong Zn-signals recorded in the light-green brushstrokes of *Landscape in Switzerland* (sample 20) and *Boat near the cliff* (sample 11) suggest an admixture of lithopone and/or barium white and zinc white.

Figure 4. Visible light image and MA-XRF maps of *Landscape in Switzerland*, showing the distribution of the major elements. The greyscale corresponds to the intensity of the signal of each element: white equals high intensity, black means low intensity. The colour maps combine distribution of Cu- and Cr-based pigments and Fe- and Sn-based pigments.

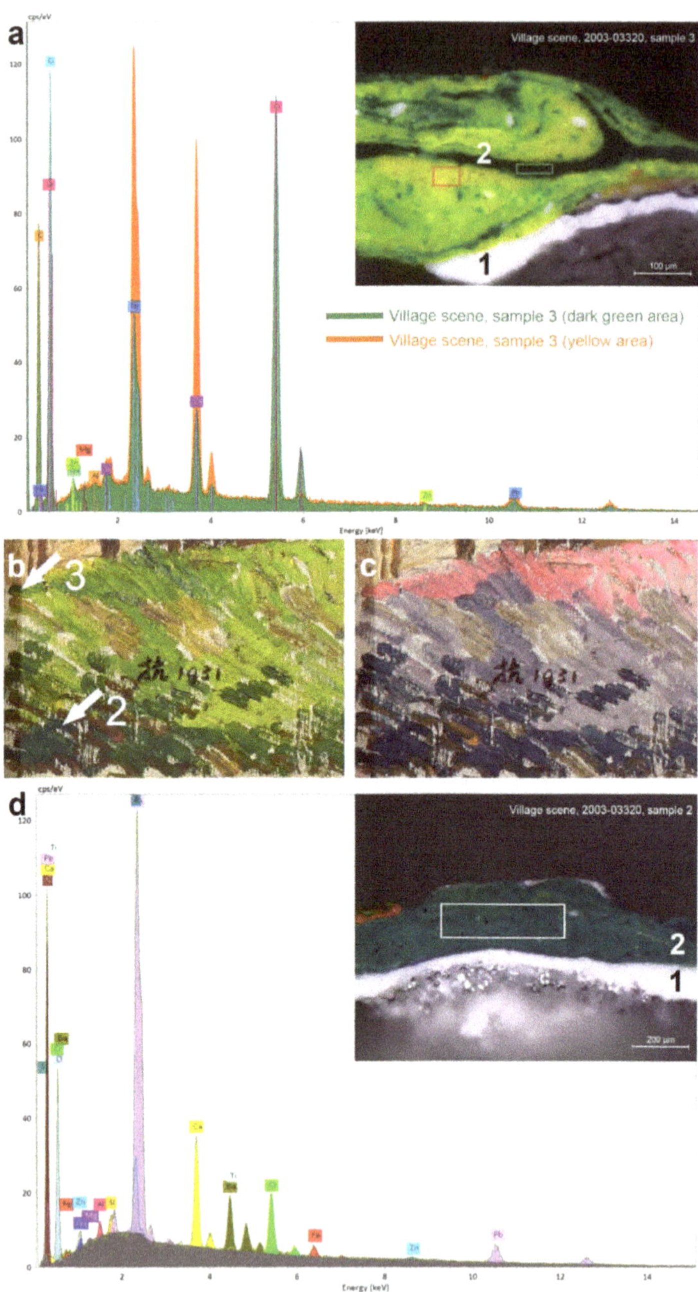

Figure 5. SEM-EDS spectra of: (**a**) green and yellow areas in sample 3 extracted from *Village scene* and inset microscopy image of the cross-section with marked areas of analyses; (**d**) dark-green paint from sample 2 extracted from the same painting and inset microscopy image of the cross-section with the marked area of analysis. The detail of the painting shows the sampling spots (**b**) and infrared false-colour image of the same areas, revealing the distribution of viridian (purple) and the mixture of Prussian blue with viridian and lead white (tints of violet) (**c**).

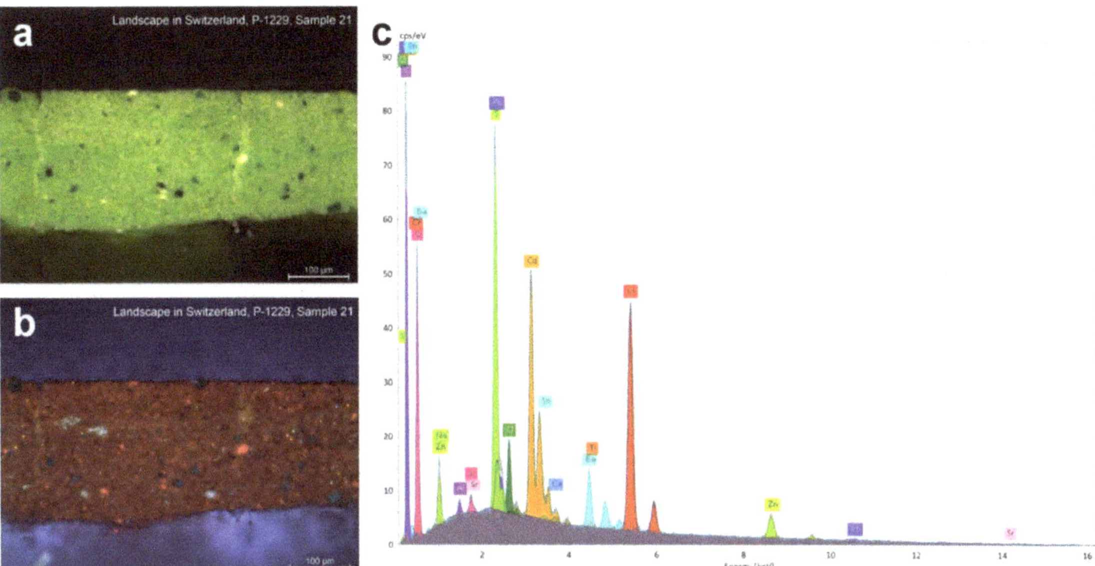

Figure 6. Microscopy images of the cross-section of sample 21, extracted from *Landscape in Switzerland*, photographed in: (**a**) VIS; (**b**) UV. The corresponding SEM-EDS spectra of the green paint are seen in (**c**). Red fluorescence of yellow particles and a strong Cd-signal suggested the presence of cadmium yellow.

Based on the SEM-EDS and PLM analyses, dark green was very often achieved by mixing viridian with Prussian blue in combination with lead white and/or chrome yellow. This can be exemplified by sample 2 in *Village scene* (Figure 5d). The SEM-EDS analysis of the dark-green paint revealed that it is rich in lead, chromium, and iron, which can be assigned to lead white, viridian, and Prussian blue. The latter was observed with PLM by dark-blue isotropic particles with a low refractive index that appear dark green with a Chelsea filter. This pigment mixture is consistent with the IRFC imaging, as the dark-violet colour is determined by the purple representation of viridian, combined with a dark-blue representation of Prussian blue (Figure 5b,c). A similar pigment mixture was identified with PLM and SEM-EDS in dark-green paint from *Countryside in France* (sample 11). However, FTIR did not depict any peaks attributable to viridian due to overlapping bands of other compounds; the most intensive peaks of this pigment fell in the range of 500–400 cm^{-1}, behind the spectral range of the measurement. Although the artist did not employ Prussian blue in the blue painted areas, he preferred it for his hue modification of the green colours. Interestingly, in *French lady* (sample 15), in addition to Prussian blue, a cobalt blue was added to the green paint to produce the desired hue.

The green colour, obtained by mixing blue and yellow, was found in *Landscape in Switzerland* (sample 20). The green sampling area appears blue in the IRFC, suggesting the presence of Prussian blue. The SEM-EDS detection of Pb, Cr, and Fe elements, combined with PLM and FTIR analyses, confirmed chrome yellow by absorption peaks at 1061, 967, 826, and 626 cm^{-1} and Prussian blue by absorption peak at 2086 cm^{-1}. The paint mixture also contains yellow iron oxide observed with PLM (anisotropic yellow particles with a high refractive index). However, its FTIR confirmation was inconclusive due to overlapping bands of chrome yellow. Moreover, there is a possibility that the artist used a commercial chrome green composed of Prussian blue and chrome yellow [26,32]. Such composite paint was available from Bourgeois Ainé in three different hues and from Lefranc in five hues under the name of *vert anglais* (Appendix A, Figures 3 and 4). It was also listed by W&N

as chrome green, available in three hues, and as cinnabar green, available in five hues (Appendix A, Figure 2).

3.1.3. Yellow

The analyses of the samples of yellow paints revealed the use of four different yellow pigments, but it is evident that chrome yellow was a prevailing pigment. It was identified as a principal component of sample 9 from *Countryside in France* and sample 8 from *Village scene*. It appears as an ingredient found together with yellow iron oxide in *Autumn colours* (sample 3) and *Breakfast* (sample 3). Nevertheless, it is difficult to ascertain if chrome yellow and iron oxide were deliberately mixed by the artist on the palette or commercially prepared. It is known that the ochres were tinted during the manufacturing process with the addition of chrome yellow [30,33], which also has its own extenders, such as barium white, gypsum, kaolin, and calcium chromate [30].

Analyses of the yellow paint mixture from *Landscape in Switzerland* (sample 23) allowed features consistent with cadmium yellow or its variants—light cadmium yellow or cadmopone—to be detected. As visualised by the MA-XRF Cd-distribution map, a usage of this pigment for painting yellow areas was very limited; however, it occurs extensively as an admixture in light-green passages (Figure 4).

The SEM-EDS detection of Co and K in the samples of yellow paint from *Landscape in Switzerland* (samples 23, 36) allowed their attribution to cobalt yellow (potassium cobaltinitrite), later confirmed with PLM by observation of isotropic, yellow, and dendritic particles with a high refractive index (Figure 7). The use of cobalt yellow by Liu Kang seems to be unusual as the pigment is known for its undesirable low hiding power in oil medium. Therefore, its main application was usually a watercolour technique [34,35]. Cobalt yellow was available from Bourgeois Ainé and Lefranc only as a dry pigment and in watercolour and gouache paints. However, it appears in W&N catalogues of oil paints (Appendix A, Figure 2).

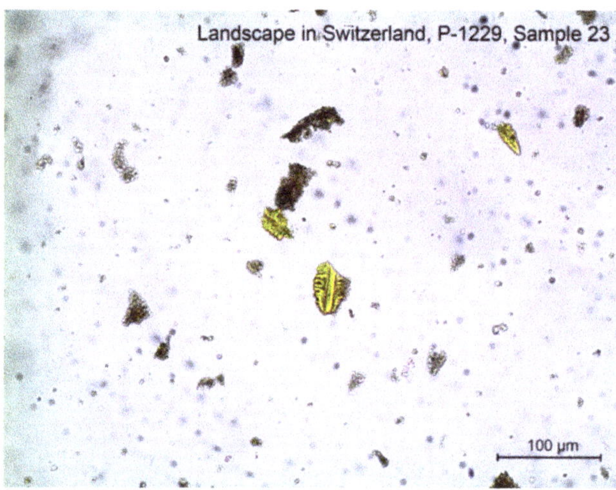

Figure 7. Cobalt yellow particles from sample 23, extracted from *Landscape in Switzerland*, photographed in plane-polarised light.

3.1.4. Brown

Brown passages were painted using predominantly brown and yellow iron oxides. Lighter tints were produced with a small addition of lead white and/or Cr-containing yellow(s). Darker-brown brushstrokes in *Boat near the cliff* (sample 14) contain Ca and P elements that can be assigned to bone black admixture, confirmed with PLM (anisotropic black and grey particles). In addition, the co-location of Fe and Mn recorded with SEM-EDS suggests the presence of umber. The SEM-EDS and PLM of the dark-brown colour from the armrest of the chair in *Breakfast* (sample 10) suggested a mixture of red iron oxide with Prussian blue and organic red pigment (unique low refractive index). FTIR completed this outcome by detecting of some absorption bands indicative of organic red at 1656, 1623, 1545, 1451, and 1270 cm^{-1}. However, more peaks typical for organic red fall in the range from 1200 to 1000 cm^{-1} and were interfered by the presence of red iron oxide. In addition, the paint mixture contains a good deal of starch grains with an extinction cross, which were visible in cross-polarised light (Figure 8a–d). However, FTIR measurements detected only one peak at 3270 cm^{-1} that is related to starch, while further identification was hampered by the intensive peak of red iron oxide at 1005 cm^{-1} and overlapping characteristic bands of starch at 1014 and 995 cm^{-1}. Starch was frequently added to the organic reds during the manufacturing process to improve its handling properties and to obtain a lighter tint [36,37]. The PLM and FTIR analyses of the paint sample correlate with the occurrence of Sn, detected with SEM-EDS, suggesting the presence of a tin-containing substrate for the organic red, which was usually cochineal lake or brazilwood lake pigment (Figure 8e) [38,39]. A composition containing alizarin crimson and brazilwood on starch and tin substrates was reported in the earlier study of Liu Kang's painting [8]. Lefranc listed a tin-containing organic red *laque anglaise* composed of *carmin* and *oxyde d'etain* (tin oxide); it was one of the most expensive, priced at 22 francs for a No. 6 tube and 9.50 francs for tube No. 2, whereas blanc d'argent (lead white) was at 3.75 francs for a tube No. 6. (Appendix A, Figure 4). *Laque anglaise* from Bourgeois Ainé had a comparable price (Appendix A, Figure 3).

3.1.5. Red

The analytical results showed that Liu Kang employed three types of organic red pigments in most of the examined paintings. They were used as primary reds or in combination with other pigments added to modify hue. An Sn-containing organic red on starch substrate was confirmed with SEM-EDS, PLM, and FTIR in *Village scene* (sample 9) and *Landscape in Switzerland* (sample 35, layer 1) and resembled the aforementioned mixture from *Breakfast* (sample 10).

The FTIR of sample 9 from the *Village scene* showed low intensity peaks at 1562 and 1545 cm^{-1}, which are indicative of organic red; however, an in-depth molecular characterisation was not possible due to overlapping signals of other compounds present in the investigated sample. A similar issue concerns the identification of starch. Although starch was well observed with PLM, only two FTIR peaks, at 3280 and 1649 cm^{-1}, were considered conclusive. Additionally, detection of Fe and co-location of Pb-, Cr-, and Ca-signals, suggested that an organic red was used in a combination with red iron oxide (absorption peaks at 1026, 1007, 935, 911, 798, 777, and 753 cm^{-1}), chrome yellow (absorption peaks at 1031, 844, 834, and 624 cm^{-1}), chalk (absorption peaks at 1410, 870, and 712 cm^{-1}), and probably lead white (1410 and 677 cm^{-1}) (Figure 9).

The sample 35 from *Landscape in Switzerland* contains two distinguishable layers. Based on the PLM, SEM-EDS, and FTIR, the top red is composed mainly of red iron oxide with an admixture of Sn-containing organic red, detected by absorption peaks at 1620, 1563, 1555, 1529, 1471, 1290, 1265, 1246, and 588 cm^{-1}, and chalk. This top layer was applied over dry paint predominantly consisting of the organic red, detected by peaks at 1621, 1562, 1552, 1531, 1310, 1265, 1247, and 604 cm^{-1}. In addition, a presence of starch was confirmed with PLM observation and FTIR by absorption peaks at 3294, 1639, 1370, 1341, 1247, 1204, 1150, 1077, 1016, 931, 861, 759, and 704 cm^{-1}, while Sn-based substrate was confirmed with the SEM-EDS. Moreover, the analysed paint contains some admixtures of lead white and probably Cr-containing yellow(s) (Figure 10). The MA-XRF map of Sn distribution in *Landscape in Switzerland* shows that some passages were initially painted with a heavy use of Sn-containing organic red (Figure 4) and finally covered with different colours as visualised on the cross-section of sample 25 (Figure 11).

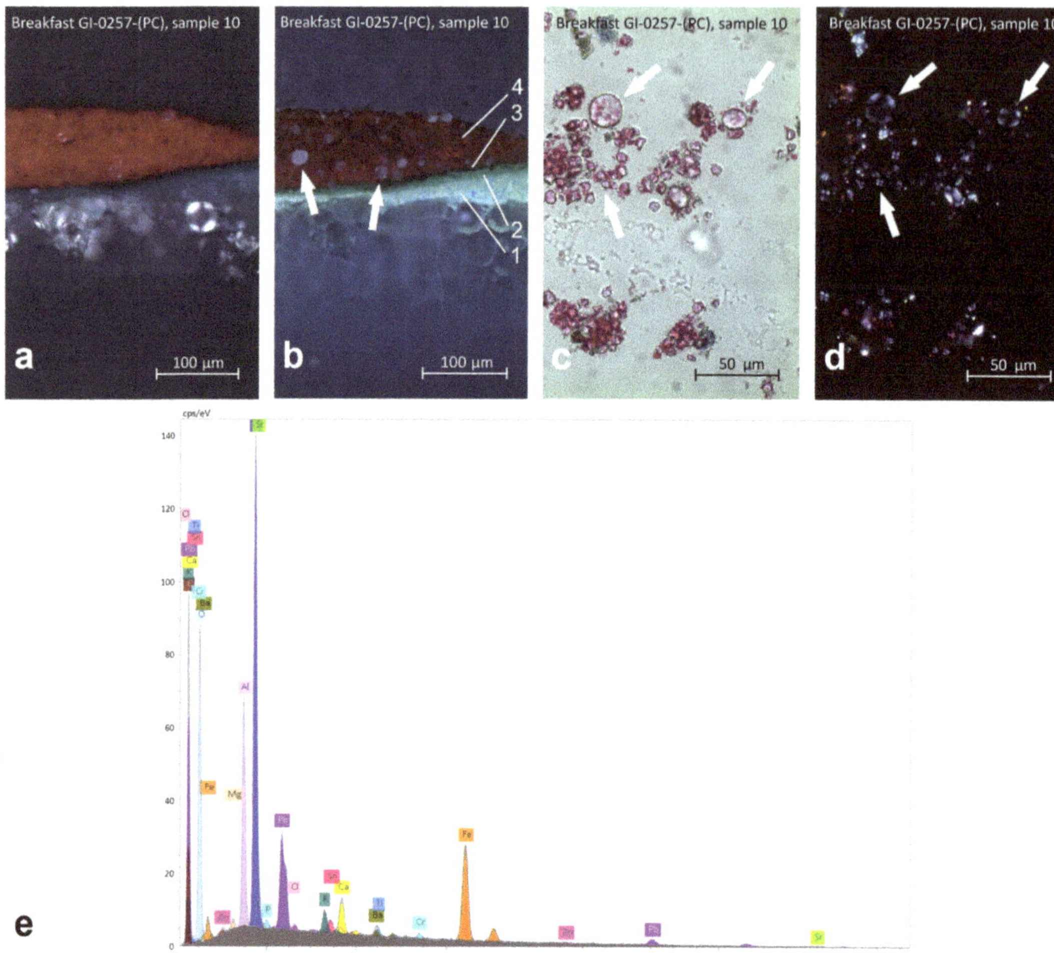

Figure 8. Microscopy images of the cross-section of sample 10, extracted from *Breakfast*, photographed in: (**a**) VIS; (**b**) UV. Circular, blue-fluorescing particles of the starch substrate are visible in layer 4 and marked with arrows (**b**). The PLM pigment dispersion from layer 4 is photographed in: (**c**) plane-polarised light; (**d**) cross-polarised light. The clumps of starch particles are marked with arrows. The corresponding SEM-EDS spectra of the red paint from layer 4 (**e**) shows a strong Sn-signal, suggesting the presence of tin substrate.

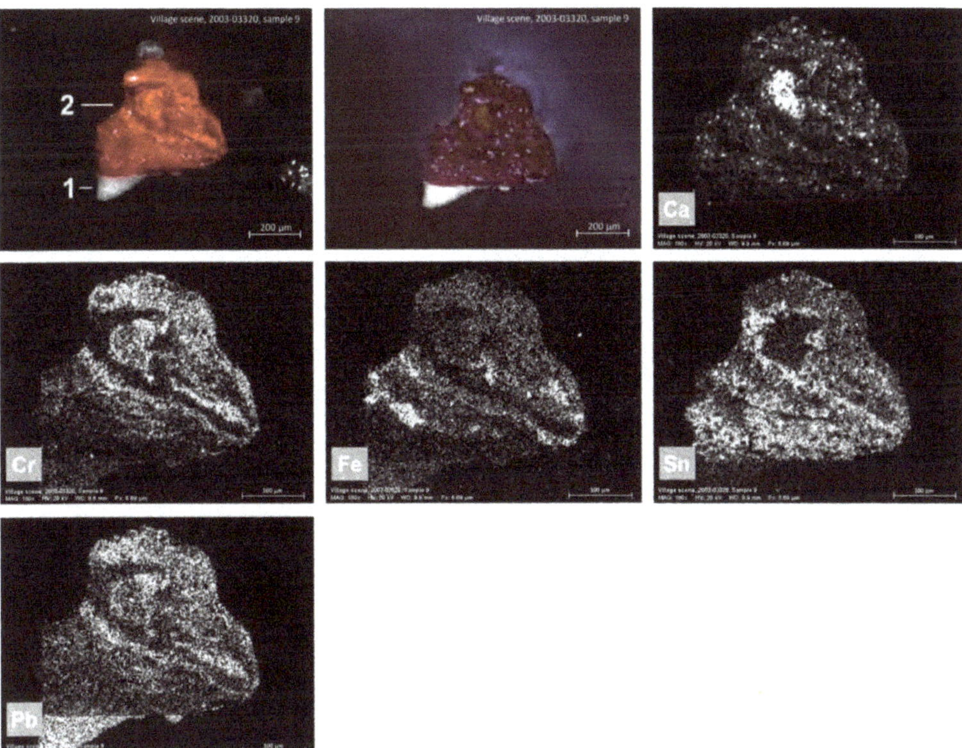

Figure 9. Microscopy image of the cross-section of sample 9 extracted from *Village scene*, photographed in VIS and UV, followed by SEM-EDS maps showing the distribution of the detected elements. The greyscale corresponds to the intensity of the signal of each element: white equals high intensity, black means low intensity. A high intensity of Sn can be assigned to the tin substrate of organic red while the co-location of Ca-, Cr-, and Pb-signals in the centre of layer 2 suggests the presence of chrome yellow.

Another organic red was detected in *Boat near the cliff* (sample 29). PLM allowed the observation of red particles with a low refractive index, which is characteristic of organic reds. An intense orange fluorescence in UV light (Figure 12b) suggests natural madder [38]; however, the SEM-EDS detection of bromine, based on Br $L\alpha_1$ and Br $K\alpha_1$ signals (Figure 12c), indicates that the red may be a compound related to eosin red—commercially known as geranium lake—which is also characterised by the orange UV fluorescence [40–42]. A comparison of the FTIR spectra of the investigated red with the reference sample of eosin Y revealed some degree of matching. Typical FTIR features consistent with the organic red were identified by absorption bands at 3335, 1561, 1455, 1345, 1221, 1174, 981, 877, 802, 766, 717, 667, and 634 cm^{-1}. However, the overlapping bands at 1455, 1174, 981, 802, 717, and 634 are also attributable to lithopone and/or barium white, oil, and acrylic resin, the latter of which is considered to be a varnish and applied during the conservation treatments in 2006 (Figure 13) [43]. Other elements present in the sample, such as Pb, Ba, Al, and S, were difficult to interpret. They can be assigned to lead white admixture and barium white extender. However, it is known that eosin was used to produce lake pigments (such as geranium lake), usually precipitated on an Al- or Pb-containing substrate [32,44–47]. Thus, Ba and S could be assigned to barium white, which is a common extender of lake pigments. Nevertheless, a presence of lithopone is suggested based on the FTIR detection of characteristic peaks at 1174, 1116, 1077, 981, 634, and 607 cm^{-1}.

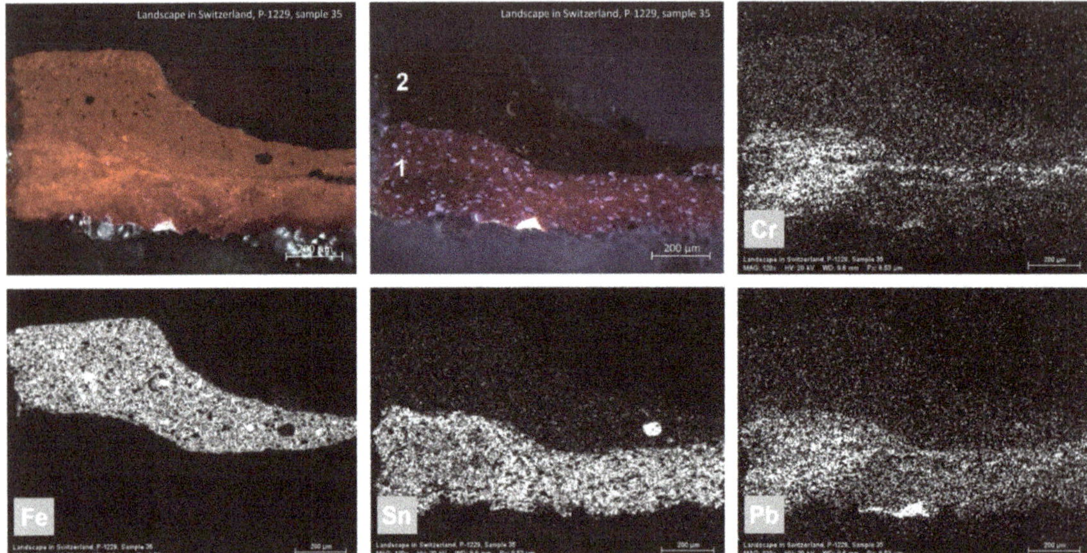

Figure 10. Microscopy image of the cross-section of sample 35 extracted from *Landscape in Switzerland*, photographed in VIS and UV, followed by SEM-EDS maps showing the distribution of the detected elements. The greyscale corresponds to the intensity of the signal of each element: white equals high intensity, black means low intensity. The high intensity of Sn in layer 1 can be assigned to the tin substrate of organic red, which is probably mixed with chrome yellow based on the co-location of Cr- and Pb-signals. Layer 2 reveals strong Fe- and weak Sn-signals, suggesting a mixture of iron oxide with organic red with tin substrate.

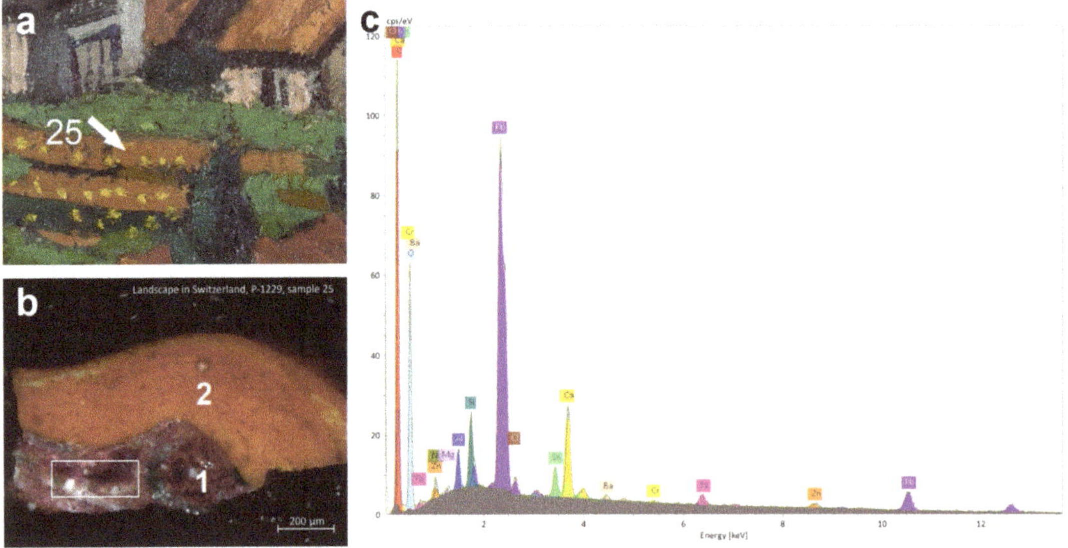

Figure 11. (**a**) Detail of *Landscape in Switzerland*, showing the sampling spot; (**b**) microscopy image of the cross-section of sample 25 with the marked area of SEM-EDS elemental analyses; (**c**) corresponding SEM-EDS spectra of the analysed area from layer 1, indicating a strong Sn-signal from the tin substrate of organic red.

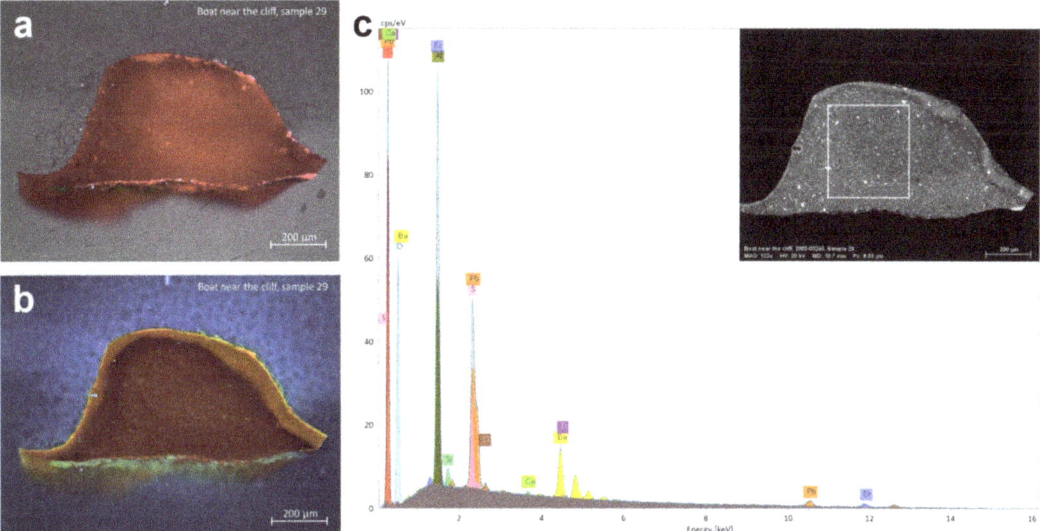

Figure 12. Microscopy image of the cross-section of sample 29 extracted from *Boat near the cliff* photographed in: (**a**) VIS; (**b**) UV. SEM-EDS spectra of the sample and inset backscattered electron (BSE) image with marked area of analyses (**c**). The spectra shows Br peaks which could be indicative of geranium lake, as well as Pb, Ba, Al, and S, which can be assigned to lead white and barium white extender and/or Pb- or Al-based substrate.

Lefranc marketed the *laque geranium* (geranium lake) with information that it is aniline-based pigment (Appendix A, Figure 4), although recent research identified eosin in its sample from 1926 [48]. Bourgeois Ainé listed *laque géranium* and *rouge géranium* (geranium red) without a chemical description, while geranium lake from W&N was derived from coal tar, according to their catalogue (Appendix A, Figures 2 and 3). Although the investigated red paint from *Boat near the cliff* exhibits features intrinsic to geranium lake, more analyses are needed to better elucidate its composition.

The organic red was also observed with PLM in the sample extracted from *Countryside in France* (sample 7). The sample does not fluoresce in UV, and the SEM-EDS detection of Al might be indicative of Al-containing substrate for the organic red detected with FTIR by peaks at 1617, 1576, 1559, 1506, 1499, 1447, 1417, 1397, 1343, 1303, 1276, and 1256 cm^{-1} [49]. However, the insufficient suit of FTIR peaks or low intensity of peaks did not allow a conclusive attribution.

In *Autumn colours* (sample 7) and *French lady* (sample 9), red paints are composed mainly of red iron oxide, modified with minor admixtures.

3.1.6. White

UVR photography is a powerful tool for a preliminary differentiation of lead white painted areas from zinc white and titanium white (titanium dioxide). Thus, lead white was observed in almost all paintings, based on its unique ability to reflect UV (Figure 14a,b). Additional SEM-EDS analyses showed that lead white occurs with a calcium carbonate, which was probably added by the manufacturer as an extender. It is worth noting that the examined white from the *Countryside in France* (sample 29) is a ground layer intentionally exposed by the artist during the painting process. It is composed predominantly of lead white with admixtures of barium, zinc, and titanium whites [9], suggesting it is of a different grade from the lead white identified in the artist's white paints [28]. Likewise, the white ground skilfully exposed by the artist in *Boat near the cliff* for describing foamy water is composed of mixture of lead and zinc whites [9]. White brushstrokes that turn dark grey and black in the UVR of *Landscape in Switzerland* and *Boat near the cliff* suggested a use of UV-absorbent titanium white or zinc white (Figure 14c,d). The latter was confirmed by the yellow-green UV fluorescence and SEM-EDS measurements of the white paint cross-sections from both paintings. Additionally, the MA-XRF of *Landscape in Switzerland* visualised a strong Zn-signal, which correlates with the white painted areas, while the Pb distribution map suggests chrome yellow and admixtures of lead white (Figure 4). The SEM-EDS detection of Ba and S in the sample 22 of *Landscape in Switzerland* suggests a common admixture of lithopone and/or barium white [50]. A minor and trace presence of Ti identified only in the colour mixtures may suggest a commercial admixture of titanium white.

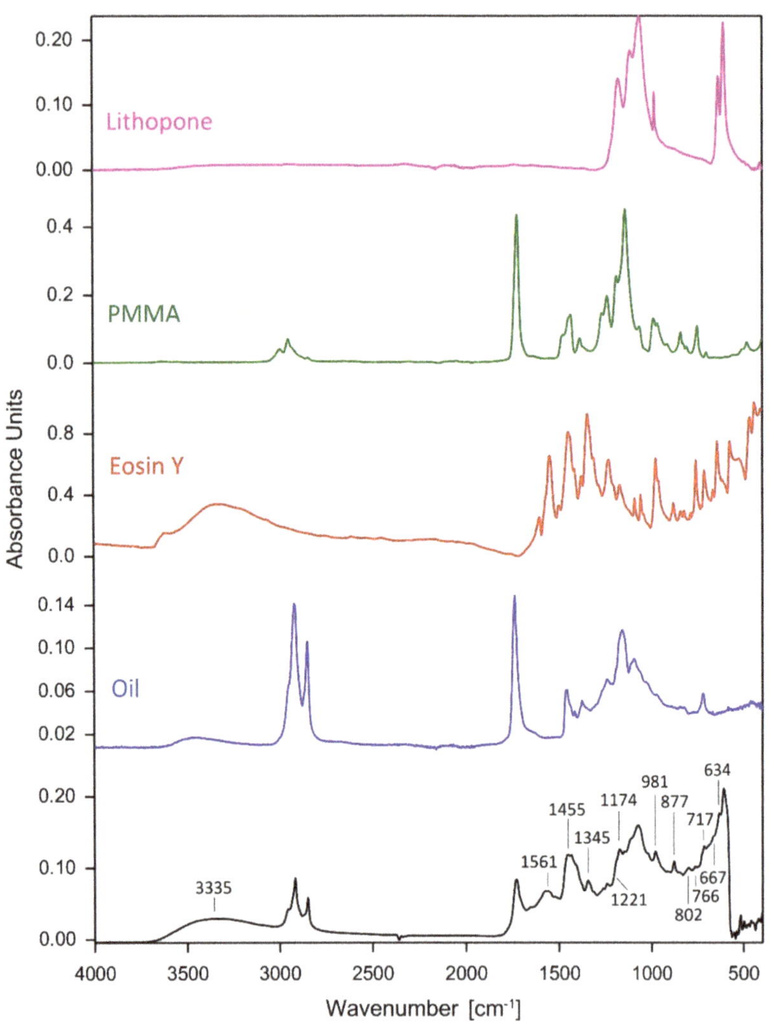

Figure 13. ATR-FTIR spectra of red paint from sample 29, taken from the *Boat near the cliff*, with labelled marker peaks of organic red and spectra of reference samples, identifying oil, eosin Y, acrylic resin, and lithopone.

Figure 14. VIS and corresponding UVR detail images of (**a**,**b**) *Countryside in France;* (**c**,**d**) *Boat near the cliff.* The UVR images indicate that white house in *Countryside in France* and foamed water in *Boat near the cliff* (marked with red arrows) show a strong UV reflectance attributable to lead white. White brushstrokes on the boat and clouds in *Boat near the cliff* appear dark grey and black in UVR (marked with yellow arrows), suggesting a use of UV absorbent zinc white, later confirmed with SEM-EDS.

3.1.7. Black

The IRFC photography determined that the black paints were very often mixtures of black with other pigments. The SEM-EDS and PLM analyses revealed that bone black is prevalent only in black brushstrokes of *Autumn colours* (sample 9). The intense black found in other paintings resulted entirely from mixing carbon blacks with ultramarine, Prussian blue, cobalt blue, and viridian. In *Village scene* (sample 13, 14) and *Landscape in Switzerland* (sample 18), it is evident that the role of the black pigment is taken by ultramarine.

3.2. Binding Media and Other Identified Compounds

The FTIR spectra from all the examined paint samples depicted characteristic bands at around 2925, 2850, 1730, 1460, 1160, and 720 cm^{-1}, confirming a consistent use of drying oil. Furthermore, the presence of zinc soap was confirmed by an absorption peak at 1539 cm^{-1} in *Countryside in France* (sample 7) and *Landscape in Switzerland* (sample 20), while lead soap was found in French lady (sample 16) by an absorption peak at 1514 cm^{-1}. The formation of metal soaps can be explained by the probable reaction of free fatty acids from the oil binder with metals present in the lead- and zinc-containing pigments [51,52]. The detection of the acrylic resin in three of the investigated samples may correspond to the varnish applied during the conservation treatments [43,53,54].

3.3. Painting Process

It is noteworthy that the VIS and NIR photography of the paintings did not detect the presence of the preparatory underdrawings. However, Liu Kang probably studied the subject matter by small-scale sketching prior to painting the composition on the canvas. For instance, a watercolour sketch of *Breakfast* from the Liu collection reveals an early idea of the composition for the canvas painting *Breakfast*, which was executed in the same year (Figure 15). Technical evidence from the *Countryside in France*, *Village scene*, *Breakfast*, and *Self-portrait* indicated that the artist approached the canvas with a clear concept of the general composition, which was established with rough brushstrokes usually of a black paint (Figure 1b,d,g and Figure 2e).

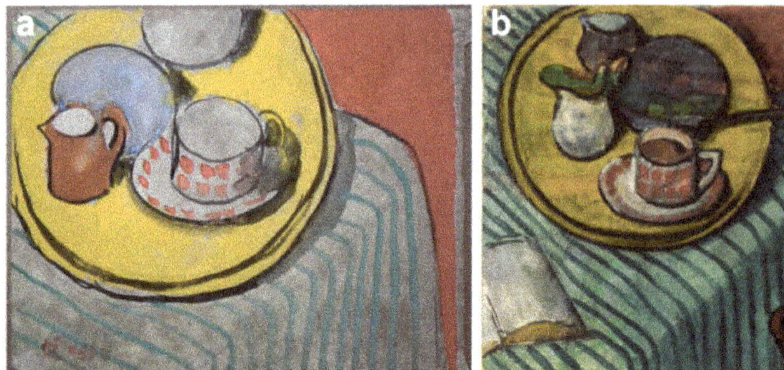

Figure 15. (**a**) Liu Kang, *Breakfast*, 1932, watercolour, 23 × 29 cm. Liu Kang Family Collection. Image courtesy of Liu family. (**b**) Liu Kang, *Breakfast*, 1932, oil on canvas, 46 × 54 cm.

The next step of the painting process involved a gradual colouring in of the outlines of the forms, providing the base for further work, as seen in *Countryside in France* and *Landscape* (Figures 1b and 2b). This method, known as *ébauche* (coloured sketch), was a common practice among Impressionists [55]. Analyses of the brushstrokes revealed that the middle-ground buildings of *Countryside in France* and the foreground house and a tree of *Landscape* were painted first. Then, the artist continued building the colouristic structure around these main subjects. The sky in *Countryside in France* was probably sketched at the end, after the background greenery was completed. Both sketches were conducted in local, vivid colours. A minimal suggestion of light effects observed in the sky of *Countryside in France* was achieved by the increased transparency of the colour, allowing the white ground to show through. The rapid development of the composition during the initial painting phase was facilitated by a few factors observed in *Countryside in France*. The small size of the painting support (number 10) reduced the time required to cover the surface with colours. The use of the absorbent or semi-absorbent ground with the ability to draw the oil from the paint accelerated the drying of the paint layer [9,56]. A thin application of colours mixed with lead white promoted the rapid drying of the paint layer. The presence of the signatures and dates indicates that the artist envisaged the coloured sketches as a completed exercise.

The further painting process can be observed in *My landlady, Madame Normand* (Figure 2f). Here, the artist intensified the colours of the main subject and built-up details with thicker paint. The advanced light effects were achieved by the increased opacity of different tints of white paint. Although the green background is sketchy and transparent, the artist's signature and date on the painting indicate that he considered the artwork as completed.

Liu Kang's variety of methods of handling the paint show constant self-development. In *Autumn colours*, his application of the paint using a small brush in short, vigorous, and descriptive paint touches reflects his attention to detail (Figures 1a and 16a). A daring adoption of parallel brushstrokes in *St Gingolph, Lac Leman, Switzerland* is reminiscent of van Gogh's style (Figures 2a and 16b). The structure of the paint layer with attractive touches is achieved by contrasting juxtaposition of greens with reds and yellows with blues. However, another painting, *Landscape in Switzerland*, shows some modification to this painting method. While the background depicting lake, mountains, and sky was conducted with directional touches, the foreground buildings and fields are depicted by coloured patches (Figures 1c and 16c,d).

Figure 16. Details showing different types of brushwork in: (**a**) *Autumn colours*; (**b**) *St Gingolph, Lac Leman, Switzerland*; and (**c**,**d**) upper and bottom parts of *Landscape in Switzerland*.

The analyses of the paint structure of *Landscape in Switzerland* revealed the hardened brushwork beneath the upper layers, suggesting that the painting, including the signature and date, was executed in a wet-on-dry technique, as illustrated by the microscopic images (Figures 10 and 11b). Hence, two distinct painting sessions can be identified, where the second session, according to the artist's date and signature, could have been in 1930, at least six months after Liu Kang's trip to Switzerland. A combination of wet-on-wet and wet-on-dry paint applications was detected in other examined paintings, suggesting that the artist did not attempt to complete the work at one sitting and sometimes worked further on the composition after the initial painting was dry. This tendency could be a result of Liu Kang working on several paintings during one session as documented in the photograph from 1929 taken in Saint-Gingolph, Switzerland. Interestingly, one of the paintings seen in the photograph appears to be unfinished; hence, it can be hypothesised that the artist tended to apply finishing touches later (Figure 17).

Figure 17. Archival photograph of Liu Kang during a painting session in Saint-Gingolph, Switzerland, in 1929: (**a**) Detail of the photograph showing a painting, probably unfinished; (**b**) Liu Kang Family Collection. Images courtesy of Liu family.

A good example of a painting executed rapidly in a single sitting is *Village scene*. In this artwork, there are significantly few details while the general artistic expression is achieved by directional and highly textured brushwork combined with solid colours (Figures 1d, 5b and 18c,d). Meanwhile, *Boat near the cliff* is characterised by a synthesis of colours and forms achieved by the broad and flat application of the paint (Figure 1f). Although the painting appears to be spontaneously and rapidly executed, the analyses proved a wet-on-dry execution, suggesting that later modifications were conducted over previous partially hardened touches.

Figure 18. Details showing the incorporation of the colour of ground in the painting process in: (**a**) *Boat near the cliff*; (**b**) *Still life with books, Paris*; and (**c**) *Village scene*. Initial compositional lines are exposed in: (**c**,**d**) *Village scene*.

During the painting process, Liu Kang experimented with incorporating the white colour of the ground layer with other paint colours reflecting inspiration from Modernists' techniques [9,57]. A good example is *Countryside in France*, in which the colour of the exposed ground was utilised for depicting the main building (Figure 14a). In contrast, the ground in *Landscape* remains exposed between the patches of applied paint, giving the artist a chance to make adjustments of colour and form (Figure 2b). In *Boat near the cliff*, the ground was used as a highlight and colour in its own right to describe foamy water (Figure 18a). However, *Still life with books, Paris*, and *Village scene* exhibit a new experiment with a white ground. Liu Kang enhanced the brilliance of the painted scenes by exposing bright accents of the white ground through the spaces between directional brushstrokes (Figures 5b and 18b,c). This painting method seems to support the notion that the artist sometimes skipped the colour sketching and confidently applied the colours after establishing the general composition with dark outlines. Moreover, to moderate the monotony of a methodical brushwork, the artist creatively produced effects of the broken flow of the paint, by dragging a loaded brush across the textured ground layer (Figures 5b and 18a–c) [9].

An interesting example of the artist's inventive way of utilising the colour and texture of the painting support is *Self-portrait* (Figure 19). As it was executed on the reverse side of an earlier composition, the brownish canvas made it particularly well suited for providing colouristic unity among the mixed yellows and browns. Moreover, the colour and texture of an un-primed canvas were skilfully exposed to give an impression of the back side of a depicted painting.

Figure 19. Detail of *Self-portrait*, showing the incorporation of the colour and texture of the painting support in depicting the reverse side of the painting in his artwork. The inset detail shows the structure of the canvas.

Another feature common in the examined paintings is the presence of a strong contour. The lines accompany the painting process from the sketching until the final stage, and they play a crucial role in the aesthetic of the paintings giving greater definition to the forms of the subjects. In *Village scene,* some of the initial compositional lines describing the highest distant hilltops as well as the shapes of the houses are still visible through the paint layer (Figure 18c,d). In the final stage, the artist usually reinforced the outlines of the subjects with a dark paint composed mainly of bone black with ultramarine and/or Prussian blue, as identified in *Village scene* (sample 13, 14), *Landscape in Switzerland* (sample 18), and *Breakfast* (sample 14).

3.4. Reusing Earlier Paintings

The evidence collected from five examined paintings revealed that Liu Kang had a practice of reusing earlier, unwanted compositions or utilising their reverse sides. For example, a visual inspection of the painted edges of the *St Gingolph, Lac Leman, Switzerland* revealed the presence of colours to be unrelated to the final image. The raking light and RTI examination of *Autumn colours* pointed out the surface brushstrokes that skip over the more complex texture of the underlying paint scheme. Subsequent transmitted NIR photography conducted with a camera facing the back of the painting revealed a presence of a still life composition that had been created in the vertical orientation. Further XRR analysis confirmed that the underlying painting depicts plants in flowerpots (Figure 20).

A visual examination of *Breakfast* provided some indications of another composition underneath. For instance, a different paint scheme was observed in the areas that were not completely covered by the current painting. Moreover, the final paint layer is characterised by several dark paint strokes that do not correspond to the present composition. They are visible on the green table top and red background, probably due to decreasing hiding power of thinly applied upper paint layer (Figure 21) [58]. This feature is similar to the reported case study of *Seafood* by Liu Kang, in which his hidden self-portrait was discovered beneath [8]. The NIR photography of *Breakfast* rotated 180° revealed a view that could be interpreted as a riverbank with trees.

Besides the aforementioned examples of paintings over discarded compositions, Liu Kang also utilised the reverse sides of earlier paintings. *Self-portrait* and *Portrait of a man with his pipe, Paris* are examples of artworks created directly on un-primed canvases. The artist's practice of reusing unsatisfactory compositions or utilising their reverse sides could have been motivated by a temporary shortage of materials or by financial constraints.

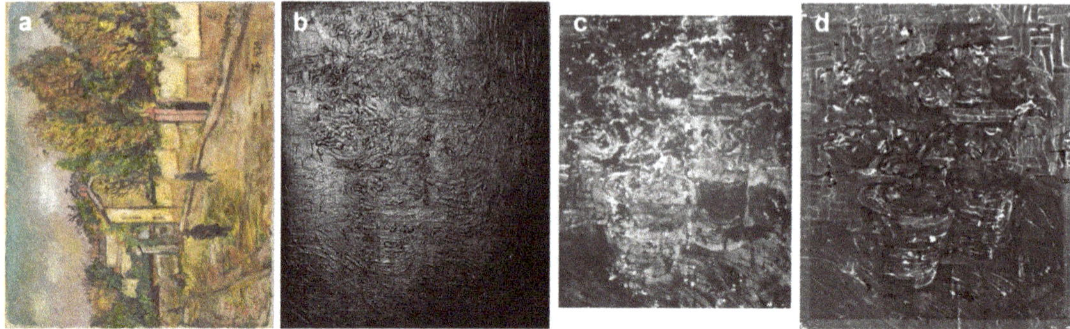

Figure 20. Images of *Autumn colours* rotated at 90° anticlockwise and photographed in: (**a**) VIS; (**b**) RTI; and (**c**) transmitted NIR executed with camera facing the back of the painting, then cropped to remove the strainer bars, inverted horizontally, and reproduced with the same relative scale; (**d**) XRR. The images revealed paint features of hidden still life painting with plants in the flowerpots.

Figure 21. VIS (**a**) and NIR (**b**) images of *Breakfast* rotated at 180°. NIR reveals the hidden view of a riverbank with trees.

4. Conclusions

The study of Liu Kang's paintings from the Paris phase provides information about the artist's choice of pigments and advances our knowledge of his working practice. Selecting artworks from both the NGS and Liu family collections enabled the tracking of the development of the artist's painting technique during the said period. In addition, the artist's family archives provided insights into his painting process. The imaging techniques, such as IRFC and UVR, played a crucial role in the tentative identification of pigments and guiding sampling of the material for further detailed analyses. RTI, XRR, and NIR provided valuable information about artist's painting process by visualising the hidden compositions underlying *Autumn colours* and *Breakfast*. The MA-XRF scanning of *Landscape in Switzerland* highlighted the distribution of elements that are indicative of certain pigments and their role in the evolution of the painting. Additional PLM, SEM-EDS, and FTIR analyses led to the identification of the paint mixtures. The contemporary colourmen catalogues are a precious resource that supplemented the interpretation of certain materials found in the course of paints' analyses.

The interpretation of the collected data unveiled a restricted palette of colours and the preferential use of ultramarine, viridian, chrome yellow, iron oxides, organic reds, lead white, and bone black. These major pigments were accompanied with the artist's admixtures. For instance, cobalt blue was hardly used and was recorded in some blue, green, and black paint mixtures. Prussian blue was used as a tint for green and black colours. Emerald green had a similar role as an admixture of greens. Additions of cadmium yellow and cobalt yellow are interesting as they were found exclusively in *Landscape in Switzerland*. PLM, SEM-EDS, and FTIR were particularly effective with the detection of three types of organic reds. One could be an organic red on tin and starch substrates. The second is probably a compound related to eosin red on Al- or Pb-containing substrate—known as a geranium lake. The third is probably an organic red on an Al-containing substrate. However, more analyses are needed to better characterise the organic red pigments. Although lead white is a predominant white pigment, its role seems to have been reduced in favour of the lithopone and/or barium white and zinc white in *Landscape in Switzerland* and *Boat near the cliff*. Artist also liked to incorporate the white of the ground into the final effect. Carbon black was added to other colours to modify their shade. However, deep blacks usually appear in bold outlines and were achieved by mixing carbon black with ultramarine, Prussian blue, cobalt blue, and viridian.

FTIR analyses additionally confirmed the presence of an oil binder in all investigated paintings from the NGS collection. Moreover, the identification of formation of metal soaps and geranium lake pigment will have some relevance for future conservation diagnostics. This is because metal soaps may contribute to the deterioration process of the paint layers, while geranium lake has a strong fading tendency [59,60].

Studies of the artist's technique revealed that he adopted coloured sketching; however, he was also able to skip this stage and effectively create compositions with rapid and decisive brushstrokes. It can be speculated that he had conducted small-scale compositional studies in different techniques prior to painting on the canvas. The subsequent progression of painting process can be characterised by constant and generally successful experimentation with brushwork. Ultimately, Liu Kang's paintings from the Paris phase are defined by continual exploration and learning of different Modernist painting styles. This very mature artistic approach benefitted him upon his return to China, when he felt a strong need to develop his own painting style.

Author Contributions: Conceptualization, D.L.; methodology, D.L.; investigation, D.L., T.K., and B.S.; data curation, D.L.; writing—original draft preparation, D.L.; writing—review and editing, D.L. and T.K; visualisation, D.L. and B.S. All authors have read and agreed to the published version of the manuscript.

Funding: This research received no external funding.

Institutional Review Board Statement: Not applicable.

Informed Consent Statement: Not applicable.

Data Availability Statement: The data presented in this study are available on request from the corresponding author.

Acknowledgments: The authors would like to acknowledge the National Gallery Singapore for allowing us to investigate the paintings and the Heritage Conservation Centre for supporting this study. The authors are indebted to Gretchen Liu for sharing Liu Kang's family painting collection and the artist's archival photographs and Hanna Szczepanowska at West Virginia University for her critical review, which was essential for further improvement of the content. They also wish to thank Roger Lee (Assistant Painting Conservator from the Heritage Conservation Centre) for his assistance at RTI; Kenneth Yeo Chye Whatt (Principal Radiographer from the Division of Radiological Sciences at the Singapore General Hospital); and Steven Wong Bak Siew (Head and Senior Consultant from the Department of Radiology at the Sengkang General Hospital) for enabling the X-ray radiography. The authors wish to thank Bruker Nano GmbH for lending us the MA-XRF (M6 Jetstream) and facilitating the scanning.

Conflicts of Interest: The authors declare no conflict of interest.

Appendix A

Table A1. Overview of the materials identified in the paint samples extracted from the investigated paintings from the NGS collection.

Title and Inventory Number	Date	Colour	Sample	SEM-EDS * and XRF ** Detected Elements	PLM, SEM-EDS, XRF Assignment	FTIR Identification
Autumn colours, GI-0255 (PC)	1930	Blue	19	**C**, **O**, **Pb**, Cr, Ca, Si, Al, Na, Ba, S, As, (Sr, Fe, Cu, Zn)	Lead white, lithopone and/or barium white and zinc white, chalk, viridian, ultramarine, emerald green or Sheele's green, Prussian blue	
		Green	5	**C**, **O**, **Pb**, Ba, Sr, S, As, Cr, Cu, (Si, Ca, Al, Fe, Na)	Lead white, barium white, chalk, viridian, emerald green or Sheele's green, Prussian blue	
		Green	18	**C**, **O**, **As**, **Cu**, Ba, Pb, S, Ca, (Ti, Na, Cr, Cl, Si, Fe, Mn)	Emerald green, barium white, lead white, chalk, viridian, umber, possible Cr-containing yellow(s) and titanium white	Emerald green, lithopone and/or barium white, lead white, possible chrome yellow, oil
		Yellow	3	**Pb**, **C**, **O**, Ca, (Cl, Cr, Si, Al, Fe, Ba)	Lead white, chrome yellow, chalk, barium white, yellow iron oxide	
		Brown	4	**Pb**, **C**, **O**, Ca, Fe, Al, Si, (Cr, Ba, Sr, Ti, Na)	Lead white, chalk, yellow iron oxide, barium white, possible Cr-containing yellow(s), and titanium white	
		Red	7	**O**, **C**, **Fe**, Pb, Ca, Ti, Si, Al, Mg, (P, Na, Sr, Cl, Zn, K, Cr, S)	Red iron oxide, lead white, chalk, titanium white, bone black, possible Cr-containing yellow(s)	
		Black	9	**C**, **O**, **Ca**, Pb, P, (Si, Al, Na, Fe, Ba, Mg, Cr, Cl, Zn, As)	Bone black, lead white, lithopone and/or barium white and zinc white	
		White	2	**Pb**, **C**, **O**, Ca, (Al, Cl, Si, Zn)	Lead white, chalk, zinc white	
Countryside in France, 2003-03365	1930	Blue	5	**C**, **Pb**, **O**, Si, Al, Ca, Ba, Na, S, Zn, (Ni, K, Cr, Ti, Sr, Mg)	Lead white, carbon black, ultramarine, lithopone and/or barium white and zinc white, possible titanium white	
		Green	11	**O**, **Pb**, **C**, **Ba**, Cr, Fe, S, Ca, Si, Al, Ti, (Na, K, Zn, Cl)	Lead white, chalk, lithopone and/or barium white and zinc white, viridian, Prussian blue, possible Cr-containing yellow(s), titanium white	Lead white, chalk, lithopone and/or barium white and zinc white, Prussian blue, possible chrome yellow, oil, acrylic resin assigned to the conservation varnish
		Green	4	**Pb**, **C**, **O**, Cr, Ca, Ba, Si, Zn, Al, (S, Fe, Ti, Na)	Lead white, chrome yellow, possible other Cr-containing yellow(s), chalk, viridian, lithopone and/or barium white and zinc white, Prussian blue	

Table A1. *Cont.*

Title and Inventory Number	Date	Colour	Sample	SEM-EDS * and XRF ** Detected Elements	PLM, SEM-EDS, XRF Assignment	FTIR Identification
		Yellow	9	**Pb**, **C**, **O**, Cr, (S, Cl, Na, Zn, Al)	Chrome yellow, zinc white	
		Brown	41	*Pb, Fe, K, Si, Ca*	Lead white, iron oxide, chalk	
		Brown	49	*Fe, Ca*	Iron oxide, chalk	
		Red	7	**C**, **Pb**, **Ba**, **O**, S, Cr, Al, Si, (Zn, Ni, Sr, Ca, P)	Lead white, chrome yellow, lithopone and/or barium white and zinc white, organic red, bone black	Lead white, chrome yellow, lithopone and/or barium white and zinc white, organic red, oil, zinc soap
		Black	16	**O**, **C**, **Pb**, Si, Al, Na, Ba, S, Zn, Ca, (K, Cr, Ti, Sr)	Lead white, ultramarine, carbon black, lithopone and/or barium white and zinc white	
		White	29	**C**, **Pb**, **O**, **Ba**, Ti, Zn, S, Na, (Fe, Si, Al)	Lead white, barium white, zinc white, titanium white	Lead white, barium white, zinc white, oil, proteins
Landscape in Switzerland, P-1229	1930	Blue	5	**Zn**, **C**, **O**, **Na**, Al, Ca, Si, (S, Pb, Sr, Mg)	Zinc white, ultramarine, lead white	
		Green	20	**Pb**, **C**, **O**, **Zn**, Cr, Fe, Na, (Ba, Si, Al, Ca, Cl)	Chrome yellow, yellow iron oxide, Prussian blue, lead white, lithopone and/or barium white and zinc white	Chrome yellow, possible iron oxide, Prussian blue, lead white, lithopone and/or barium white and zinc white, oil, zinc soap, acrylic resin assigned to the conservation varnish
		Green	21	**C**, **O**, **Cd**, **Cr**, S, Ba, Zn, Cl, (Na, Pb, Al, Si, Sn, Ti, Ca, Sr)	Cadmium yellow, lithopone and/or barium white and zinc white, viridian, chalk, lead white, chrome yellow, possible titanium white	
		Green	24	**C**, **O**, **Pb**, Ba, Zn, Cr, Al, Cl, Na, S, Fe, (Cu, Mg, Si, Ti, Ca, As, Sr, P)	Chrome yellow, viridian, Prussian blue, lithopone and/or barium white and zinc white, emerald green or Sheele's green, possible titanium white	
		Yellow	23	**Pb**, **O**, **Zn**, C, K, Na, Ba, Co, Cd, (Mg, Ca, Fe, Si, S, Al, Cr, Cl)	Chrome yellow, possible other Cr-containing yellow(s), lithopone and/or barium white and zinc white, cobalt yellow, cadmium yellow, yellow iron oxide	
		Yellow	36	**Zn**, **C**, **O**, Pb, Na, Ca, Fe, K, Ba, Co, (Si, As, S, Mg, Al, Cr, Cl)	Lithopone and/or barium white and zinc white, lead white, chrome yellow, possible other Cr-containing yellow(s), yellow iron oxide, cobalt yellow	

Table A1. *Cont.*

Title and Inventory Number	Date	Colour	Sample	SEM-EDS * and XRF ** Detected Elements	PLM, SEM-EDS, XRF Assignment	FTIR Identification
		Brown	25, layer 2	**O**, **Zn**, **C**, Ba, Na, Pb, Fe, Ca, S, Al, Si, K, (Ti, Co, Mg, As, Sr, Cu)	Lithopone and/or barium white and zinc white, lead white, yellow iron oxide, cobalt yellow, possible emerald green or Sheele's green and titanium white	
		Red	25, layer 1	**C**, **O**, **Pb**, Ca, Sn, Si, Fe, (Al, Ba, Zn, Na, Cl, Cr, Mg)	Lead white, organic red, red iron oxide, starch	
		Red	35, layer 2	**O**, **Fe**, **C**, **Si**, Al, Ca, (K, Sn, Mg, Ti, S, Pb, Zn)	Red iron oxide, organic red, chalk	Red iron oxide, organic red, chalk, oil
		Red	35, layer 1	**C**, **O**, **Pb**, Sn, (Cr, Fe, Ca, Si, Al, Cl)	Lead white, organic red, iron oxide, possible Cr-containing yellow(s)	Lead white, organic red, iron oxide, starch, oil
		Black	18	**C**, **O**, **Zn**, Al, Ca, Ba, S, Na, (Si, Fe, Pb, P, Sr, Mg, Cu, Cl)	Lithopone and/or barium white and zinc white, ultramarine, bone black, Prussian blue	
		White	22	**Zn**, **C**, **O**, **Na**, (S, Ba, Mg, Ca, Al)	Lithopone and/or barium white and zinc white, chalk	
Village scene, 2003-03320	1931	Blue	20	**Pb**, **C**, **O**, **Ca**, (Na, Si, Al, Cr, Cl, Mg, Sr, Zn)	Lead white, ultramarine, viridian, chalk, possible zinc white	
		Green	2	**C**, **Pb**, **O**, Ba, Ca, Cr, S, Fe, (Zn, Al, Na, Si, Ti, Mg)	Lead white, lithopone and/or barium white and zinc white, chalk, viridian, Prussian blue	
		Green	3	**O**, **Cr**, **C**, **Pb**, Ca, (Zn, Na, Si, Mg, Al)	Viridian, lead white, chrome yellow, possible other Cr-containing yellow(s), chalk	
		Yellow	8	**Pb**, **C**, **O**, **Cr**, (Ca, S, Na, Si, Cl, Al)	Chrome yellow, chalk	
		Brown	5	**Pb**, **C**, **O**, Ca, Fe, Si, Al, (Zn, Sr, Mg, Cl, Ba, Cr, P)	Lead white, lithopone and/or barium white and zinc white, yellow iron oxide, bone black, possible Cr-containing yellow(s)	
		Brown	10	**O**, **Fe**, **C**, **Ba**, S, Ca, Si, (Pb, Al, Ti, Sr, Cl)	Yellow iron oxide, barium white, chalk, lead white	
		Red	9	**C**, **O**, **Pb**, Sn, Ca, Cr, Si, Fe, (Al, Zn, Ba, P)	Chrome yellow, possible lead white, organic red, starch, chalk, lithopone and/or barium white and zinc white, red iron oxide, bone black	Chrome yellow, possible lead white, organic red, iron oxide, starch, chalk, oil
		Black	13	**O**, **C**, Ca, Si, Pb, Na, Al, S, P, Zn, (Ba, K, Fe, Sr, Cr, Cl)	Ultramarine, lead white, lithopone and/or barium white and zinc white, Prussian blue, bone black	

Table A1. *Cont.*

Title and Inventory Number	Date	Colour	Sample	SEM-EDS * and XRF ** Detected Elements	PLM, SEM-EDS, XRF Assignment	FTIR Identification
		Black	14	**O**, **Pb**, **C**, **Si**, Al, Na, S, Ca, (K, As, Zn, Ba, Fe, Sr, Cl)	Lead white, ultramarine, lithopone and/or barium white and zinc white, red iron oxide, carbon black	
		White	17	**Pb**, **C**, **O**, Ca, (Cl, Al, Mg)	Lead white, chalk	
French lady, 1993-00996	1931	Blue	20	**C**, **O**, **Pb**, Al, Ca, Ba, Si, Co, Na, (S, Cr, Mg, Ti, Sr, K, P, Fe, Zn, Cl)	Lead white, chalk, lithopone and/or barium white and zinc white, cobalt blue, viridian, ultramarine, bone black	
		Green	15	**C**, **O**, **Pb**, **Cr**, Ba, Ca, (As, Si, Al, Na, Cu, Ti, Fe, Co, Cl, S)	Lead white, viridian, chalk, barium white, cobalt blue, Prussian blue, possible emerald green or Sheele's green and titanium white	
		Green	16	**Pb**, **C**, **O**, Ba, Ca, As, Cr, S, Cu, (Fe, Ti, Na, Al, Si, Cl)	Lead white, barium white, chalk, emerald green or Sheele's green, Prussian blue, possible Cr-containing yellow(s) and titanium white	Lead white, lithopone and/or barium white, chalk, possible emerald green or Sheele's green, Prussian blue, chrome yellow, oil, lead soap
		Brown	17	**Pb**, **O**, **C**, Ca, Si, Fe, Al, (Zn, Ba, Cr, Cl, Mg, K, Na, Ti)	Lead white, yellow iron oxide, Cr-containing yellow(s), lithopone and/or barium white and zinc white	
		Red	9	**Pb**, **C**, **O**, Ca, Ba, Fe, Si, (Al, Cl, Cr, Ti, Mg)	Lead white, chalk, Cr-containing yellow(s), barium white, iron oxide	
		Black	19	**O**, **C**, **Fe**, **Ba**, Pb, Al, S, Co, Na, K, Si, As, (Ca, P, Ti, Zn, Mg, Cr, Sr, Cl)	Prussian blue, lithopone and/or barium white and zinc white, lead white, cobalt blue, bone black, viridian, possible titanium white	
		Black	6	**O**, **C**, **Pb**, Si, Al, S, Na, Ca, As, K, Fe, (Sr, Ba, Cl, Mg, Ti, Cr)	Lead white, ultramarine, carbon black, Prussian blue, barium white, viridian, possible titanium white	
Boat near the cliff, 2003-03249	1931	Blue	3	**C**, **Zn**, **O**, **Ba**, Na, S, Cr, (Al, Si, Sr)	Lithopone and/or barium white and zinc white ultramarine, viridian	
		Blue	21	**C**, **O**, Na, Zn, Al, Si, S, Ba, (K, Sr, Cr, Ca, Cl)	Ultramarine, viridian, lithopone and/or barium white and zinc white	
		Blue	26	**C**, **O**, **Ba**, **Zn**, S, Cr, Na, Pb, (Ti, Ca, Al, Si, Co, Mg)	Lithopone and/or barium white and zinc white, lead white, viridian, ultramarine, cobalt blue	

Table A1. *Cont.*

Title and Inventory Number	Date	Colour	Sample	SEM-EDS * and XRF ** Detected Elements	PLM, SEM-EDS, XRF Assignment	FTIR Identification
		Green	11	**O**, **C**, **Ba**, **Zn**, Pb, S, Fe, Cr, Ca, Na, Si, (Al, Ti, Mg, P, Cl, K)	Lithopone and/or barium white and zinc white, chalk, chrome yellow, yellow iron oxide, Prussian blue, viridian, bone black, possible titanium white	
		Green	12	**O**, **C**, **Ba**, Ca, Pb, S, Cr, Fe, (Ti, Sr, Si, Al, Na, Zn)	Lithopone and/or barium white and zinc white, viridian, Prussian blue, chrome yellow	
		Yellow	25	**O**, **C**, **Ba**, Ca, Pb, S, Zn, Cr, (Sr, Na, Ti, Mg, Si, Al, Cd, Fe)	Lithopone and/or barium white and zinc white, chrome yellow, cadmium yellow, iron oxide, possible titanium white	
		Brown	14	**C**, **O**, **Zn**, Fe, Na, Al, Ba, (Ca, P, S, Si, Mn, Mg, Cr, Pb, Sr)	Lithopone and/or barium white and zinc white, umber, bone black, chrome yellow, possible other Cr-containing yellow(s)	
		Red	29	**C**, **O**, Pb, Ba, Br, Al, S, (Si, Ti, Cl, Ca)	Br-containing organic red on Pb- or Al-based substrate, barium white, possible titanium white	Lithopone and/or barium white, organic red, oil, acrylic resin assigned to the conservation varnish
		Black	17	**C**, **O**, **Zn**, Pb, Na, Mg, (Si, Al, Fe, S, Ba, Ca, K)	Lithopone and/or barium white and zinc white, lead white, carbon black, Prussian blue	
		White	22	**Zn**, **C**, **O**, Na, (Si, Mg, Pb, Al)	Zinc white, lead white	
Breakfast, GI-0257 (PC)	1932	Green	4	**O**, **Cr**, **C**, Pb, Ba, S, Ca, (Na, Al, Si, Mg, Ti, Sr)	Viridian, lead white, barium white, chalk, possible titanium white	
		Yellow	3	**O**, **C**, **Fe**, Si, Pb, Al, Ca, (K, As, Cr, Sr, P, Ba, Cl, Ti, Zn)	Yellow iron oxide, chrome yellow, lead white, lithopone and/or barium white and zinc white, bone black, possible titanium white	
		Brown	10	**O**, **C**, **Fe**, **Si**, Pb, Al, Sn, Ca, Sr, Ba, K, (Ti, Cr, Mg, Cl, Zn, P)	Red iron oxide, organic red, lead white, chalk, Prussian blue, starch, Cr-containing yellow(s), lithopone and/or barium white and zinc white, bone black, possible titanium white	Red iron oxide, organic red, lead white, chalk, Prussian blue, starch, oil
		Black	14	**O**, **Ca**, **Pb**, **C**, **Cr**, P, Ba, Si, (Fe, Al, Na, Ti, Mg, K, Cl, Sr)	Lead white, bone black, viridian, barium white, Prussian blue, possible titanium white	
		White	7	**Pb**, **C**, **O**, **Ca**, (Cl, Si, Na, Al, Mg)	Lead white, chalk	

* Major elements are given in bold, minor elements in plain type and trace elements in brackets. ** Elements detected with XRF are given in italics.

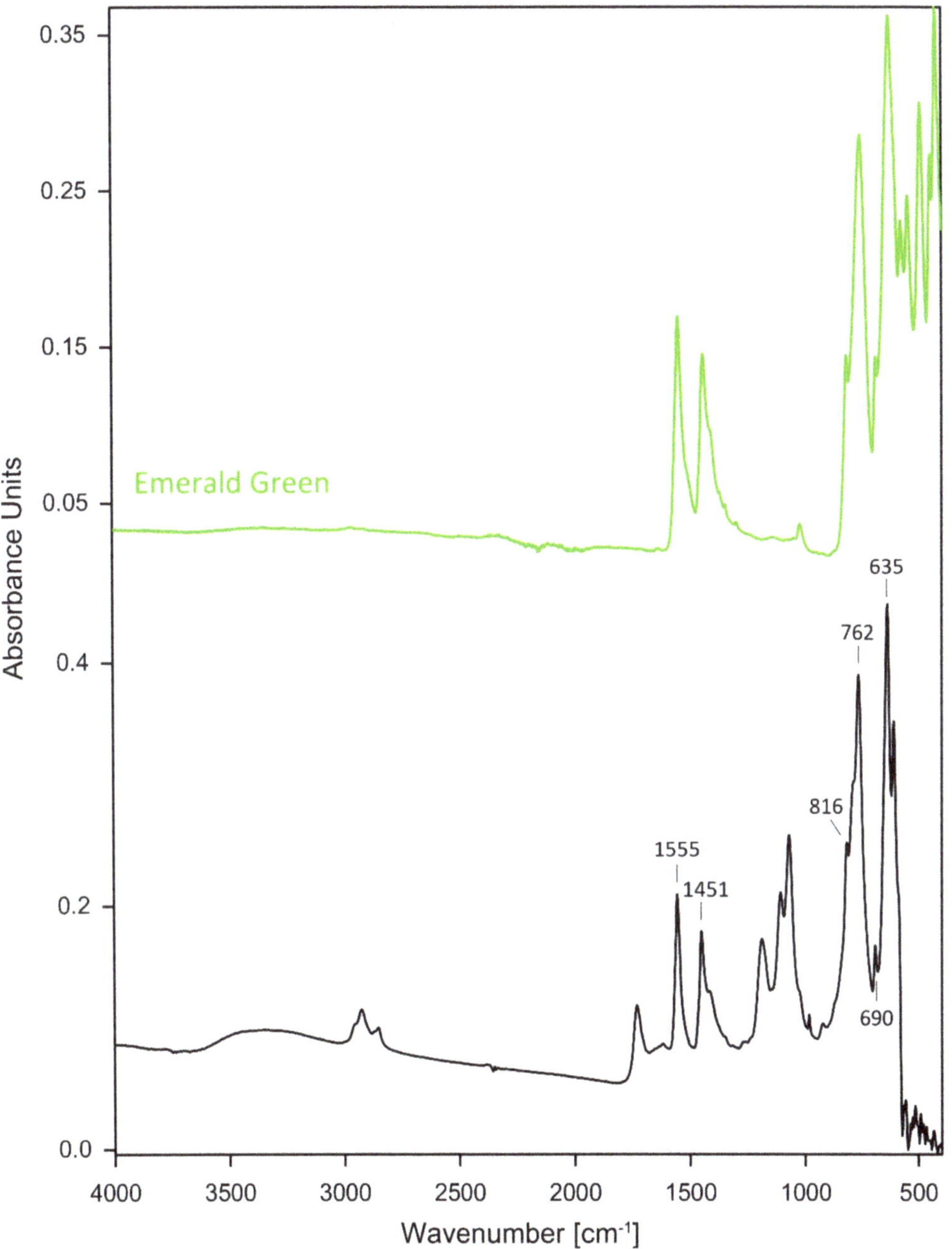

Figure 1. ATR-FTIR spectra of green paint from sample 18, taken from *Autumn colours*, with labelled marker peaks of emerald green and spectra of a reference sample of emerald green.

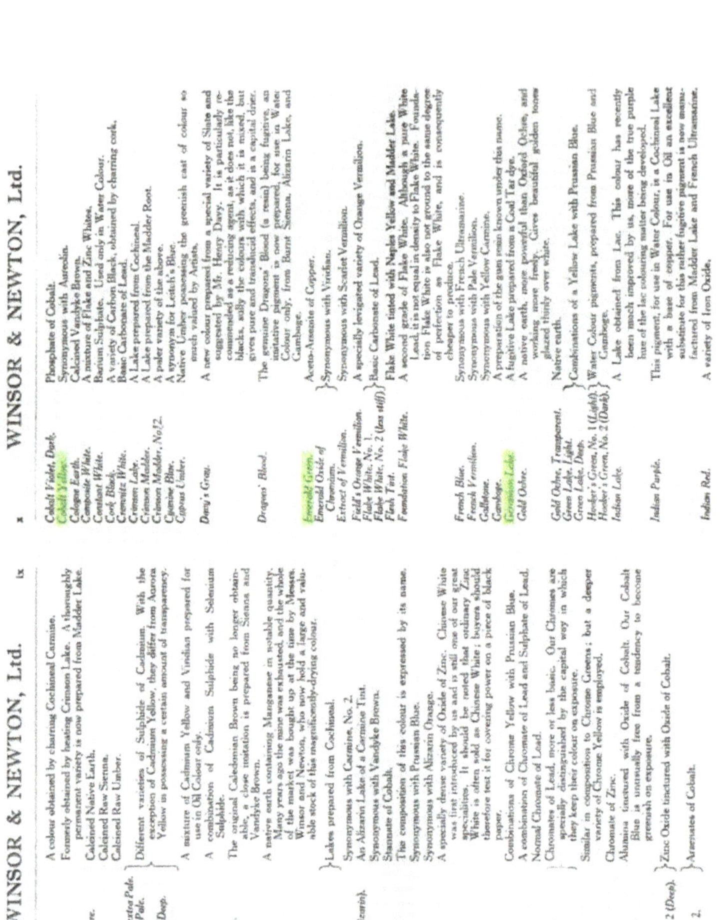

Figure 2. Compositions of pigments, listed in the catalogue of W&N from 1928, used for the manufacturing of oil and watercolours. The highlighted details show carmines, varieties of chrome greens, cinnabar green, cobalt yellow, emerald green, and geranium lake.

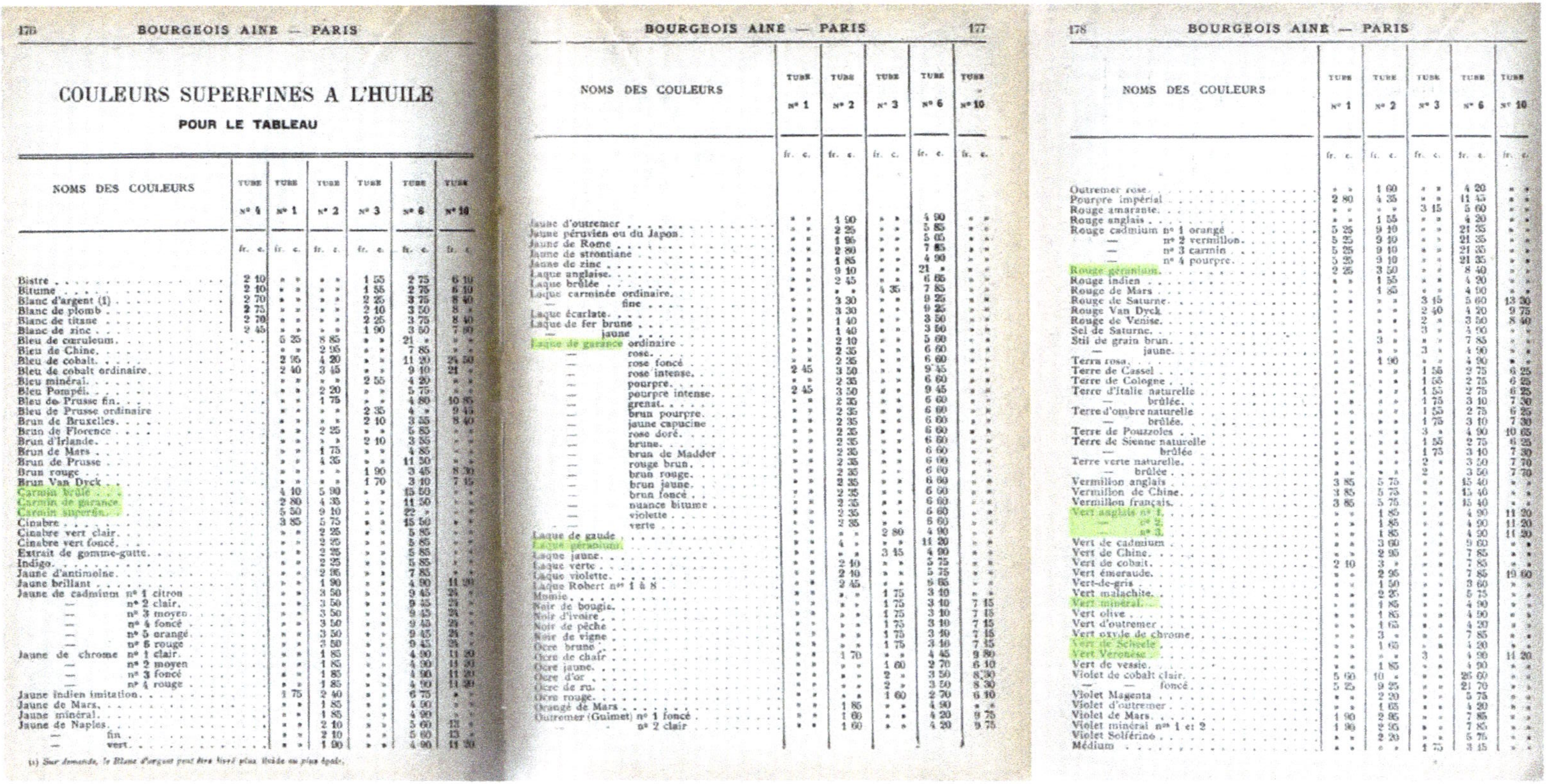

COULEURS SUPERFINES A L'HUILE

POUR LE TABLEAU

NOMS DES COULEURS	TUBE N° 4 (fr. c.)	TUBE N° 1 (fr. c.)	TUBE N° 2 (fr. c.)	TUBE N° 3 (fr. c.)	TUBE N° 6 (fr. c.)	TUBE N° 10 (fr. c.)
Bistre	2 10	» »	» »	1 55	2 75	6 10
Bitume	2 10	» »	» »	1 55	2 75	6 10
Blanc d'argent (1)	2 70	» »	» »	2 25	3 75	8 40
Blanc de plomb	2 75	» »	» »	2 10	3 50	8 »
Blanc de titane	2 70	» »	» »	2 25	3 75	8 40
Blanc de zinc	2 45	» »	» »	1 90	3 50	7 80
Bleu de cœruleum		5 25	8 85	» »	21 »	» »
Bleu de Chine		» »	2 95	» »	7 85	» »
Bleu de cobalt		2 95	4 20	» »	11 20	24 50
Bleu de cobalt ordinaire		2 40	3 45	» »	9 10	21 »
Bleu minéral		» »	» »	2 55	4 20	» »
Bleu Pompéi		» »	2 20	» »	5 75	» »
Bleu de Prusse fin		» »	1 75	» »	4 80	10 85
Bleu de Prusse ordinaire		» »	» »	2 35	4 »	9 45
Brun de Bruxelles		» »	» »	2 10	3 55	8 40
Brun de Florence		» »	2 25	» »	5 85	» »
Brun d'Irlande		» »	» »	2 10	3 55	» »
Brun de Mars		» »	1 75	» »	4 85	» »
Brun de Prusse		» »	4 35	» »	11 30	» »
Brun rouge		» »	» »	1 90	3 45	8 30
Brun Van Dyck		» »	» »	1 70	3 10	7 15
Carmin brûlé		4 10	5 90	» »	15 50	» »
Carmin de garance		2 80	4 35	» »	11 50	» »
Carmin superfin		5 50	9 10	» »	22 »	» »
Cinabre		3 85	5 75	» »	15 50	» »
Cinabre vert clair		» »	2 25	» »	5 85	» »
Cinabre vert foncé		» »	2 25	» »	5 85	» »
Extrait de gomme-gutte		» »	2 25	» »	5 85	» »
Indigo		» »	2 25	» »	5 85	» »
Jaune d'antimoine		» »	2 85	» »	7 85	» »
Jaune brillant		» »	1 90	» »	4 90	11 20
Jaune de cadmium n° 1 citron		» »	3 50	» »	9 45	24 »
— n° 2 clair		» »	3 50	» »	9 45	24 »
— n° 3 moyen		» »	3 50	» »	9 45	24 »
— n° 4 foncé		» »	3 50	» »	9 45	24 »
— n° 5 orangé		» »	3 50	» »	9 45	24 »
— n° 6 rouge		» »	3 50	» »	9 45	24 »
Jaune de chrome n° 1 clair		» »	1 85	» »	4 90	11 20
— n° 2 moyen		» »	1 85	» »	4 90	11 20
— n° 3 foncé		» »	1 85	» »	4 90	11 20
— n° 4 rouge		» »	1 85	» »	4 90	11 20
Jaune indien imitation		1 75	2 40	» »	6 75	» »
Jaune de Mars		» »	1 85	» »	4 90	» »
Jaune minéral		» »	1 85	» »	4 90	» »
Jaune de Naples		» »	2 10	» »	5 60	13 »
— fin		» »	2 10	» »	5 60	13 »
— vert		» »	1 90	» »	4 90	11 20

NOMS DES COULEURS	TUBE N° 1 (fr. c.)	TUBE N° 2 (fr. c.)	TUBE N° 3 (fr. c.)	TUBE N° 6 (fr. c.)	TUBE N° 10 (fr. c.)
Jaune d'outremer	» »	1 90	» »	4 90	» »
Jaune péruvien ou du Japon	» »	2 25	» »	5 85	» »
Jaune de Rome	» »	1 95	» »	5 05	» »
Jaune de strontiane	» »	2 80	» »	7 85	» »
Jaune de zinc	» »	1 85	» »	4 90	» »
Laque anglaise	» »	9 10	» »	21 »	» »
Laque brûlée	» »	2 45	» »	6 65	» »
Laque carminée ordinaire	» »	» »	4 35	7 85	» »
— fine	» »	3 30	» »	9 25	» »
Laque écarlate	» »	3 30	» »	9 25	» »
Laque de fer brune	» »	1 40	» »	3 50	» »
— jaune	» »	1 40	» »	3 50	» »
Laque de garance ordinaire	» »	2 10	» »	5 60	» »
— rose	» »	2 35	» »	6 60	» »
— rose foncé	» »	2 35	» »	6 60	» »
— rose intense	2 45	3 50	» »	9 45	» »
— pourpre	» »	2 35	» »	6 60	» »
— pourpre intense	2 45	3 50	» »	9 45	» »
— grenat	» »	2 35	» »	6 60	» »
— brun pourpre	» »	2 35	» »	6 60	» »
— jaune capucine	» »	2 35	» »	6 60	» »
— rose doré	» »	2 35	» »	6 60	» »
— brune	» »	2 35	» »	6 60	» »
— brun de Madder	» »	2 35	» »	6 60	» »
— rouge brun	» »	2 35	» »	6 60	» »
— brun rouge	» »	2 35	» »	6 60	» »
— brun jaune	» »	2 35	» »	6 60	» »
— brun foncé	» »	2 35	» »	6 60	» »
— nuance bitume	» »	2 35	» »	6 60	» »
— violette	» »	2 35	» »	6 60	» »
— verte	» »	2 35	» »	6 60	» »
Laque de gaude	» »	» »	2 80	4 90	» »
Laque géranium	» »	4 »	» »	11 20	» »
Laque jaune	» »	» »	3 45	4 90	» »
Laque verte	» »	2 10	» »	5 75	» »
Laque violette	» »	2 10	» »	5 75	» »
Laque Robert n°s 1 à 8	» »	2 45	» »	6 65	» »
Momie	» »	» »	1 75	3 10	» »
Noir de bougie	» »	» »	1 75	3 10	7 15
Noir d'ivoire	» »	» »	1 75	3 10	7 15
Noir de pêche	» »	» »	1 75	3 10	7 15
Noir de vigne	» »	» »	1 75	3 10	7 15
Ocre brune	» »	» »	1 75	3 10	7 15
Ocre de chair	» »	1 70	» »	4 45	9 80
Ocre jaune	» »	» »	1 60	2 70	6 10
Ocre d'or	» »	» »	2 »	3 50	8 30
Ocre de ru	» »	» »	2 »	3 50	8 30
Ocre rouge	» »	» »	1 60	2 70	6 10
Orangé de Mars	» »	1 85	» »	4 90	» »
Outremer (Guimet) n° 1 foncé	» »	1 60	» »	4 20	9 75
n° 2 clair	» »	1 60	» »	4 20	9 75

NOMS DES COULEURS	TUBE N° 1 (fr. c.)	TUBE N° 2 (fr. c.)	TUBE N° 3 (fr. c.)	TUBE N° 6 (fr. c.)	TUBE N° 10 (fr. c.)
Outremer rose	» »	1 60	» »	4 20	» »
Pourpre impérial	2 80	4 35	» »	11 45	» »
Rouge amarante	» »	» »	3 15	5 60	» »
Rouge anglais	» »	1 55	» »	4 20	» »
Rouge cadmium n° 1 orangé	5 25	9 10	» »	21 35	» »
— n° 2 vermillon	5 25	9 10	» »	21 35	» »
— n° 3 carmin	5 25	9 10	» »	21 35	» »
— n° 4 pourpre	5 25	9 10	» »	21 35	» »
Rouge géranium	2 25	3 50	» »	8 40	» »
Rouge indien	» »	1 55	» »	4 20	» »
Rouge de Mars	» »	1 85	» »	4 90	» »
Rouge de Saturne	» »	» »	3 15	5 60	13 30
Rouge Van Dyck	» »	» »	2 40	4 20	9 75
Rouge de Venise	» »	» »	2 »	3 50	8 40
Sel de Saturne	» »	» »	3 »	4 90	» »
Stil de grain brun	» »	3 »	» »	7 85	» »
— jaune	» »	» »	3 »	4 90	» »
Terra rosa	» »	1 90	» »	4 90	» »
Terre de Cassel	» »	» »	1 55	2 75	6 25
Terre de Cologne	» »	» »	1 55	2 75	6 25
Terre d'Italie naturelle	» »	» »	1 55	2 75	6 25
— brûlée	» »	» »	1 75	3 10	7 30
Terre d'ombre naturelle	» »	» »	1 55	2 75	6 25
— brûlée	» »	» »	1 75	3 10	7 30
Terre de Pouzzoles	» »	» »	3 »	4 90	10 65
Terre de Sienne naturelle	» »	» »	1 55	2 75	6 25
— brûlée	» »	» »	1 75	3 10	7 30
Terre verte naturelle	» »	» »	2 »	3 50	7 70
— brûlée	» »	» »	2 »	3 50	7 70
Vermillon anglais	3 85	5 75	» »	15 40	» »
Vermillon de Chine	3 85	5 75	» »	15 40	» »
Vermillon français	3 85	5 75	» »	15 40	» »
Vert anglais n° 1	» »	1 85	» »	4 90	11 20
— n° 2	» »	1 85	» »	4 90	11 20
— n° 3	» »	1 85	» »	4 90	11 20
Vert de cadmium	» »	3 60	» »	9 60	» »
Vert de Chine	» »	2 95	» »	7 85	» »
Vert de cobalt	2 10	3 »	» »	7 85	» »
Vert émeraude	» »	2 95	» »	7 85	19 60
Vert-de-gris	» »	1 50	» »	3 60	» »
Vert malachite	» »	2 25	» »	5 75	» »
Vert minéral	» »	1 85	» »	4 90	» »
Vert olive	» »	1 85	» »	4 90	» »
Vert d'outremer	» »	1 65	» »	4 20	» »
Vert oxyde de chrome	» »	3 »	» »	7 85	» »
Vert de Scheele	» »	1 65	» »	4 20	» »
Vert Véronèse	» »	» »	3 »	4 90	11 20
Vert de vessie	» »	1 85	» »	4 90	» »
Violet de cobalt clair	5 60	10 »	» »	26 60	» »
— — foncé	5 25	9 25	» »	21 70	» »
Violet Magenta	» »	2 20	» »	5 75	» »
Violet d'outremer	» »	1 65	» »	4 20	» »
Violet de Mars	1 90	2 95	» »	7 85	» »
Violet minéral n°s 1 et 2	1 90	2 95	» »	7 85	» »
Violet Solférino	» »	2 20	» »	5 75	» »
Médium	» »	» »	1 75	3 15	» »

Figure 3. Oil colours, listed in the catalogue of Bourgeois Ainé from 1929. The highlighted details show carmines, a range of madder lakes, geranium lake, geranium red, varieties of chrome greens, emerald green, Scheele's green, and its variant mineral green.

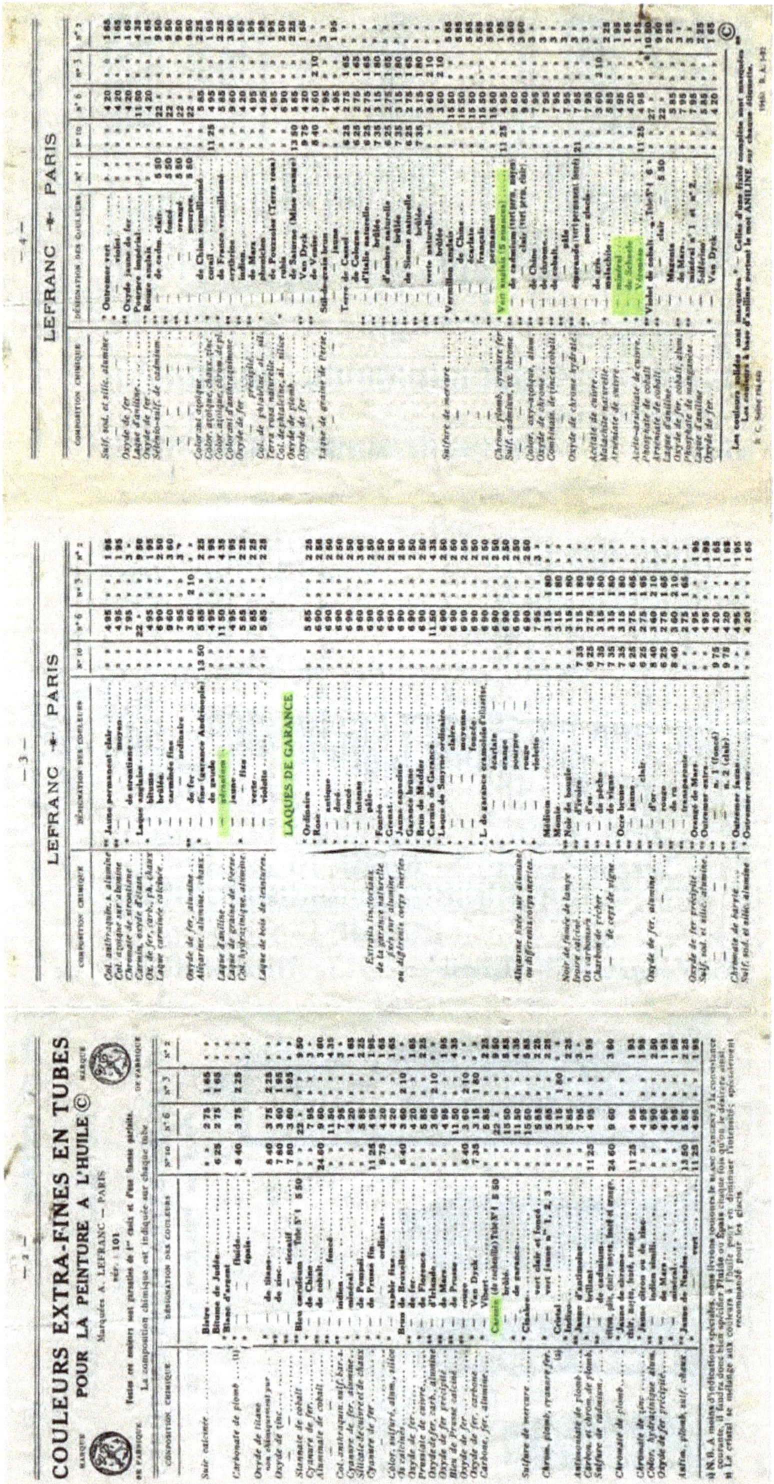
COULEURS EXTRA-FINES EN TUBES
POUR LA PEINTURE A L'HUILE
LEFRANC ✦ PARIS
LAQUES DE GARANCE
LEFRANC ✦ PARIS

Figure 4. Oil colours, listed in the catalogue of Lefranc from 1931. The highlighted details show carmines, geranium lake, a range of madder lakes, varieties of chrome greens, emerald green, Scheele's green, and its variant mineral green.

References

1. Croizier, R. Post-Impressionists in Pre-War Shanghai: The Juelanshe (Storm Society) and the Fate of Modernism in Republican Shanghai. In *Modernity in Asian Art*; Clark, J., Ed.; Wild Peony: Broadway, NSW, Australia, 1993; p. 135.
2. Zheng, J. *The Modernization of Chinese Art: The Shanghai Art College, 1913–1937*; Leuven University Press: Leuven, Belgium, 2016; pp. 92, 96.
3. Kwok, K.C. *Journeys: Liu Kang and His Art = Yi Cheng: Liu Kang Qi Ren Qi Yi*; National Arts Council: Singapore, 2000; p. 49.
4. Liu, K. Liu Haisu and Contemporary Chinese Art. In *Liu Kang: Essays on Art & Culture*; Siew, S., Ed.; National Art Gallery: Singapore, 2011; p. 153.
5. Liu, K. *Liu Kang Ba Shi Ba Sui = Liu Kang at 88F*; Singapore Soka Association: Singapore, 1998; p. 22.
6. Ong, Z.H.; Wong, S.; Ong, P.; Tng, P. *Life of Love, Life of Art*; Yeo, W.W., Ed.; National Art, Gallery: Singapore, 2011; p. 34.
7. Lizun, D. A preliminary study of Liu Kang's palette and the discovery and interpretation of hidden paint layers. *Herit. Sci.* **2020**, *8*, 21. [CrossRef]
8. Lizun, D.; Szroeder, P.; Kurkiewicz, T.; Szczupak, B. Examination of painting technique and materials of Liu Kang's Seafood and hidden self-portrait. *Int. J. Conserv. Sci.* **2021**, *12*, 3–26.
9. Lizun, D.; Kurkiewicz, T.; Szczupak, B. Technical examination of Liu Kang's Paris and Shanghai painting supports (1929–1937). *Herit. Sci.* **2021**, *9*, 37. [CrossRef]
10. Liu, K.; Fu, L.; Fu, T.O. *Liu Kang: Essays on Art & Culture*; Siew, S., Ed.; National Art Gallery: Singapore, 2011; p. 126.
11. Cosentino, A. Identification of pigments by multispectral imaging; a flowchart method. *Herit. Sci.* **2014**, *2*, 12. [CrossRef]
12. Cosentino, A. Practical notes on ultraviolet technical photography for art examination. *Conserv. Património* **2015**, 53–62. [CrossRef]
13. Cosentino, A. Infrared technical photography for art examination. *e-Preserv. Sci.* **2016**, *13*, 1–6.
14. Warda, J.; Frey, F.; Heller, D.; Kushel, D.; Vitale, T.; Weaver, G. *The AIC Guide to Digital Photography and Conservation Documentation*; American Institute for Conservation of Historic and Artistic Works: Washington, DC, USA, 2011.
15. Schroer, C.; Bogart, J.; Mudge, B.; Lum, M. Guide to Highlight Image Capture. Available online: http://culturalheritageimaging.org/What_We_Offer/Downloads/RTI_Hlt_Capture_Guide_v2_0.pdf (accessed on 27 March 2021).
16. Schroer, C.; Bogart, J.; Mudge, B.; Lum, M. Guide to Highlight Image Processing. Available online: http://culturalheritageimaging.org/What_We_Offer/Downloads/rtibuilder/RTI_hlt_Processing_Guide_v14_beta.pdf (accessed on 27 March 2021).
17. Schroer, C.; Bogart, J.; Mudge, B.; Lum, M. Guide to RTIViewer. Available online: http://culturalheritageimaging.org/What_We_Offer/Downloads/rtiviewer/RTIViewer_Guide_v1_1.pdf (accessed on 27 March 2021).
18. Alfeld, M.; Pedroso, J.V.; van Eikema Hommes, M.; van der Snickt, G.; Tauber, G.; Blaas, J.; Haschke, M.; Erler, K.; Dik, J.; Janssens, K. A mobile instrument for in situ scanning macro-XRF investigation of historical paintings. *J. Anal. At. Spectrom.* **2013**, *28*, 760–767. [CrossRef]
19. Mactaggart, P.; Mactaggart, A. *A Pigment Microscopist's Notebook*; Chard: Mactaggart, AB, Canada, 1998.
20. Le Salon 1930. *Société Des Artistes Français*; Imprimerie Georges Lang: Paris, France, 1930.
21. Le Salon 1932. *Société Des Artistes Français*; Imprimerie Georges Lang: Paris, France, 1932.
22. *Catalogue De La 41e Exposition Au Grand Palais Des Champs-Elysées Du 17 Janvier Au 2 Mars 1930*; Société Des Artistes Indépendants: Paris, France, 1930.
23. Berrie, B.H. Prussian Blue. In *Artists' Pigments: A Handbook of Their History and Characteristics*; West FitzHugh, E., Ed.; National Gallery of Art: Washington, DC, USA, 1997; Volume 3, p. 195.
24. Kirby, J.; Saunders, D. Fading and colour change of Prussian blue: Methods of manufacture and the influence of extenders. *Natl. Gallery Tech. Bull.* **2004**, *25*, 73–99.
25. Feller, R.L. Barium Sulfate–Natural and Synthetic. In *Artists' Pigments: A Handbook of Their History and Characteristics*; Feller, R.L., Ed.; National Gallery of Art: Washington, DC, USA, 1986; Volume 1, p. 47.
26. Newman, R. Chromium Oxide and Hydrated Chromium Oxide. In *Artists' Pigments: A Handbook of Their History and Characteristics*; West FitzHugh, E., Ed.; National Gallery of Art: Washington, DC, USA, 1997; Volume 3, p. 281.
27. Buti, D.; Rosi, F.; Brunetti, B.G.; Miliani, C. In-situ identification of copper-based green pigments on paintings and manuscripts by reflection FTIR. *Anal. Bioanal. Chem.* **2013**, *405*, 2699–2711. [CrossRef] [PubMed]
28. Bomford, D.; Kirby, J.; Leighton, J.; Roy, A. *Art in the Making: Impressionism*; National Gallery: London, UK, 1990; pp. 48, 61.
29. Fiedler, I.; Bayeard, M.A. Emerald green and Sheele's green. In *Artists' Pigments: A Handbook of Their History and Characteristics*; West FitzHugh, E., Ed.; National Gallery of Art: Washington, DC, USA, 1997; Volume 3, p. 238.
30. Kühn, H.; Curran, M. Chrome Yellow and Other Chromate Pigments. In *Artists' Pigments: A Handbook of Their History and Characteristics*; Feller, R.L., Ed.; National Gallery of Art: Washington, DC, USA, 1986; Volume 1, pp. 190, 194, 196, 207.
31. Fiedler, I.; Bayard, M.A. Cadmium Yellows, Oranges and Reds. In *Artists' Pigments: A Handbook of Their History and Characteristics*; Feller, R.L., Ed.; National Gallery of Art: Washington, DC, USA, 1986; Volume 1, pp. 65, 74, 80.
32. Eastaugh, N.; Walsh, V.; Chaplin, T.; Siddall, R. *Pigment Compendium: A Dictionary and Optical Microscopy of Historical Pigments*; Butterworth-Heinemann: London, UK, 2008; pp. 103, 157, 173.
33. Helwig, L. Iron Oxide Pigments: Natural and Synthetic. In *Artists' Pigments: A Handbook of Their History and Characteristics*; Berrie, B.H., Ed.; National Gallery of Art: Washington, DC, USA, 2007; Volume 4, p. 73.
34. Gettens, R.J.; Stout, G.L. *Painting Materials: A Short Encyclopaedia*; Dover Publications: New York, NY, USA, 2015; p. 110.

35. Corman, M. Cobalt Yellow (Aureolin). In *Artists' Pigments: A Handbook of Their History and Characteristics*; Feller, R.L., Ed.; National Gallery of Art: Washington, DC, USA, 1986; Volume 1, p. 37.
36. Kirby, J.; van Bommel, M.; Verhecken, A.; Spring, M.; Vanden Berghe, I.; Stege, H.; Richter, M. *Natural Colorants for Dyeing and Lake Pigments: Practical Recipes and Their Historical Sources*; Archetype Publications: London, UK, 2014; pp. 78, 101.
37. Schweppe, H.; Roosen-Runge, H. Carmine–Cochineal Carmine and Kermes Carmine. In *Artists' Pigments: A Handbook of Their History and Characteristics*; Feller, R.L., Ed.; National Gallery of Art: Washington, DC, USA, 1986; Volume 1, pp. 263, 272.
38. Schweppe, H.; Winter, J. Madder and Alizarin. In *Artists' Pigments: A Handbook of Their History and Characteristics;* West FitzHugh, E., Ed.; National Gallery of Art: Washington, DC, USA, 1997; Volume 3, pp. 112, 124, 131.
39. van Bommel, M.; Geldof, M.; Hendriks, E. An Investigation of Organic Red Pigments Used in Paintings by Vincent Van Gogh (November 1885 to February 1888). In *ArtMatters-Netherlands Technical Studies in art*; Hermens, E., Ed.; Waanders Publishers: Zwolle, The Netherlands, 2005; Volume 3, pp. 128, 129.
40. Centeno, S.A.; Hale, C.; Carò, F.; Cesaratto, A.; Shibayama, N.; Delaney, J.; Dooley, K.; van der Snickt, G.; Janssens, K.; Stein, S.A. Van Gogh's Irises and roses: The contribution of chemical analyses and imaging to the assessment of color changes in the red lake pigments. *Herit. Sci.* **2017**, *5*, 18. [CrossRef]
41. Geldof, M.; de Keijzer, M.; van Bommel, M.; Pilz, K.; Salvant, J.; van Keulen, H.; Megens, L. Van Gogh's Geranium Lake. In *Van Gogh's Studio Practice;* Jansen, L., Geldof, M., Haswell, R., Hendriks, E., van Heugten, S., Eds.; Yale University Press: New Haven, CT, USA, 2013; pp. 268–289.
42. Ghelardi, E.; Degano, I.; Colombini, M.P.; Mazurek, J.; Schilling, M.; Learner, T. Py-GC/MS applied to the analysis of synthetic organic pigments: Characterization and identification in paint samples. *Anal. Bioanal. Chem.* **2015**, *407*, 1415–1431. [CrossRef] [PubMed]
43. *Painting Examination and Treatment Report. Boat Near the Cliff. Conservation ID:101356; Heritage Conservation Centre: Singapore, 2006.*
44. Fieberg, J.E.; Knutås, P.; Hostettler, K.; Smith, G.D. "Paintings fade like flowers": Pigment analysis and digital reconstruction of a faded pink lake pigment in Vincent van Gogh's Undergrowth with two figures. *Appl. Spectrosc.* **2017**, *71*, 794–808. [CrossRef] [PubMed]
45. Kirby, J. The Reconstruction of Late 19th-Century French Red Lake Pigments. In *Art of the Past: Sources and Reconstruction: Proceedings of the First Symposium of the Art Technological Source Research Study Group*; Clarke, M., Townsend, J., Stijnman, A., Eds.; Archetype Publications, ICN Amsterdam: London, UK, 2005; p. 75.
46. de Keijzer, M.; van Bommel, M. Bright New Colours: The History and Analysis of Fluorescein, Eosin, Erythrosine, Rhodamine and Some of Their Derivatives. In *The Diversity of Dyes in History and Archaeology;* Kirby, J., Ed.; Archetype Publications: London, UK, 2017; pp. 326–338.
47. Pozzi, F.; Basso, E.; Centeno, S.A.; Smieska, L.M.; Shibayama, N.; Berns, R.; Fontanella, M.; Stringari, L. Altered identity: Fleeting colors and obscured surfaces in Van Gogh's Landscapes in Paris, Arles, and Saint-Rémy. *Herit. Sci.* **2021**, *9*, 15. [CrossRef]
48. Degano, I.; Tognotti, P.; Kunzelman, D.; Modugno, F. HPLC-DAD and HPLC-ESI-Q-ToF characterisation of early 20th century lake and organic pigments from Lefranc archives. *Herit. Sci.* **2017**, *5*, 7. [CrossRef]
49. Kirby, J.; Spring, M.; Higgitt, C. The technology of red lake pigment manufacture: Study of the dyestuff substrate. *Natl. Gallery Tech. Bull.* **2005**, *26*, 71–87.
50. Kühn, H. Zinc white. In *Artists' Pigments: A Handbook of Their History and Characteristics;* Feller, R.L., Ed.; National Gallery of Art: Washington, DC, USA, 1986; Volume 1, p. 178.
51. Duffy, M.; Martins, A.; Boon, J.J. Metal Soaps and Visual Changes in a Painting by René Magritte–The Menaced Assassin, 1927. In *Issues in Contemporary Oil Paint;* van den Berg, K.J., Burnstock, A., de Keijzer, M., Krueger, J., Learner, T., Tagle, d.A., Heydenreich, G., Eds.; Springer International Publishing: Cham, Switzerland, 2014; pp. 197–203.
52. Hermans, J.J.; Keune, K.; Van Loon, A.; Iedema, P.D. Toward a Complete Molecular Model for the Formation of Metal Soaps in Oil Paints. In *Metal Soaps in Art: Conservation and Research;* Casadio, F., Keune, K., Noble, P., Van Loon, A., Hendriks, E., Centeno, S.A., Osmond, G., Eds.; Springer International Publishing: Cham, Switzerland, 2019; pp. 47–67.
53. *Painting Examination and Treatment Report. Countryside in France. Conservation ID: 113162; Heritage Conservation Centre: Singapore, 2006.*
54. *Painting Examination and Treatment Report. Landscape in Switzerland. Conservation ID: 102968;* Heritage Conservation Centre: Singapore, 1993.
55. Callen, A. *The Art of Impressionism: Painting Technique & the Making of Modernity;* Yale University Press: New Haven, CT, USA, 2000; pp. 163–169.
56. Jirat-Wasiutynski, V.; Newton, H.T. Absorbent grounds and the matt aesthetic in Post-Impressionist painting. *Stud. Conserv.* **1998**, *43*, 235–239. [CrossRef]
57. Callen, A. *Techniques of the Impressionists;* Orbis: London, UK, 1982; pp. 155, 178.
58. Brill, T.B. *Light: Its Interaction with Art and Antiquities;* Plenum Press: New York, NY, USA, 1980; p. 89.
59. Burnstock, A.; Lanfear, I.; van den Berg, K.J.; Carlyle, L.; Clarke, M.; Hendriks, E.; Kirby, J. Comparison of the Fading and Surface Deterioration of Red Lake Pigments in Six Paintings by Vincent Van Gogh with Artificially Aged Paint Reconstructions. In Proceedings of the ICOM-CC 14th Triennial Meeting, The Hague, The Netherlands, 12–15 March 2020; Verger, I., Ed.; James & James: London, UK, 2005; pp. 459–466.
60. Chieli, A.; Miliani, C.; Degano, I.; Sabatini, F.; Tognotti, P.; Romani, A. New insights into the fading mechanism of geranium lake in painting matrix. *Dye. Pigment.* **2020**, *181*, 108600. [CrossRef]

heritage

MDPI

Article

Jazz Colors: Pigment Identification in the Gouaches Used by Henri Matisse

Ana Martins [1,*], Anne Catherine Prud'hom [2], Maroussia Duranton [3], Abed Haddad [1,4,5], Celine Daher [6], Anne Genachte-Le Bail [3] and Tiffany Tang [1]

1 The David Booth Conservation Department, The Museum of Modern Art, 11 W53rd Street, New York, NY 10019, USA; abed_haddad@moma.org (A.H.); tiffany.h.tang@gmail.com (T.T.)
2 Service de la Restauration, Cabinet d'Art Graphique, Musée National d'Art Moderne, Centre Pompidou, CEDEX 04, 75191 Paris, France; Anne-catherine.PrudHom@centrepompidou.fr
3 Institut National du Patrimoine (INP), Département des Restaurateurs, Laboratoire de Recherche, 124 Rue Henri Barbusse, 93300 Aubervilliers, France; maroussia.duranton@culture.gouv.fr (M.D.); anne.genachte-le-bail@culture.gouv.fr (A.G.-L.B.)
4 Ph.D. Program in Chemistry, City University of New York Graduate School and University Center, 365 5th Avenue, New York, NY 10016, USA
5 Department of Chemistry, City University of New York, City College of New York, 160 Convent Avenue, New York, NY 10031, USA
6 Centre de Recherche sur la Conservation (CRC, USR 3224), Musée National d'Histoire Naturelle, Ministère de la Culture, CNRS, CP21, 36 Rue Geoffroy-Saint-Hilaire, 75005 Paris, France; celine.daher@mnhn.fr
* Correspondence: a.martins@vangoghmuseum.nl

Abstract: *Jazz*, the illustrated book by Henri Matisse, is a testament to the vitality of the artist in the last decade of his career. The book consists of twenty illustrations reproduced in 370 copies using a stencil-based printing technique and the same Linel gouaches the artist had used for the original maquettes. This study reports on the comprehensive analysis carried out to identify the pigments in the gouaches used in *Jazz* by transmitted and reflectance infrared, Raman, SERS, and X-ray fluorescence spectroscopies, and describes the lightfastness of these gouaches as evaluated by microfaedometry. This study also highlights the necessity of a multi-analytical approach for comprehensive identification of artist materials and investigates the suitability of portable and non-invasive techniques. The results were consistent across the three copies investigated: a portfolio in the collection of the Museum of Modern Art in New York and two books in the collection of the Musée National d'Art Moderne in Paris. In total, 39 distinct colors were characterized, with the magenta, pinks, and blues being the most fugitive.

Keywords: Henri Matisse; cut-outs; gouache; Pigment identification; light sensitivity; Raman spectroscopy; X-ray fluorescence spectroscopy (XRF); Fourier transform infrared spectroscopy (FTIR); microfaedometry (MFT)

Citation: Martins, A.; Prud'hom, A.C.; Duranton, M.; Haddad, A.; Daher, C.; Genachte-Le Bail, A.; Tang, T. *Jazz* Colors: Pigment Identification in the Gouaches Used by Henri Matisse. *Heritage* **2021**, *4*, 4205–4221. https://doi.org/10.3390/heritage4040231

Academic Editor: Diego Tamburini

Received: 26 August 2021
Accepted: 29 October 2021
Published: 4 November 2021

1. Introduction

Jazz, the artist book illustrated by Henri Matisse and published by Tériade in 1947, is considered a defining moment in the life of the artist, when his *papiers découpés*, or paper cut-outs, took a life of their own as a new medium in his practice. Matisse's first cut-outs were maquettes for the cover of Christian Zervos' *Cahiers d'Art* nos. 3–5 in 1936 and for the first cover of Tériade's rival publication *Verve* in 1937 [1]. Encouraged to create a *livre d'artiste* in a similar style by Tériade [2], Matisse composed a series of twenty maquettes between August 1943 and March 1944, with " ... figures in vivid and violent tones, resulting from crystallizations of memories of the circus, popular tales or travel" [3]. In 1947, after several trials and sustained insistence from Tériade, the maquettes were ultimately reproduced and printed by Edmond Vairel and Draeger Frères, Paris. As stated in the last page of the *Jazz* book, the plates were reproduced using a stencil method also known as pochoir [4], and using the same Linel gouaches Matisse had used for the maquettes. While he approved of the production technique and proofed all plates, Matisse initially disliked the final appearance of *Jazz*, lamenting to a friend " ... It is absolutely a failure" [5]. Nevertheless, the process itself influenced his future work, where the cut-outs became an end to themselves rather than preliminary models for future work. The maquettes are now part of the collection at the Musée National d'Art Moderne in Paris (MNAM), and a total of 250 books, 100 portfolios, and 20 non-circulated copies are in collections and repositories all over the world [6]. The fifth plate in *Jazz* is reproduced in Figure 1 to illustrate the vitality of the work.

Figure 1. Henri Matisse *Horse, Rider and Clown*, from *Jazz* 1947. One from a portfolio of twenty pochoirs (42.3 × 65.5 cm). Published by Tériade, Paris and printed by Edmond Vairel, Paris, Draeger Frères, Paris. Edition of 100. MoMA object number: 291.1948.5. Copyright © 2021 Succession H. Matisse/Artists Rights Society (ARS), New York. All reproductions of this work are excluded from the CC: BY License.

A full understanding of the materials used in the reproduction of *Jazz* is important and long overdue because of its significance and reach. It is vital that institutions continue taking the necessary precautions when exhibiting and storing *Jazz* to ensure the integrity and vibrancy of its colors.

While systematic studies on the composition of commercial paints used by artists abound: Bocour [7], HKS, W&N, and Maimeri [8], Talens [8,9], and Ripolin [10] to only cite a few, no such studies appear to have been carried out for commercial gouaches. Information in the literature on the Linel gouaches used by Matisse, for instance, is essentially limited to light sensitivity, and no studies were found in the literature concerning the Sennelier gouaches which he also used in his early cut-outs [11]. Matisse was aware of the fading of some of these colors, in particular, the organic-based violets and pinks [12]. Fading of the blues, on the other hand, has been established visually and by spectrophotometry in *Blue Nude I* and *The Swimming* Pool [11,13], while the colors in *Acanthes* appear to be relatively stable based on microfadeometry [14]. The same technique established that of the colors present in *The Snail*, the green, orange, white, and magenta are light-sensitive [15]. A set of cut-out scraps from Matisse's studio, donated to MoMA by the artist's family in 2013, was also recently tested and showed a range of light sensitivities between fugitive and stable [16]. Until now, and to the best of the authors' knowledge, neither chemical nor light stability analyses have been extended to the gouaches used in the printing of *Jazz*. Therefore, any information on the composition and stability of these gouaches is expected to be of interest to the institutions and collectors holding a copy of *Jazz*.

This article reports on a comprehensive analysis of the colors used in three reproductions: a portfolio in the collection of the Museum of Modern Art in New York (MoMA) and two books in the Musée National d'Art Moderne de Paris (MNAM), one in the collection and the other one in the study collection. The objective of the study was twofold: on the one hand to gain knowledge on the nature of the pigments present in the gouaches and their light sensitivity, and on the other hand, to support the assertion that all the copies of *Jazz* were made with the same gouaches. This is also part of a larger study that aims at comparing the materials used in the original cut-out maquettes made in 1943–44 and the pochoirs made in 1947, and also compare these materials with the wider set of cut-out scraps donated to the museum. The identification of the pigments relied on a multi-analytical approach using non-invasive techniques including portable X-ray fluorescence (p-XRF), reflectance spectroscopy (RS), reflectance Fourier transform infrared spectroscopy (r-FTIR), portable Raman spectroscopy (p-Raman), and microfadeometry (MFT) and was complemented with analysis of micro-samples by Fourier transform infrared spectroscopy (μ-FTIR) spectroscopy, confocal Raman spectroscopy, and surface-enhanced Raman (SERS) spectroscopy. Principal component analysis (PCA) was used as a data exploratory method to identify gouaches with similar p-XRF, RS, and r-FTIR spectral signatures and help establish the number of gouaches used by Vairel to reproduce *Jazz*.

2. Methodology and Analytical Techniques

The books and portfolios contain twenty prints designated here by P1 to P20 in the same sequence they appear in the book. The analyses were performed on a *Jazz* portfolio in the MoMA collection and two *Jazz* books in the MNAM collection. A preliminary examination of the MoMA portfolio under an optical microscope (Figure 2) showed minimal overlap between colors but some cross-contamination by paint splatters at the boundary between different color fields, likely caused by the use of a brush to apply the paints [4]. Analysis and sampling were therefore carried out at least 2 cm away from these boundaries whenever possible. Some of the letters and elements in prints P2, P17, and P19, however, were painted over the colored background. This was taken into account when interpreting the data and was confirmed for the three copies of *Jazz* examined.

Figure 2. Details of *Jazz* plates examined under the optical microscope (×10 magnification) showing minimal overlap between color fields and some cross-contamination of gouache particles.

The first phase of the study focused on determining the total number of gouaches used to reproduce the MoMA portfolio using non-invasive techniques. For each of the twenty prints, p-XRF analysis and RS were carried on two to three distinct spots for each color and paper support leading to a total of 228 p-XRF spectra and 396 reflectance spectra. The analysis established that for every color in a print, the composition of the two or three spots analyzed was the same. Based on that conclusion, the number of spots for r-FTIR analysis was reduced to one spot per color for each print (142 spectra). Principal component analysis (PCA) was then used as a data exploratory method to identify gouaches with the same p-XRF, RS, and r-FTIR spectral signatures and help establish the number of gouaches used by Vairel to reproduce the MoMA *Jazz* copy. MFT was also carried out on representative spots for each color to evaluate their sensitivity to light exposure. In the second phase of the project, p-XRF, RS, and r-FTIR analyses were carried out on one of the MNAM copies (one spot for each color) to confirm that the same gouaches had been used as in the MoMA copy.

Pigment identification proved challenging for some of the colors using non-invasive analysis exclusively. Further analysis using p-Raman clarified a few of the unknowns but ultimately sampling and analyses by μ-FTIR, Raman, and SERS were carried out on micro-samples taken from the *Jazz* book in the MNAM study collection to provide a more comprehensive overview of the gouache palette and confirm or complement the non-invasive analysis. A representative micro-sample was taken using a tungsten needle for each of the 39 colors identified by non-invasive analysis and whenever possible from the edge of the plate (the plates that were sampled are indicated in Table S1 in the Supplemental Information document SI).

2.1. Infrared Spectroscopies (μ-FTIR and r-FTIR)

Micro-FTIR (μ-FTIR) analysis was carried out in transmission mode using a Nicolet iS50-FTIR coupled with a Thermo Nicolet Continuum infrared microscope equipped with an MCT detector. Spectra were collected in the 4000–600 cm^{-1} range with a 4 cm^{-1} resolution and 128 scans and using the Thermo Scientific OMNIC 9.0 software package.

Reflectance FTIR (r-FTIR) was carried out using a portable Bruker ALPHA spectrometer equipped with an external reflection module and a DTGS detector. Spectra were collected in the 7000–400 cm^{-1} range and a 4 cm^{-1} resolution over 128 scans. Measurements were made in non-contact mode over a spot size of approximately 4 mm in diameter and a 1 to 1.5 cm distance from the object (the focal distance from the surface was adjusted using the instrument's built-in camera). The background was periodically acquired using a built-in flat gold mirror. The spectra were converted to pseudo-absorbance.

Spectra were examined using the Spectral Search and Multicomponent Search tools available in the Thermo Scientific OMNIC Specta 2.0 software, as well as the IRUG FTIR spectral database [17]. Both original and Kramers–Kronig (KK) transformed r-FTIR spectra were evaluated.

2.2. Raman Spectroscopies (Confocal Raman, p-Raman, SERS)

Confocal Raman spectra (Raman) were collected using a Renishaw In-via Raman system equipped with a 785 nm diode laser at powers between 0.3 to 3 mW, a 1200 lines/mm grating, and a Leica confocal microscope with a 100x objective. Final spectra represent an average of five acquisitions of 10 s. AgNPs for SERS were synthesized using the Lee-Meisel method [18] and measurements were obtained before and after using acid pretreatments for hydrolyzing colorants into the colloidal solution [19] (see supplementary information (SI) for a complete description). SERS spectra were collected using a 532 nm diode laser at powers between 0.5 to 2.5 mW, 1800 lines/mm grating, and a Leica confocal microscope with a $\times 100$ objective. Final spectra represent an average of five acquisitions of 10 s. Portable Raman (p-Raman) spectra were acquired using a Bruker BRAVO Handheld Raman spectrometer equipped with 785 nm and 853 nm diode lasers (DuoLASER^{TM}), a resolution of 10–12 cm^{-1} and 400 ms exposure time over 30 acquisitions. The output laser is $\approx$ 50 mW and spot size diameter 1 mm) measurements were carried in non-contact mode at 1–2 mm to prevent thermal degradation or damage to the object. All Raman spectra were evaluated using Omnic Specta.

Reagents for surface-enhanced Raman (SERS) spectroscopy: silver nitrate ($AgNO_3$) (99.9999% trace metals basis) and sodium citrate ($\geq$99%, FG) were used for silver nanoparticle (AgNPs) synthesis. Potassium nitrate (KNO_3) (BioXtra, $\geq$99.5%) was used for AgNP aggregation. Nitric acid (HNO_3) (ACS Reagent, 70%) and hydrofluoric acids (HF) (ACS Reagent 48%) were used for colorant hydrolysis into AgNP colloid. All materials were purchased from Millipore-Sigma.

2.3. Reflectance Spectrophotometry (RS)

Reflectance spectra were taken using the X-rite eXact spectrophotometer in the range of 400–700 nm with a 10 nm resolution and a 4 mm diameter spot size. The purpose of these measurements was to identify colors with the same signature spectra across multiple plates and not for pigment identification. It was thus deemed acceptable to perform the measurement through a clear mylar (PET) film (3.8 microns) to avoid direct contact with the surface (preliminary test indicated a 90% reduction in the reflectance).

2.4. X-ray Fluorescence (p-XRF)

Portable X-ray fluorescence (p-XRF) analysis was performed at MoMA with a Bruker Tracer III-SDD handheld XRF instrument with an Rh excitation source and silicon drift detector (5 mm diameter approximate spot size) under He purge. The instrument was operated at 40 kV and 3 μA, and spectra were acquired for 120 s. An Elio spectrometer (XGLab) equipped with an Rh target and 1 mm beam size was used for the analysis at MNAM. An integrated CCD camera and two laser pointers allowed precise focus on the region of interest. All analyses were performed in atmospheric conditions, with no He purge, at 40 kV and 100 μA, with a collection time of 200 s. All the spectra were examined with the Bruker Artax 7.2 software.

2.5. Microfadeometry (MFT)

Illumination was achieved with an Oriel Instruments LIK-LMP Light Intensity Controller Kit equipped with a Si Detector Head and Xe-arc lamp controlled by a Spectra-Physics power supply coupled to a thermo-Oriel Digital Flux controller. Spectra were registered using a Control Development spectrometer in the range of 380–900 nm. Light contact to the object's surface was achieved using a 100 μm diameter bifurcated fiber optic, which ensures a narrow beam radius and high light intensity. Microfading was performed on three distinct spots for each color (in the same plate and color field) for a duration of 15 minutes. ISO Blue Wool (BW) 1, 2, and 3 were used as internal standards for comparison [20]. The average ΔE76 curves were obtained using the Spectral Viewer software developed at the Getty Conservation Institute [21]. A BW equivalent scale (BW_{eq}) was established by fitting a power function to the ΔE values at t = 15 min for BW 1, 2, and 3 [22,23]. A BW_{eq} value was then calculated for each color by interpolation of that curve for the ΔE values at t = 15 min for each color. the value obtained is rounded to the nearest category: $BW_{eq} < 2$ is highly light-sensitive; $BW_{eq} = 2$, 2.5, and 3 are light-sensitive, and $BW_{eq} > 3$ is moderately light-sensitive to stable. It is important to note the limitation of MFT for elucidating lightfastness beyond $BW_{eq} = 3$ [24].

2.6. PCA Analysis

Principal component analysis (PCA) was used as an exploratory data analysis method to group colors with very similar p-XRF, RS, and r-FTIR spectral signatures and in this way help determine the number of paints used to reproduce *Jazz*. PCA was carried out separately on the p-XRF, RS, and r-FTIR spectra using the SOLO-MIA (Eigenvector) software package. Preprocessing of the spectra included Poisson scaling (for the p-XRF spectra in the 1–15 KeV range), first derivative (for RS), and SNV (for r-FTIR), followed by mean centering.

3. Results and Discussion

3.1. Pigment Identification

Jazz was printed on Velin d'ARCHES paper as specified on the last page of the *Jazz* books. Analysis of the paper support with r-FTIR and p-Raman only produced bands related to cellulose. No additives or coatings were detected with the techniques in use. Analysis with p-XRF detected small amounts of Al, Si, P, S, K, Ca, Fe, and Zn. This small mineral contribution from the paper support was taken into account when evaluating the p-XRF results obtained for the gouaches. The r-FTIR, p-Raman, and p-XRF spectra obtained for the paper support of the twenty MoMA plates and the twenty MNAM pages were very similar, confirming that the same paper was used for both reproductions.

PCA was used as an exploratory method to distinguish gouaches of a similar color but potentially different composition by comparing their p-XRF, RS, and r-FTIR spectra. The contribution of the paper support was the same for all the spots analyzed and thus had no impact on the PCA analysis. The example provided in Figure 3 illustrates how PCA was used to distinguish three groups of yellow gouaches based on their p-XRF spectra. The scores and loading plots (Figure 3a,b respectively) extracted by PCA differentiate gouaches richer in either calcium sulfate (Ca, S) or barium sulfate (Ba, S, Sr) from gouaches containing neither extender. The same procedure was repeated to examine the data obtained with r-FTIR and with RS. Ultimately, seven different yellow gouaches were distinguished based on the similarities between their XRF, r-FTIR, and RS spectral signatures. The same analysis was carried out on the rest of the colors, leading to the final conclusion that 39 different gouaches had been used to reproduce *Jazz*.

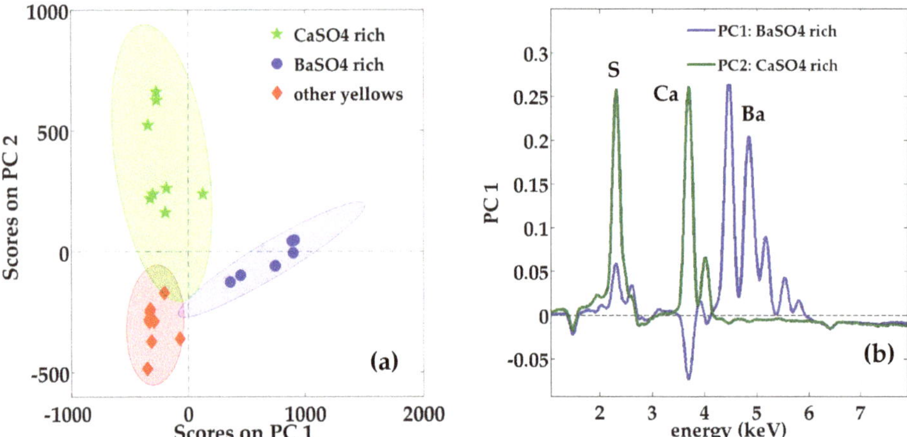

Figure 3. Results of PCA analysis carried out on the p-XRF spectra taken on 23 yellow spots across the 20 MoMA *Jazz* plates. The scores plot (**a**) distinguishes three different types of yellows. The loading plot (**b**) indicates that the distinction is based on the presence or not of $CaSO_4$ or $BaSO_4$ extenders.

This information was used to guide further non-invasive analysis by p-Raman and MFT for each color. It was also used to guide sampling for complementary analysis by µ-FTIR, Raman, and SERS. Table 1 summarizes the pigments identified in each gouache. Tables S1 and S2 in the SI provide a more detailed account of the results obtained for each gouache with the different analytical methods and the prints where each of these gouaches was used. p-XRF, IR, and Raman spectra obtained for the black gouaches are provided in Figure 4 (also included in the Supplementary Materials as Figure S1 in the Supplementary Materials), Figure 5 (Figure S2), and Figure 6 (Figure S3) to illustrate the findings. The spectra for the other colors are provided in the Supplementary Materials (Figures S1–S31).

Table 1. Pigments and auxiliary compounds detected and identified across three copies of *Jazz* using a multi-analytical approach (PR49:2: Calcium Lithol Red; PG7: Phthalocyanine green, PY3: Hansa Yellow 10G; PY5: Hansa Yellow 5G; PY6: Hansa Yellow 3G; PY10: Hansa Yellow R; PR3: Toluidine Red; PR4: Permanent Red R) * *not analyzed.*

Color		Pigment(s) Identified	MFT	Color		Pigment(s) Identified	MFT
Black 1 (Bk1)		Bone black/carbon black Prussian blue Silicates and metal oxides	>3	Green 3 (G3)		PG7 Synthetic ultramarine Barium sulfate	2.5
Black 2 (Bk2)		Bone black/carbon black Prussian blue Silicates and metal oxides	>3	Yellow 1 (Y1)		PY5	>3
Violet (V)		Crystal violet/methyl violet Rhodamine 6G Copper ferrocyanide Barium sulfate	3	Yellow 2 (Y2)		PY5 PY10	>3
Magenta (M)		Rhodamine 6G Rhodamine 3B Copper ferrocyanide Phototungstic (acid) Barium sulfate	2.5	Yellow 3 (Y3)		PY5 PY6 Bariumsulfate/zinc white/ lithopone	3
Pink 1 (Pk1)		Crystal violet/methyl violet Rhodamine 6G Copper ferrocyanide Barium sulfate	2.5	Yellow 4 (Y4)		PY3 PY6 Barium sulfate	3
Pink 2 (Pk2)		PR49:2	2	Yellow 5 (Y5)		PY3 PY10 Lead chromate Calcium sulfate	>3
Blue 1 (B1)		Synthetic ultramarine	2	Yellow 6 (Y6)		PY3 Gypsum Barium sulfate	>3
Blue 2 (B2)		Synthetic ultramarine Titanium white Calcium carbonate Zinc white	3	Yellow 7 (Y7)		PY5 PY3 Barium sulfate	>3
Blue 3 (B3)		Synthetic ultramarine Zinc white/ Barium sulfate/lithopone	3	Orange 1 (O1)		PY5 PR4	2.5
Blue 4 (B4)		Synthetic ultramarine	2	Orange 2 (O2)		PY5 PY10 Lead chromate Calcium sulfate	3
Blue 5 (B5)		Synthetic ultramarine	2	Orange 3 (O3)		PY5 PY10	3
Green 1 (G1)		PG7 PY3 Barium sulfate Aluminum Hydrate	2	Orange 4 (O4)		Iron oxide (yellow ochre) Lithopone Calcium carbonate	>3
Green 2 (G2)		PG7 PY3 Barium sulfate	>3	Orange 5 (O5)		Iron oxide (yellow ochre)	2.5
Orange 6 (O6)		PR4 PY5 Iron oxide (Mars red) Iron oxide (ochre) Calcium sulfate Lead chromate	2	Tan 2 (T2)		Iron oxide	*
Red 1 (R1)		PR3 PR49:2 Vermilion Barium sulfate	2.5	White 1 (W1)		Barium sulfate/zinc white/ lithopone	≈3
Red 2 (R2)		PR3 PR49:2 Lead chromate Barium sulfate	>3	White 2 (W2)		Zinc white Barium sulfate/lithopone	*

Table 1. *Cont.*

Color		Pigment(s) Identified	MFT	Color		Pigment(s) Identified	MFT
Red 3 (R3)		PR49:2	>3	Gray 1 (Gy1)		Zinc white Barium sulfate/zinc white/ lithopone Prussian blue	2.5
Red 4 (R4)		PR3 PR49:2	3	Gray 2 (Gy2)		Titanium white (anatase) Calcium carbonate Ultramarine blue Prussian blue Barium sulfate/zinc white/ lithopone	2.5
Brown (Br)		Iron oxide (Sienna/Mars red) Lead chromate Calcium sulfate	>3	Gray 3 (Gy3)		Titanium white (anatase) Calcium carbonate Ultramarine blue Prussian blue	*
Tan 1 (T1)		Possibly iron oxide Barium sulfate/zinc white/Lithopone	2.5	Paper		Low mineral content	2

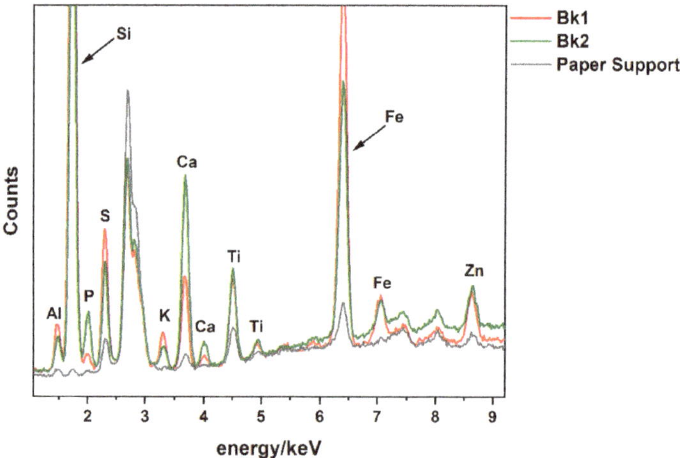

Figure 4. p-XRF spectra for the black gouaches—(red) Bk1 and (gray) paper support for plate P1 and (green) Bk2 for plate P8. Elements present in both gouaches: Al, Si, P, S, K, Ca, Ti, Cr, Mn, Fe, and Zn (small amounts of Al, Si, S, Ca, Fe, and Zn are also present in the paper).

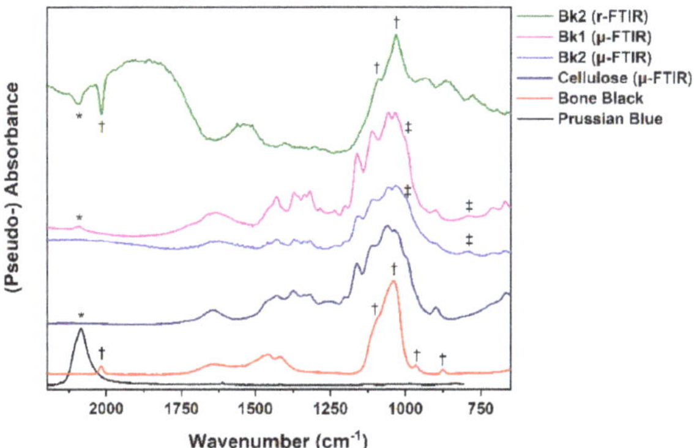

Figure 5. (Non-KK transformed) r- and μ-FTIR spectra of Black 1 (Bk1) and Black 2 (Bk2) alongside IRUG [17] reference spectra for cellulose, (†) Bone black (875, 962, 1038, 1087 and 2013 cm^{-1}) and (*) Prussian blue (2094 cm^{-1}); (‡) silicates/aluminosilicates were also observed (1001, 794 cm^{-1}).

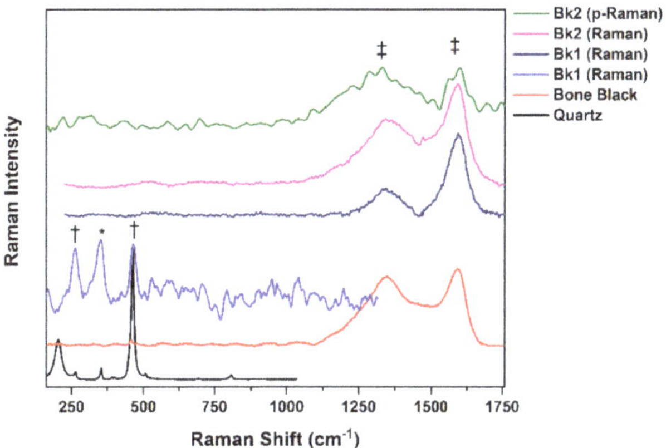

Figure 6. Confocal and p-Raman spectra of Black 1 (Bk1) and Black 2 (Bk2) alongside reference spectrum for Bone black [17,25] (D and G broad bands (‡) at 1367 and 1590 cm^{-1}—shifted to 1320 cm^{-1} in p-Raman). (†) Quartz (468 cm^{-1}) and silicates (272 cm^{-1}) [17,26] were also observed (355 cm^{-1} unassigned *).

3.1.1. Medium, Extenders, Additives, or Impurities

Traditionally, gouache paints are prepared with a gum rabic binder, white fillers for opacity, and organic or inorganic pigments [27]. A detailed medium analysis was beyond the scope of this work, but gum, though generally difficult to distinguish from cellulose was detected in several samples by μ-FTIR (Figure 7) by the presence of the characteristic O-H bending band at 1650–1620 cm^{-1} [28]. The analyses also revealed the presence of a variety of white inorganic compounds in the gouaches acting as extenders, lake substrates, and other potential additives. All are easily detected by p-XRF (key elements indicated in Table S1) and in most cases confirmed by infrared and Raman spectroscopies. Barium sulfate ($BaSO_4$) was the most prevalent extender and was identified by XRF (Ba, S, Sr) [29], FTIR [30], and Raman [31] spectroscopies. Calcium carbonate ($CaCO_3$) was identified by FTIR [32] and Raman [31] as well. Calcium sulfate was present in the gypsum form ($CaSO_4{\cdot}2H_2O$) based on the presence of sulfate vibration bands and OH stretching and deformation bands in the IR [33] and Raman [31] spectra. Alumina hydrate was identified by μ-FTIR through hydroxyl deformation vibration bands [34]. In the magenta, violet, and red gouaches, and in the yellow and orange gouaches containing PY5, a significant amount of aluminum was detected by XRF, but the source could not be identified. A small but varying amount of zinc was detected in all the gouaches by XRF, possibly as an additive or contamination.

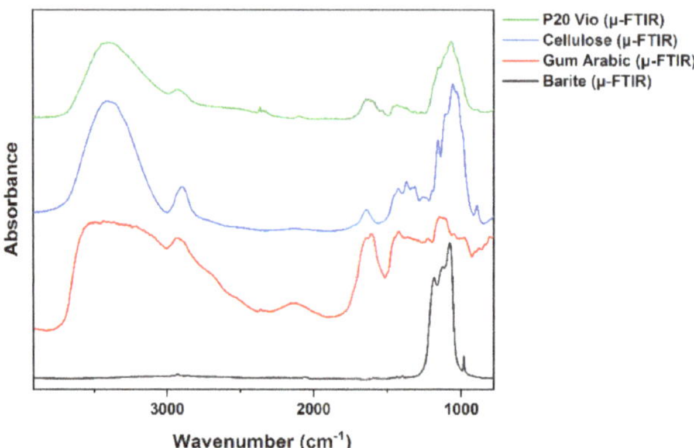

Figure 7. μ-FTIR spectrum obtained for the violet gouache (V) alongside IRUG [17] reference spectra for barium sulfate, cellulose, and gum rabic (gray region) the characteristic O-H bending mode at 1650–1620 cm^{-1} [17,28].

3.1.2. Black Gouaches

Two distinct black gouaches were identified. Both contained a mixture of bone/carbon black, Prussian blue, and silicates with lesser amounts of metal oxides, including possibly iron oxides. The relative concentrations of these pigments were however different: the more predominant Black 1 (Bk1) was slightly richer in Prussian blue and silicates; Black 2 (Bk2), found only in three prints, was richer in bone/carbon black. This pigment is an impure black carbon pigment prepared from charring animal bones containing mainly calcium hydroxyapatite ($Ca_{10}(PO_4)_6(OH)_2$) and small amounts of magnesium phosphate ($Mg_3(PO_4)_2$) and calcium carbonate ($CaCO_3$) [27], which is why it could be detected by the presence of P and Ca in the p-XRF spectra (Figure 4), especially in Bk2 where it was more concentrated. Bone black in Bk2 was also confirmed by FTIR through a characteristic band at 2013 cm^{-1} assigned to degradation products of the pigment synthesis [35]. The characteristic main phosphate band (ν_3 at 1036 cm^{-1}) associated with the hydroxyapatite overlaps with the cellulose and silicate bands in the μ-FTIR and (non-KK transformed) r-FTIR spectra [36] and neither 962 nor 875 cm^{-1} ν_2 bands were detected (Figure 5). The presence of the characteristic D and G broad bands in the Raman spectra [25] confirmed the presence of carbon black, either as a constituent of bone black and/or as a pigment (Figure 6). The additional phosphate band at 960 cm^{-1} in the Raman spectrum is generally weak [37] and was not observed.

Prussian blue is a synthetically produced ferric ferrocyanide blue pigment ($KFe[Fe(CN)_6]$) [38]. Fe and K were detected by p-XRF (these elements can also indicate the presence of natural iron oxides). Further confirmation was provided by r- and μ-FTIR with the characteristic CN asymmetric stretch [30]. The pigment was not observed in either Raman modalities, although it is detectable [31]. The p-XRF spectra also suggested the presence of silicates (Al, Si, K) and metal oxides (Ti, Cr, and possibly Fe). The presence of silicates and aluminosilicates was confirmed by μ- and r-FTIR [33], although several bands overlapped with bone black and cellulose. Confocal Raman spectroscopy confirmed the presence of quartz and silicates [26].

3.1.3. Violet Gouache

The violet gouache (V) was only used in pochoir P20. SERS (Figure S4) identified Rhodamine 6G (R6G) [39] and methyl violet and/or crystal violet (MV/CV) [40]. The weak

Cu and Fe peaks in p-XRF (Figure S5) suggested the presence of copper ferrocyanide (CF), also confirmed by the characteristic CN asymmetric stretch in the μ- and r-FTIR spectra. CF could be the precipitating anion for either R6G (PR169) or for MV (PV27) [41].

3.1.4. Magenta Gouache

The composition of the magenta gouache (M) was the same across nine prints. Reference to this gouache appears in published correspondence between Matisse, Tériade, and Gaut, the then director of the Linel company [42]. The source pigment for this gouache was manufactured in Germany and became unavailable after WWII. In the letter, Matisse mentions the importance of a "Violet Fixe" and the large quantity needed to reproduce a large work, indubitably *Jazz*. SERS spectra indicated the presence of R6G, and confocal Raman also indicated the presence of Rhodamine 3B (R3B) [43] (Figure S6). The p-XRF analysis suggested that these two pigments are precipitated with CF—also confirmed by r-FTIR—and possibly phospho-tungsto-molybdic acid (PTM) since P and W were also detected (Figure S7). The presence of these constituents could be related to the presence of PV2, an R3B-PTM pigment, and PR169, an R6G-CF pigment [41].

3.1.5. Pink Gouaches

Two distinct pink gouaches were identified: R6G and CV/MV by SERS in Pink 1 (Pk1), while the presence of CF was detected by p-XRF (Figure S8) though not confirmed by FTIR. The second pink gouache (Pk2) used in P20 is slightly darker and warmer in tone and contained Calcium Lithol Red (PR49:2) as identified by Raman and SERS (Figure S9) [44]. The p-XRF spectra showed significant peaks for Ca and S as expected, the latter relating to the sulfonated group of the beta-naphthol [41].

3.1.6. Blue Gouaches

A total of five distinct blue gouaches were identified and all included synthetic ultramarine blue. Blue 1 (B1) was the most prominent blue and contained no additional pigments or fillers while Blue 2 (B2), used in P19 and similar to B1 in tonality, also contained titanium white (TiO_2), zinc white (ZnO), and calcium carbonate. The intermediate Blue 3 (B3) used in P6 contained zinc white. Both light Blue 4 (B4), used in P12, and Blue 5 (B5), used in P15, had a light gray-blue tone but no additional pigment was detected aside from ultramarine blue. Ultramarine blue ($3Na_2O.3Al_2O_3.6SiO_2.2Na_2S$) was detected in all the blues by p-XRF (Figure S10) through the presence of Al, Si, and S (K and Ca were also present in small quantities) [45]. In μ-FTIR and r-FTIR spectra (with no KK transformation), the characteristic Si-O and Al-O stretch as well as the Si-O main bending mode were readily observed (Figure S11) [30,45]. The pigment was also detected with confocal and p-Raman spectroscopy through the Raman modes of the sulfur radical anions S_3^- and S_2^- acting as chromophores (Figure S12) [45]. Those bands were weak but still discernible for the light B4 and B5 blues. Titanium white in B2 was detected by p-XRF (Ti), more specifically anatase as confirmed by confocal and p-Raman [46]. Zinc white was detected by p-XRF but could not be confirmed by Raman or infrared spectroscopy, though detectable by the latter [47].

3.1.7. Green Gouaches

Three distinct green gouaches were used in *Jazz*, a darker (G1), a lighter (G2) yellow-green, and a teal green (G3) that appears streakier. G1 and the lighter G2 both contained Hansa Yellow 10G (PY3), a yellow pigment with a greenish tint, and phthalocyanine green (PG7). The two are often found combined in yellow-greens [48]. PG7 was observed in confocal and p-Raman [49] (Figure S13) and detected by p-XRF through the presence of Cl and Cu peaks (Figure S14), though not confirmed by FTIR [17,50]. Conversely, PY3 was more readily observed by μ- and (non-KK transformed) r-FTIR [17] (Figure S15) than by Raman [49] spectroscopies, emphasizing the advantage and necessity of complementary techniques. The green gouache was applied on top of the gray background in P17 and P19, and both p-Raman and r-FTIR detected a signal from underlying pigments. G3 contains PG7 and ultramarine blue. Both pigments were confirmed by p-XRF and Raman. Some of the PG7 bands were weakly detected by µ-FTIR.

3.1.8. Yellow Gouaches

A total of seven different yellows (Y1 to Y7) were identified across twelve prints, with tonalities varying from light green-yellow to dark orange-yellow. Several prints present both a lighter and a darker yellow. All the pigments were identified as Arylid Hansa yellows, by themselves or combined: green-yellows PY3 and PY5 (Hansa Yellow 5G); red-yellows PY6 (Hansa Yellow 3G) and PY10 (Hansa Yellow R). It is notable that PY5 and PY10 are no longer in production [41]. All were easily identified by µ-FTIR and by r-FTIR when sufficiently abundant (Figures S16 and S17) [17], and/or confocal and p-Raman (Figures S18 and S19) [49]. Cl is a constituent present in most of these monoazo yellows and was detected by p-XRF (Figure S20). A small amount of lead chromate was seen in Y5 as confirmed by p-XRF (Cr, Pb) and µ-Raman (Figure S21) [31].

3.1.9. Orange Gouaches

Overall analysis revealed the use of six different orange (O1 to O6) gouaches that contained one or more organic and/or inorganic pigments: PY5, PY10, Permanent Red R (PR4), iron oxides, and lead chromate. O1 is a bright orange that contained PY5 and PR4 both confirmed by FTIR [17] and Raman [51] analysis. The light Orange 2 and Orange 3 have similar tonalities, and both contained PY5 and PY10 as confirmed by FTIR and Raman analysis. O2, however, also contained a small amount of lead chromate and a significant amount of gypsum. The composition of O3 was the same as Y2, but the ratio of the PY10 (reddish yellow) to PY5 (greenish-yellow) was reversed based on the intensity of their respective bands in the Raman spectra. Orange 4 and Orange 5 contained ochre, a natural iron oxide pigment rich in kaolinite [52], as confirmed by p-XRF (Al, Si, K, Fe) (Figure S22), r- and µ-FTIR [34], and confocal Raman [31] (Figure S23a). Orange 6 had a more complex composition and contained PY5, PR4 detected by FTIR and Raman analysis of the samples (Figures S24–S26), possibly a small amount of lead chromate based on the p-XRF analysis, Mars Red [31] (Figure S23b) and a second iron oxide pigment (214, 252, 304, 402, 463, and 560 cm^{-1}), the latter two detected by confocal Raman.

3.1.10. Red Gouaches

The FTIR (Figures S27 and S28) and Raman analysis suggested that four different reds (R1 to R4) were used. R1, R2, and R4 contained mainly Toluidine Red (PR3) [17,50] and PR49:2—also found in Pk2. The p-XRF detected the presence of small amounts of lead chromate (Pb, Cr) in R2, while vermilion (HgS) was detected in R1 (Figure S29), as confirmed by Raman (Figure S28) [30]. PR49:2 was the only pigment detected in the less orange-hued R3.

3.1.11. Brown Gouache

The brown gouache (Br) was only used in two prints, P2 and P20. The main pigment was a natural iron oxide with a high content of aluminosilicates (possibly Sienna) and was confirmed by the hematite bands in r-FTIR [33] and ochre bands in Raman [31]. The p-XRF analysis also detected a small amount of lead chromate aside from the iron oxides.

3.1.12. An Gouaches

Darker (T1) and lighter (T2) tan gouaches were used in print P10 and P11, respectively. Based on the p-XRF analysis, both gouaches contained a small amount of a natural iron oxide pigment rich in aluminosilicates.

3.1.13. White Gouaches

The white areas in the prints were generally the paper background left in reserve except in prints P12 and P17 where white gouaches were used. Both gouaches appeared to have the same composition based on the p-XRF analysis (S, Zn, Sr, Ba) although White 2 (W2) in P17 was much richer in Zn (Figure S30). The presence of these elements can indicate the presence or combination of barium sulfate ($BaSO_4$), zinc white (ZnO), and lithopone (30%:70% $ZnS:BaSO_4$) [27]. Further distinction was not possible using Raman or infrared spectroscopies as neither zinc white nor zinc sulfide in lithopone [53] are easily detected. The Zn to Ba peak ratio in the p-XRF spectrum for White 1 (W1) however was characteristic of lithopone [54] while W2 had a much higher content of Zn, suggesting that it contained zinc white and barium sulfate or lithopone.

3.1.14. Gray Gouaches

Three different grays (Gy1 to Gy3) were distinguished. Gy1, used only in print P18, contained possibly zinc white and barium sulfate (based on the high Zn to Ba ratio detected by p-XRF—Figure S31) and/or lithopone, and a small amount of Prussian blue (confirmed by a very weak CN band in r-FTIR). Gy2, on the other hand, was mainly composed of titanium white in anatase form according to Raman analysis, and calcium carbonate confirmed by Raman and FTIR (and a small amount of zinc white and barium sulfate or lithopone). The gray tonality was due to the presence of ultramarine blue and Prussian blue in (Figure S23c), both confirmed by p-XRF and by Raman [31]. Gy3 was present in P9 and was similar to Gy2 in tonality and composition except for the presence of the zinc-based pigment.

3.2. Color and Color Stability

Microfaedometry indicated that no color is alarmingly light-sensitive, or in any case more sensitive than the paper support (Figure 8). It is important to note that although the tonality of two colors might be similar, like for example Blue 1 and 2, their light sensitivity can be different as a result of their different composition. This highlights the importance of confirming the composition of all the gouaches across this work and other cut-outs by Matisse.

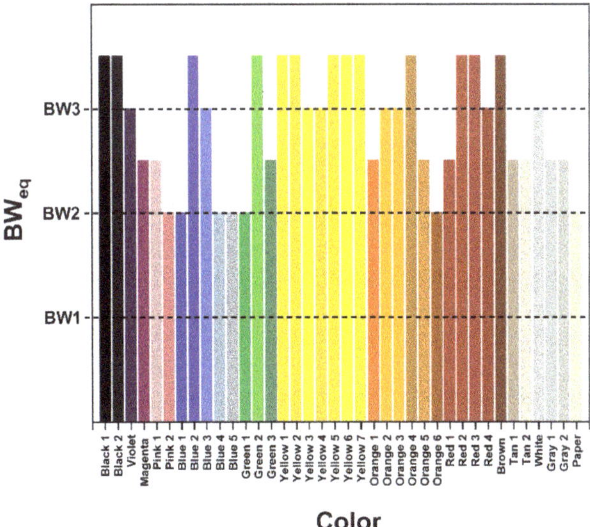

Figure 8. The light sensitivity of all colors represented in *Jazz* on a range from BW1 (most sensitive) to BW3 (moderate sensitivity).

The most sensitive colors appeared to be Blue 1, 4, and 5, Pink 2, and Orange 1. The light blue and pink colors in this group are more transparent. Consequently, the microfaedometry testing also reflects the light sensitivity of the paper support. Fading of Matisse's blue gouaches was reported in the past [11] for *The Swimming Pool* (1952) and linked to the acidic environment created by the burlap support [45]. Despite being a lightfast pigment [53], studies have shown that fading of this pigment can be promoted by the paint medium [55]. It appears, however, that the presence of white pigments and fillers in B2 and B3 increases the lightfastness of the gouaches. Orange 6 and the slightly less light-sensitive Orange 1 are the only gouaches that contain PR4, and their sensitivity was unexpectedly high considering the good lightfastness reported for PR4 in the literature (though they are reported to darken overexposure) [41]. Conversely, the sensitivity of Green 1 and to a lesser extent of Green 3, both richer in PG7 than the more light-stable Green 2, was also unexpected [41]. In contrast, although aniline (CV/MV) and xanthene dyes (rhodamines) have reputedly poor lightfastness [41,56], neither the magenta nor the violet gouaches were exceedingly light-sensitive.

The results of the microfaedometry testing led to new recommendations for the exhibition of the *Jazz* plates at MoMA. The prints have been grouped based on their vulnerability to color change and prioritized for upcoming displays depending on their exhibition history. Prints P14, P15, P16, and P20 are considered the most susceptible to fading as they contain at least two of the most vulnerable colors (BW_{eq} = 2), while prints P6, P7, P10, P11, and P18 are slightly more stable but still contain three or more moderately sensitive colors (BW_{eq} = 2.5).

4. Conclusions

The identification of the pigments present in the gouaches used by Henri Matisse was important to understand the current condition of the *Jazz* prints and support preventive conservation strategies. Based on the analysis carried out at MoMA and at the MNAM, a total of 39 distinct colored gouaches were used to reproduce the *Jazz* prints. Some of the colors, though similar in tonality, have a different composition which has implications for their light stability and display recommendations. The composition of the gouaches was consistent across the three copies of *Jazz* investigated and it is reasonable to assume that this applies to the rest of the existing copies since they were printed at the same time [6].

This study also illustrates the value and limitations and challenges of using a non-invasive approach that combines p-XRF, r-FTIR, and p-Raman to characterize the Linel gouaches used by Matisse in his cut-outs. More than twenty different pigments and extenders were detected and characterized across the twenty *Jazz* plates. All the pigments were identified using a non-invasive approach that combines p-XRF, r-FTIR, and p-Raman analysis with the exception of the methyl violet and/or crystal violet (MV/CV) and the Rhodamine 3B and 6G pigments that could only be identified by confocal Raman or SERS. The organic pigments PR3, PR4, PR49:2, PY3, PY5, PY6, and PY10 were identified by r-FTIR and p-Raman when present as major constituents, while PG7 was better identified by Raman. p-XRF was particularly suited to detect the presence of mineral pigments, especially when in low concentration, as it was the case of lead chromate and vermilion. It also hinted at the presence of the triarylcarbonium pigments (R3B and R6G) precipitated with complex ions (copper ferrocyanide or phosphotungstic acid). It was also helpful to confirm the presence of inorganic pigments and extenders identified by r-FTIR and/or p-Raman namely Prussian Blue, synthetic ultramarine, bone black, iron oxides, titanium white, barium sulfate and/or lithopone, calcium sulfate, and carbonate. While most of the colors were at least as light-stable as the paper background, adhering to lighting and exhibition guidelines for sensitive works on paper is important to sustain the vibrancy of the most sensitive colors and preserve the harmony and energy of the compositions so vital to Matisse.

Matisse was known to have used gouaches straight out of the tube and unmixed for his cut-outs [11], but the gouaches used in *Jazz* might have been made-to-order as

Linel colorists worked closely with Matisse, Tériade, and Vairel to reproduce this large number of prints. Some of the gouaches have very similar tonalities but slightly different compositions, like the reds, some of the yellows and blues, as well as the blacks. This might have been deliberate to produce slight nuances across the plates or stemmed from the need to prepare multiple batches of gouaches. Although the gouaches identified in *Jazz* might have a unique composition, the analytical findings should still be relevant for the study of Matisse's cut-outs, including for dating and authentication purposes.

Supplementary Materials: The following are available online at https://www.mdpi.com/article/10.3390/heritage4040231/s1, Figure S1: p-XRF spectra for the Black 1 (Bk1) and Black 2 (Bk2), Figure S2: r- and μ-FTIR spectra of Black 1 (Bk1) and Black 2 (Bk2), Figure S3: Confocal and p-Raman spectra of Black 1 (Bk1) and Black 2 (Bk2), Figure S4: SERS spectra of Violet (V), Figure S5: p-XRF spectra for the Violet (V), Figure S6: Normal (confocal) Raman and SERS spectra of Magenta (M), Figure S7: p-XRF spectra for the Magenta (M), Figure S8: p-XRF spectra for the Pink 1 (Pk1) and Pink 2(Pk2), Figure S9: Confocal Raman spectrum of Pink 2 (Pk2), Figure S10: p-XRF spectra for the Blue 1 (B1), Blue 2 (B2) and Blue 3 (B3), Figure S11: r- and μ-FTIR spectra of Blue 2 (B2), Figure S12: Confocal and p-Raman spectra of Blue 2 (B2), Figure S13: Confocal and p-Raman spectra of Green 2 (G2), Figure S14: p-XRF spectra for the Green 1 (G1) and Green 3 (G3), Figure S15: r- and μ-FTIR spectra of Green 2 (G2), Figure S16: r- and μ-FTIR spectra of Yellow 5 (Y5), Figure S17: r- FTIR spectrum of Yellow 3 (Y3), Figure S18: Confocal and p-Raman spectra of Yellow 4 (Y4), Figure S19: Confocal and p-Raman spectra of Yellow 1 (Y1) and Yellow 2 (Y2), Figure S20: p-XRF spectra for the Yellow 1 (Y1), Yellow 4 (Y4) and Yellow 5 (Y5), Figure S21: Confocal Raman spectrum of Yellow 5 (Y5), Figure S22: p-XRF spectra for the Orange 4 (O4) and Orange 5 (O5), Figure S23: (a) Confocal Raman spectrum of Orange 5 (O5); confocal Raman spectrum of Orange 6 (O6); confocal and p-Raman spectrum of Gray 2 (Gy2), Figure S24: r- and μ-FTIR spectra of Orange 6 (O6), Figure S25: r- and μ-FTIR spectra of Orange 6 (O6), Figure S26: Confocal Raman spectrum of Orange 6 (O6), Figure S27: r- and μ-FTIR spectra of Red 4 (R4), Figure S28: Confocal Raman spectrum of Red 1 (R1), Figure S29: p-XRF spectra for the Red 1 (R1) and Red 2 (R2), Figure S30: p-XRF spectra for White 1 (W1) and White 2 (W2), Figure S31: p-XRF spectra for the Gray 1 (Gy1), Gray 2 (Gy2) and Gray 3 (Gy3), Table S1: All pigments and auxiliary compounds detected and identified across three copies of Jazz using a multi-analytical approach, Table S2: Gouaches identified in each Jazz plate.

Author Contributions: Conceptualization, A.M. and M.D.; Data curation, A.M.; Formal analysis, A.M., M.D., A.H., C.D., A.G.-L.B. and T.T.; Investigation, A.M. and A.C.P.; Methodology, A.M., A.C.P. and M.D.; Project administration, A.M. and M.D.; Resources, A.C.P.; Writing—original draft, A.M.; Writing—review & editing, A.C.P., M.D., A.H., C.D., A.G.-L.B. and T.T. All authors have read and agreed to the published version of the manuscript.

Funding: This research was funded by Fondation des Sciences du Patrimoine and the Cooperation and International Mobility 2018 program; National Science Foundation (CHE-1402750) and HRD-1547830 (IDEALS CREST); City University of New York PSC-CUNY Faculty Research Award Program, Grant No. 69079.

Data Availability Statement: Data available upon request to authors.

Acknowledgments: Martins, Prud'hom, Duranton, Daher, and Geanatche acknowledge the financial support of the Fondation des Sciences du Patrimoine and the Cooperation and International Mobility 2018 program, as well as the support from Veronique Serano-Stedman and Jonas Strove at the Musée National D'Art Moderne de Paris. Haddad is indebted to the National Science Foundation and the City University of New York. Tang acknowledges the MoMA Summer Internship Program. Martins is grateful for the support of MoMA colleagues Karl Buchberg, Laura Neufeld, and Jodi Hauptman. Martins would also like to thank Georges Matisse and the Matisse family for their generous support. The authors acknowledge the support of Federica Pozzi, Anna Cesaratto, Marco Leona, and The Network Initiative for Conservation Science (NICS) for lending the Bruker Bravo p-Raman instrument. The authors would like to thank Xavier Gallet, Matthieu Lebon and Olivier Tombret (UMR 7194 CNRS–Histoire Naturelle de l'Homme Préhistorique) for providing access their Elio portable XRF.

Conflicts of Interest: The authors declare no conflict of interest.

References

1. Cullinan, N. Chromatic Composition. In *Henri Matisse: The Cut-Outs*; Buchberg, K., Cullinan, N., Hauptman, J., Serota, N., Eds.; Tate Publishing: London, UK, 2014.
2. Rabinow, R.A. *The Legacy of La Rue Férou: Livres D'Artiste Created for Tériade by Rouault, Bonnard, Matisse, Léger, Le Corbusier, Chagall, Giacometti, and Miró*; New York University, Graduate School of Arts and Science: New York, NY, USA, 1995.
3. Wheeler, M. *The Last Works of Henri Matisse: Large Cut Gouaches*; Museum of Modern Art: New York, NY, USA, 1962.
4. Saudé, J. *Traité D'enluminure D'art au Pochoir*; Éditions de L'IBIS: Paris, France, 1925.
5. Matisse, H.; Rouveyre, A.; Finsen, H. *Matisse—Rouveyre: Correspondance*; Flammarion: Paris, France, 2001.
6. Lavarini, B. *Henri Matisse: Jazz [1943–1947]: Ein Malerbuch als Selbstbekenntnis*; Scaneg Verlag: München, Germany, 2000.
7. Rogge, C.E.; Epley, B.A. Behind the Bocour Label: Identification of Pigments and Binders in Historic Bocour Oil and Acrylic Paints. *J. Am. Inst. Conserv.* **2017**, *56*, 15–42. [CrossRef]
8. Izzo, F.C.; van den Berg, K.J.; van Keulen, H.; Ferriani, B.; Zendri, E. Modern oil paints—Formulations, organic additives and degradation: Some case studies. In *Issues in Contemporary Oil Paint*; Burnstock, A., de Keijzer, M., Krueger, J., Learner, T., de Tagle, A., Heydenreich, G., van den Berg, K.J., Eds.; Springer International Publishing: Berlin/Heidelberg, Germany, 2014; pp. 75–104.
9. Pause, R.; van der Werf, I.D.; van den Berg, K.J. Identification of Pre-1950 Synthetic Organic Pigments in Artists' Paints. A Non-Invasive Approach Using Handheld Raman Spectroscopy. *Heritage* **2021**, *4*, 1348–1365. [CrossRef]
10. Gautier, G.; Bezur, A.; Muir, K.; Casadio, F.; Fiedler, I. Chemical Fingerprinting of Ready-Mixed House Paints of Relevance to Artistic Production in the First Half of the Twentieth Century. Part I: Inorganic and Organic Pigments. *Appl. Spectrosc.* **2009**, *63*, 597–603.
11. Buchberg, K.; Gross, M.; Lohrengel, S. Materials and Techniques. In *Henri Matisse: The Cut-Outs*; Buchberg, K., Cullinan, N., Hauptman, J., Serota, N., Eds.; Tate Publishing: London, UK, 2014.
12. King, A. *Henri Matisse: Paper Cut-Outs*; St. Louis Art Museum: St. Louis, MO, USA, 1977.
13. Mochon, R. *Matisse's "Swimming Pool": A Microfading Study*; Internal Report; Museum of Modern Art: New York, NY, USA, 2018.
14. Liang, H.; Lange, R.; Andrei, L. *Examination of the Light Sensitivity of Acanthes by Henri Matisse Using Microfade Spectroscopy*; Internal Report; ISAAC Mobile Lab, Nottingham Trent University: Nottingham, UK, 2012.
15. Townsend, J. *Microfading Report on the Snail*; Report T00540; Tate: London, UK, 2013.
16. Martins, A. *Microfading Report on the Matisse Gouaches Historical Set*; The Museum of Modern Art: New York, NY, USA, 2018.
17. Price, B.A.; Boris, P.; Lomax, S.Q. (Eds.) *Infrared and Raman Users Group Spectral Database*, 2007th ed.; IRUG: Philadelphia, PA, USA, 2009; Volumes 1–2.
18. Lee, P.C.; Meisel, D.J. Adsorption and Surface-Enhanced Raman of Dyes on Silver and Gold Sols. *Phys. Chem.* **1982**, *86*, 3391–3395. [CrossRef]
19. Pozzi, F.; Lombardi, J.R.; Bruni, S.; Leona, M. Sample treatment considerations in the analysis of organic colorants by surface-enhanced Raman scattering. *Anal. Chem.* **2012**, *84*, 3751–3757. [CrossRef]
20. Druzik, J. Oriel Microfading Tester (MFT): A Brief Description. Research and Technical Studies Specialty Group Postprints. In Proceedings of the 38th Annual Meeting of the American Institute for Conservation of Historic and Artistic Works, Milwaukee, WI, USA, 11–14 May 2010.
21. Keene, L. Getty Spectral Viewer™—Program Written for Getty Conservation Institute [GCI] for Processing Newport-Oriel Microfade Tester SPEC32 Software Results. 2007. Available online: https://www.getty.edu/conservation/our_projects/science/lighting/partners.html (accessed on 19 October 2021).
22. Ford, B. A microfading survey of the lightfastness of blue, black and red ballpoint pen inks in ambient and modified environments. *AICCM Bull.* **2018**, *39*, 26–34. [CrossRef]
23. Mecklenburg, M.F.; del Hoyo-Meléndez, J.M. Development and application of a mathematical model to explain fading rate inconsistencies observed in light-sensitive materials. *Color. Technol.* **2012**, *128*, 139–146. [CrossRef]
24. Prestel, T. A classification system to enhance lightfastness data interpretation based on microfading tests and rate of colour change. *Color. Technol.* **2017**, *133*, 506–512. [CrossRef]
25. Daly, N.S.; Sullivan, M.; Lee, L.; Trentelman, K.J. Multivariate analysis of Raman spectra of carbonaceous black drawing media for the in situ identification of historic artist materials. *Raman Spectrosc.* **2018**, *49*, 1497–1506. [CrossRef]
26. Kingma, K.J.; Hemley, R.J. Raman spectroscopics tudy of microcrystalline silica. *Am. Mineral.* **1994**, *79*, 269–273.
27. Gettens, R.J.; Stout, G.L. *Painting Materials: A Short Encyclopedia*; Courier Corporation: North Chelmsford, MA, USA, 1966.
28. Derrick, M.R.; Stulik, D.; Landry, J.M. *Infrared Spectroscopy in Conservation Science*; Getty Conservation Institute: Los Angeles, CA, USA, 1999.
29. Feller, R.L. Barium Sulphate. In *Artists' Pigments: A Handbook of Their History and Characteristics*; Feller, R.L., Ed.; National Gallery of Art: Washington, DC, USA, 1986; Volume 1.
30. Vetter, W.; Schreiner, M. Characterization of pigment-binding media systems - comparison of non-invasive in-situ reflection FTIR with transmission technical FTIR microscopy E-Preserv. *Sci* **2011**, *8*, 10–22.
31. Bell, I.M.; Clark, R.J.H.; Gibbs, P.J. Raman spectroscopic library of natural and synthetic pigments (Pre- ~ 1850 AD). *Spectrochim. Acta A Mol. Biomol. Spectrosc.* **1997**, *53*, 2159–2179. [CrossRef]
32. Socrates, G. *Infrared and Raman Characteristic Group Frequencies: Tables and Charts*, 3rd ed.; John Wiley & Sons: West Sussex, UK, 2004.

33. Genestar, C.; Pons, C. Earth pigments in painting: Characterisation and differentiation by means of FTIR spectroscopy and SEM-EDS microanalysis. *Anal. Bioanal. Chem.* **2005**, *382*, 269–274. [CrossRef] [PubMed]
34. Frederickson, L.D. Characterization of hydrated alumina by infrared spectroscopy. Application to study of bauxite ores. *Anal. Chem.* **1954**, *26*, 1883–1885. [CrossRef]
35. Vila, A.; Ferrer, N.; García, J.F. Chemical composition of contemporary black printing inks based on infrared spectroscopy: Basic information for the characterization and discrimination of artistic prints. *Anal. Chim. Acta* **2007**, *591*, 97–105. [CrossRef] [PubMed]
36. Daveri, A.; Malagodi, M.; Vagnini, M.J. The Bone Black Pigment Identification by Noninvasive, In Situ Infrared Reflection Spectroscopy. *J. Anal. Methods Chem.* **2018**, *2018*, 6595643. [CrossRef] [PubMed]
37. Coccato, A.; Jehlicka, J.; Moens, L.; Vandenabeele, P.J. Raman spectroscopy for the investigation of carbon-based black pigments. *Raman Spectrosc.* **2015**, *46*, 1003–1015. [CrossRef]
38. Berrie, B. Prussian Blue. In *Artists' Pigments: A Handbook of Their History and Characteristics*; West FitzHugh, E., Ed.; National Gallery of Art: Washington, DC, USA, 1997; Volume 3.
39. Jensen, L.; Schatz, G.C.J. Resonance Raman Scattering of Rhodamine 6G as Calculated Using Time-Dependent Density Functional Theory. *J. Phys. Chem. A* **2006**, *110*, 5973–5977. [CrossRef] [PubMed]
40. Cañamares, M.V.; Chenal, C.; Birke, R.L.; Lombardi, J.R.J. DFT, SERS, and Single-Molecule SERS of Crystal Violet. *Phys. Chem. C* **2008**, *112*, 20295–20300. [CrossRef]
41. Herbst, W.; Hunger, K. *Industrial Organic Pigments: Production, Properties, Applications*, 3rd ed.; John Wiley & Sons: Weinheim, Germany, 2006.
42. Matisse, H.; Flam, J.D.; Tériade, E.; Szymusiak, D. *Matisse et Tériade*; Anthèse: Arcueil, France, 1996.
43. Steger, S.; Stege, H.; Bretz, S.; Hahn, O.J. A complementary spectroscopic approach for the non-invasive in-situ identification of synthetic organic pigments in modern reverse paintings on glass (1913–1946). *Cult. Herit.* **2019**, *38*, 20–28. [CrossRef]
44. Kennedy, A.; Stewart, H.; Stenger, J. Lithol Red: A Systematic Structural Study on Salts of a Sulfonated Azo Pigment. *Chem. Eur. J.* **2012**, *18*, 3064–3069. [CrossRef]
45. Desnica, V.; Furić, K.; Schreiner, M. Notice "Multianalytical characterisation of a variety of ultramarine pigments". *E-Preserv. Sci.* **2004**, *1*, 15–25.
46. Burgio, L.; Clark, R.J.H. Library of FT-Raman spectra of pigments, minerals, pigment media and varnishes, and supplement to existing library of Raman spectra of pigments with visible excitation. *Spectrochim. Acta Part A Mol. Biomol. Spectrosc.* **2001**, *57*, 1491–1521. [CrossRef]
47. Silva, C.E.; Silva, L.P.; Edwards, H.G.M.; de Oliveira, L.F.C. Diffuse reflection FTIR spectral database of dyes and pigments. *Anal. Bioanal. Chem.* **2006**, *386*, 2183–2191. [CrossRef] [PubMed]
48. Lewis, P.A. (Ed.) *Pigment Handbook, Volume 1: Properties and Economics*, 2nd ed.; Wiley-Interscience: New York, NY, USA, 1988.
49. Scherrer, N.C.; Stefan, Z.; Francoise, D.; Annette, F.; Renate, K. Synthetic organic pigments of the 20th and 21st century relevant to artist's paints: Raman spectra reference collection. *Spectrochim. Acta Part A Mol. Biomol. Spectrosc.* **2009**, *73*, 505–524. [CrossRef] [PubMed]
50. Buti, D.; Rosi, F.; Brunetti, B.G.; Miliani, C. In-situ identification of copper-based green pigments on paintings and manuscripts by reflection FTIR. *Anal. Bioanal. Chem.* **2013**, *405*, 2699–2711. [CrossRef] [PubMed]
51. Fremout, W.; Saverwyns, S.J. Identification of synthetic organic pigments: The role of a comprehensive digital Raman spectral library. *Raman Spectrosc.* **2012**, *43*, 1536–1544. [CrossRef]
52. Helwig, K. Iron Ocide. In *Artists' Pigments: A Handbook of Their History and Characteristics*; Berrie, B., Ed.; National Gallery of Art: Washington, DC, USA, 2007; Volume 4.
53. Buxbaum, G.; Pfaff, G. (Eds.) *Industrial Inorganic Pigments*; John Wiley & Sons: Hoboken, NJ, USA, 2006.
54. Capua, R.J. The obscure history of a ubiquitous pigment: Phosphorescent lithopone and its appearance on drawings by John la Farge. *J. Am. Inst. Conserv.* **2014**, *53*, 75–88. [CrossRef]
55. de la Rie, E.R.; Michelin, A.; Ngako, M.; Del Federico, E.; Del Grosso, C. Photo-catalytic degradation of binding media of ultramarine blue containing paint layers: A new perspective on the phenomenon of "ultramarine disease" in paintings. *Polym. Degrad. Stab.* **2017**, *144*, 43–52. [CrossRef]
56. de Keijzer, M.; van Bommel, M.R. Bright new colours: The history and analysis of fluorescein, eosin, erythrosine, Rhodamin B and some of their derivatives. In *The Diversity of Dyes in History and Archaeology*; Atkinson, J.K., Ed.; Archetype Books: London, UK, 2017.

 heritage

MDPI

Article

Reviving Alexander Calder's *Man-Eater with Pennants*: A Technical Examination of the Original Paint Palette

Abed Haddad *, Megan Randall, Lynda Zycherman and Ana Martins

The David Booth Conservation Department, The Museum of Modern Art, 11W 53rd Street, New York, NY 10019, USA; megan_randall@moma.org (M.R.); lynda_zycherman@moma.org (L.Z.); 2021amartins@gmail.com (A.M.)
* Correspondence: abed_haddad@moma.org or abedhaddad123@gmail.com

Abstract: *Man-Eater with Pennants*, a rarely exhibited sculpture in Alexander Calder's oeuvre, was commissioned by The Museum of Modern Art (MoMA) and installed in 1945. To exhibit the large standing mobile in *Alexander Calder: Modern from the Start* (2021), the derelict sculpture had to be remediated; this initiated a collaborative investigation with conservation scientists, conservators, curators, and the Calder Foundation into the original paint colors hidden beneath layers of repaint. XRF analysis was carried out to elucidate the paints' composition, followed by sampling for analysis to assess the paint stratigraphy and binders. Scrapings were analyzed by µ-FTIR and Raman spectroscopies; cross sections were examined with optical microscopy and analyzed with SEM-EDS. Analysis differentiated between the original paints, which contain Prussian blue, parachlor red, chrome yellow, and the many layers of overpaint, which contain titanium white, molybdate orange, a variety of β-Naphthol reds, red lead, and ultramarine. A model for *Man-Eater*, *Mobile with 14 Flags*, is also part of the museum's collection, and was first considered as a point of reference for the original colors. Similar analysis, however, indicates that the maquette was painted after the *Man-Eater* was first installed, therefore is not representative of the original colors. In addition to investigating an early primary palette for Calder's outdoor sculptures, this study helped develop the plan for the restoration of the original color scheme of *Man-Eater*.

Keywords: pigment identification; Raman spectroscopy; X-ray fluorescence spectroscopy; scanning electron microscopy-energy dispersive spectroscopy; microstratigraphic analysis

Citation: Haddad, A.; Randall, M.; Zycherman, L.; Martins, A. Reviving Alexander Calder's *Man-Eater with Pennants*: A Technical Examination of the Original Paint Palette. *Heritage* **2021**, *4*, 1920–1937. https://doi.org/10.3390/heritage4030109

Academic Editor: Diego Tamburini

Received: 1 August 2021
Accepted: 16 August 2021
Published: 19 August 2021
Corrected: 1 March 2022

1. Introduction

Alexander Calder (1898–1976), one of America's best-known sculptors, is renowned for developing two new idioms in modern art: mobiles, which hang from the ceiling and whose shapes move in response to air currents, and stabiles, large scale, stationary abstract sculpture, characterized by simple forms executed in sheet metal. Born into a family of artists, Calder showed facility in handling metals from a young age [1]. Although he trained as an engineer at the Stevens Institute of Technology in New Jersey and held several engineering jobs after graduation, at the age of 25, he committed himself as an artist, relocated to New York, and took art classes. In 1926, he moved to Paris where he mingled with established artists and writers of the time, including figures such as Picasso, Miró, Léger, and Duchamp. During a visit to Mondrian's studio in 1930, Calder saw, tacked to the wall, colored cardboard rectangles that Mondrian used to work out compositions [2]. Calder proposed that the rectangles could be made "to oscillate in different directions, and at different amplitudes." The visit proved to be the "shock that started things," as he wrote later [1].

In 1931, Calder began constructing compositions of metal wire and wood and made them move via inherent kinetics, activating their motion with motors and other means [2], and then later allowing free form movement with the wind. The moving sculptures were dubbed "Mobiles" by artist and friend Marcel Duchamp. These sculptures of "detached bodies floating in space" embodied Calder's fascination with the dynamism of the universe, which he sought to describe through a bold palette of primary and secondary colors [2]. Calder also created abstract stationary constructions for which Jean Arp then coined the term "Stabiles" to differentiate them from the moving sculptures.

Man-Eater with Pennants, abbreviated as *Man-Eater* in this paper, (Figure 1) was proposed by Alfred Barr and commissioned by the Board of Trustees for The Abby Aldrich Rockefeller Sculpture Garden in 1944. A maquette, *Mobile with 14 Flags* (Figure 2), was shown to museum leadership for approval before the final work was fabricated. *Man-Eater* was Calder's largest mobile sculpture to date, towering more than 9 m with a wingspan of more than 14 m. Seven painted metal pennants perch atop long vertical rods; heavy flat black plates counterbalance them at the bottom. The wind- or human-propelled elements bounce, sway, and rotate around a central post. The maquette is nearly identical in form to the final expression of the massive *Man-Eater with Pennants.* Naturally, a full-scale sculpture designed for outdoor installation must be structurally different from a table-top model. For example, on the actual mobile, two of the uppermost polygonal shapes have oval-shaped cutouts, whereas the maquette does not. The main horizontal crossbar on the mobile is heavily reinforced to bear the weight of the vertical rods and the pennants, but on the maquette, all the wires of the structure are the same delicate gauge. However, the shapes and colors of the pennants are the same on both the maquette and the *Man-Eater*; they are black, red, blue, and yellow. In addition, the upright post is a crucial grounding aspect for the large-scale sculpture, whereas the moving parts of the maquette sit on a triangular base, allowing table-top display.

Figure 1. Alexander Calder. *Man-Eater with Pennants*, 1945. Painted steel rods and sheet iron 14′ (425 cm) × approx. 30′ (915 cm) in diameter. Purchase.

Figure 2. Alexander Calder. *Mobile with 14 Flags* (maquette for *Man-Eater with Pennants*), 1945. Painted steel 53″ (134.6 cm) × approximately 50″ (127 cm) diameter. Gift of the artist.

Man-Eater with Pennants was installed in MoMA's garden from 1945–1949. From the outset, the merits and deficiencies of the mobile were debated within the museum. The curator Dorothy Miller expressed concern about the lack of movement due to the weight of the structure and the absence of ball bearings [3]. Philip Goodwin, Head of the Department of Architecture and Design, had deep reservations about the ironwork, which he called "clumsy," the "far from pleasing" colored forms, and the faulty engineering [4]. Of the opposing opinion, Curator James Johnson Sweeney thought the *Man-Eater* was one of Calder's most successful ventures in this field [5]. In addition, fears about the public injuring themselves on the mobile began right away; in the interest of public safety, a small fence was set around it. Sweeney, a fan of the sculpture, strongly advocated for their removal. "I would like to suggest in my last will and testament that the Calder mobile should move. I would recommend that the chains be taken off the monster" [6]. After four years on view, the mobile was taken off view, ostensibly because reconstruction of the garden space was underway. *Man-Eater* traveled to outdoor venues in London and Houston in the 1950s. After its last showing at MoMA from 1969–1970, it was housed in storage, where it has been since.

As for the maquette, with its purpose fulfilled, it was put away in storage with other Calder sculpture proposals. There it languished for ten years until a storage reorganization and clean-up effort was undertaken in 1959. Internal records called it an extended loan and it was returned to Calder at that time. The Registrar's carefully observed description specifies that the model was made of "unpainted iron" and also notes surface accretions and defects. Ten years later, the artist gifted the maquette to the museum. Photographs from that acquisition suggest that at some time while it was in Calder's possession between 1959 and 1969 it must have received its color coat. Because the maquette was painted sometime after the *Man-Eater* installation, the color scheme on the model cannot be considered a trial for the full-scale iteration; it is the exact opposite: the larger was the example for the smaller.

For MoMA's 2021 exhibition, *Alexander Calder: Modern from the Start*, the exciting work of reviving this forgotten sculpture began with conservation, curatorial, and registrar teams reviewing the documentation in MoMA's files and examining the work in storage. The sculpture had been repainted several times in its history. However, the only record of a repainting campaign is correspondence between the artist and curator Dorothy Canning Miller in 1948, where she describes the colors as not resembling the original paint applied by Calder even though the museum used the Ronan Japan Color paints preferred by the artist [7]. After its last exhibition between 1969 and 1970, extensive paint loss and rusting due to outdoor exposure went unaddressed. In addition, the metal rods were likely bent by painters holding them down to repaint them, or during handling [8]. This project provided a unique opportunity to study the painting history and the original colors of the large mobile and the associated maquette.

While the origins of Calder's affinity for a palette of primary colors are explored in the art historical literature, research in the scientific literature on Calder's paint choices is limited. A recent publication has investigated the red, black, and white paints used on a motorized work from 1932, including extensive overpainting from past treatments [9]. The work outlined here contributes to the discussion on the evolution of Calder's color choices and subsequent attempts by The Museum of Modern Art and others to determine the hues of the early primary palette. To our knowledge, this is the first study to delve into Calder's yellows and blues. As part of planning the restoration treatment for Man-Eater, the conservation team took cross section samples of the paint layers and mounted them for microscopy. These "slices of time" open a window into the history of the painting campaigns.

2. Materials and Methods

Scientific analysis was undertaken to better understand the pigments and binders used in *Man-Eater* and the associated maquette. Both invasive and non-invasive techniques were employed, with the latter involving both scrapings and cross sections. Initial analysis with portable X-Ray Fluorescence (p-XRF) yielded information about possible pigments and extenders as well as the composition of the metals. However, in the case of *Man-Eater*, the extensive overpainting, coupled with the lack of historical narrative about said campaigns necessitated micro-invasive sampling for more fingerprint spectroscopic techniques. Samples for analysis by micro-Fourier transform infra-red (μ-FTIR) and Raman spectroscopies were taken from areas that could be representative of different layers. While μ-FTIR also gave some indication of the binder used in the paint formulation, the pigments in each layer of the cross section were conclusively identified by Raman spectroscopy in addition to scanning electron microscopy (SEM), coupled with electron dispersive spectroscopy (EDS). Cross sections from the maquette showed a single campaign of colored paint, which makes the results more definitive than those taken from *Man-Eater*, especially as it relates to binder analysis by μ-FTIR.

The complementary nature of these methods made for a rich visualization and interpretation of the original colors of Man-Eater. These results helped determine the final colors to be used in the restoration campaign. The following is a description of the techniques used.

XRF analysis was performed with a Bruker Tracer III-SDD handheld XRF instrument with a Rh excitation source and silicon drift detector, with a 5 mm diameter approximate spot size. The instrument was operated at 40 kV and 3 μA, and spectra were acquired for 120 s. Additional XRF was performed on the maquette using a Bruker Tracer 5i at 40 kV and 4.5 μA and spectra were acquired for 120 s. All the spectra were examined with the Bruker Artax 8.0 software.

Micro-μ-FTIR (μ-FTIR) analysis was carried out in transmission mode using a Nicolet iS50-μ-FTIR coupled with a Thermo Nicolet Continuum infrared microscope equipped with an MCT detector. Spectra were collected in the 4000–600 cm^{-1} range with a 4 cm^{-1} resolution and 128 scans and using the Thermo Scientific OMNIC 9.12 software package. Spectra were examined using the Spectral Search and Multicomponent Search tools available in the Thermo Scientific OMNIC Specta 2.0 software and IRUG spectral databases [10].

Raman spectra were collected using a Renishaw In-via Raman system, equipped with a 785 nm diode laser at powers between 0.3 to 3 mW, a 1200 lines/mm grating, and a Leica confocal microscope with a 50× LWD or 100× objective. Final spectra represent an average of five acquisitions of 10 s. Raman spectra were evaluated Spectral Search and Multicomponent Search tools available in the Thermo Scientific OMNIC Specta 2.0 software and SOPRANO [11] and UCL [12]. spectral databases.

SEM-EDS was carried out under low vacuum and using a Hitachi TM3000 microscope (Tokyo, Japan) fitted with a Bruker Xflash MIN SVE and Quantax 70 software. Images were acquired using analysis mode at 15 kV and a four-segment backscatter electron (BSE) detector. The cross sections were not coated as imaging was carried out under low vacuum; remaining gas molecules in the chamber ionize the negative buildup up on the surface of uncoated, organic samples.

Optical Microscopy was carried out using a Leica DM IRM microscope using 5× and 10× objectives.

Cross sections were embedded in BioPlastic ®, [Aldon Corp., Avon, NY, USA] a blend of polyester and methacrylate monomers in a styrene solvent, trimmed with a jeweler's saw, and dry polished with Micro-Mesh® [Micro-surface Finishing Products, Wilton, IA, USA] silicon carbide or aluminum oxide abrasives.

3. Results and Discussion

3.1. Metals

Based on XRF analysis, *Man-Eater* is made of a steel of containing manganese. Areas denuded of paint were severely rusted. In most cross sections, there are two metallic layers, a reddish layer followed by a silvery one, on top of which the first colored paint layer has been applied. SEM-EDS analysis confirmed what could be concluded visually from microscopy: a layer of rusted steel rich in iron coated with a silvery, flaky layer of aluminum. Aluminum paint functions as an anti-corrosion coat on top of a steel substrate. Due to its rapid oxidation when exposed to air, an aluminum flake paint can confer a high degree of corrosion protection on a steel substrate by forming a thin, transparent layer of aluminum oxide film. This impervious layer is self-adherent and reaches a maximum thickness of 100 Å [13]. The effect has been understood since the early 1900s and was popular in the 1920s and 1930s for automotive and structural parts. In line with Calder's practice, the original color layer is applied directly atop the protective coating of aluminum [14]. The steel rust penetrated upwards through both the aluminum paint and the first colored paint layers in many cases, suggesting that the first repaint campaign was necessitated by surface paint loss due to rusting of the steel substrate.

The maquette's metal structure is unusual because the colored flags are a different metal than the remaining black elements [15]. XRF analysis showed the colored pennants atop the vertical rods to be made of an aluminum alloy containing lead, iron, silicon, manganese, and zinc. This could suggest an effort to lighten the load on the vertical rods. The lower black pennants are an iron-based alloy whose weight serves to keep the rods vertical. The XRF spectrum of the rods showed an intense peak for copper in addition to iron. A closer examination of the dull brown patches on the black-painted rods shows that they appear to be copper-clad steel, which has application in the electric and automotive industries, for example [16,17]. This metal has been observed on rare occasions in other sculptures by Calder [15]. The rivets connecting the black rods to the colored aluminum pennants are made of a brass alloy of copper, zinc, arsenic, and possibly titanium.

3.2. Yellow Paint

The color of the sole yellow pennant in *Man-Eater* had bleached considerably, rendering it remarkably light in comparison with underlying layers visible in chips on the edges. The surface was also visibly scratched and scuffed but had the brush marks of hand application. Calder is known for hand painting many of his early painted outdoor works, and retaining that quality would have been paramount to any repainting campaign.

In the yellow cross section (Figure 3), nine individual layers of paint are clearly delineated, varying in shades from intense, to pale, to greenish. Chrome yellow (P.Y. 34; C.I. 77600), chemically a lead chromate ($PbCrO_4$), was observed in the Raman spectra of the first eight layers. This was also confirmed by SEM-EDS (Figure 3), where Pb and Cr were recorded across those layers. Chrome yellow pigment was first synthesized in 1804 but only came into prominence when more abundant sources of chrome minerals were available [18]. Chemically, chrome yellow is available as a pure $PbCrO_4$ or as solid solutions, with $PbCrO_4$ and lead sulfate ($PbCr_{1-x}S_xO_4$), in shades that range from yellow to orange ($x < 0.1$) to lemon-yellow ($0.2 \leq x \leq 0.4$) and pale yellow ($x > 0.5$) with increasing sulfate concentration (C.I. 77603 when coprecipitated with $PbSO_4$) [19]. Partial replacement of the chromate in solid solutions brings about a reduction in tinctorial strength with increasing sulfate concentration but allows for the manufacture of yellows with a greenish hue. In terms of crystallography, $PbCrO_4$ presents as monoclinic and $PbSO_4$ as orthorhombic, and substitution of chromate ions by smaller sulfate ions leads to a compression of the monoclinic unit cell at low sulfate concentration and a change from monoclinic to orthorhombic when x exceeds 0.4 in a solid solution.

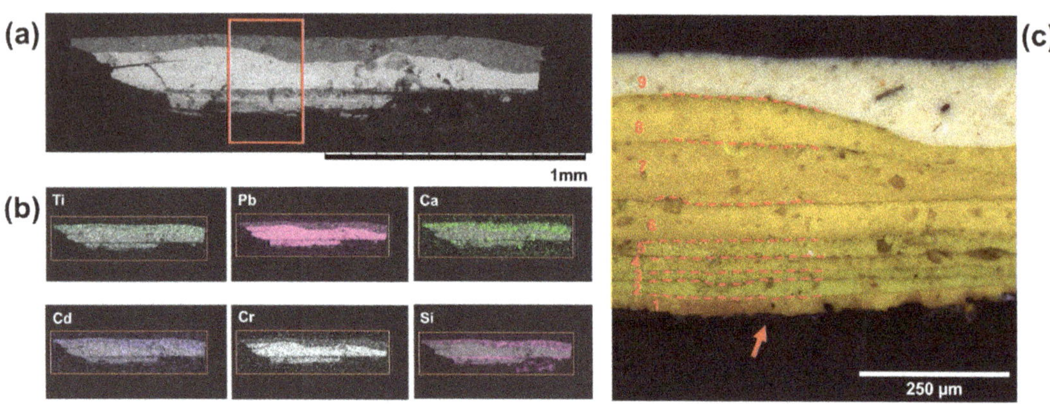

Figure 3. Cross section imaging of a sample taken from the yellow pennant in *Man-Eater*. (**a**) BSE image of the cross section and (**b**) associated EDS mapping of relevant chemical elements, of particular note is the Cd and Ti in Layer 9 and Pb and Cr in the remaining layers. (**c**) Light microscope image at 20× magnification showing all 9 layers of paint and aluminum anticorrosive paint, indicated by an arrow. © 2021 MoMA, New York, NY, USA.

This phenomenon is also observed by Raman spectroscopy (Figure 4), where shifts to higher wavenumbers of some chromate bands indicate the presence of lead sulfate [19]. The $\nu_1(CrO_4{}^{2-})$ symmetric stretching mode shifts to higher energy with increasing substitution of sulfate ions into the lattice. Discrete shifts between the spectra of the yellow layers point to at least two different yellows used in the overpainting. This is evident in Layers 1, 2, and 3 in the cross section, where $\nu_1(CrO_4{}^{2-})$ is at 841 cm^{-1} in comparison with 839 cm^{-1} in the remaining layers. The $\nu_4/\nu_2(CrO_4{}^{2-})$ bending multiplet is also influenced by sulfate substitution and cell compression; $\nu_4(CrO_4{}^{2-})$ modes for pure chrome yellow are located at 400, 376, and 357 cm^{-1}, while those at 336 and 323 cm^{-1} are attributable to $\nu_2(CrO_4{}^{2-})$ modes. Further pointing to the presence of sulfate in the first three layers, the mode at 400 cm^{-1} is shifted to 403 cm^{-1} and that at 357 cm^{-1} to 360 cm^{-1}. A band at 970 cm^{-1} attributed to a $\nu_1(SO_4{}^{2-})$ mode [19]. The pair of Layers 2 and 3 cements the presence of a solid solution of chromates and sulfates, perhaps one that is still monoclinic with few sulfates. While this peak was not seen in layer 1 due to noise, it can be said with some certainty that a paint containing a $PbCr_{1-x}S_xO_4$ pigment was the original color of the yellow pennant, one that was possibly lemony in hue at one time. Similarly, Layer 5 exhibits the presence of $\nu_1(SO_4{}^{2-})$ from lead, barium, and calcium sulfates, based on

Raman analysis (Figure 3). Layer 5 was also rich in silicates but contained less pigment material, as indicated by the elemental distribution observed in SEM-EDS. Layer 4 consists only of $PbCrO_4$ and $BaSO_4$, Similarly, the color and spectra of Layers 6, 7, and 8 consist of $PbCrO_4$ and $BaSO_4$ based on Raman analysis (Figure 3). They could have been applied successively in one repainting campaign with drying time between coats.

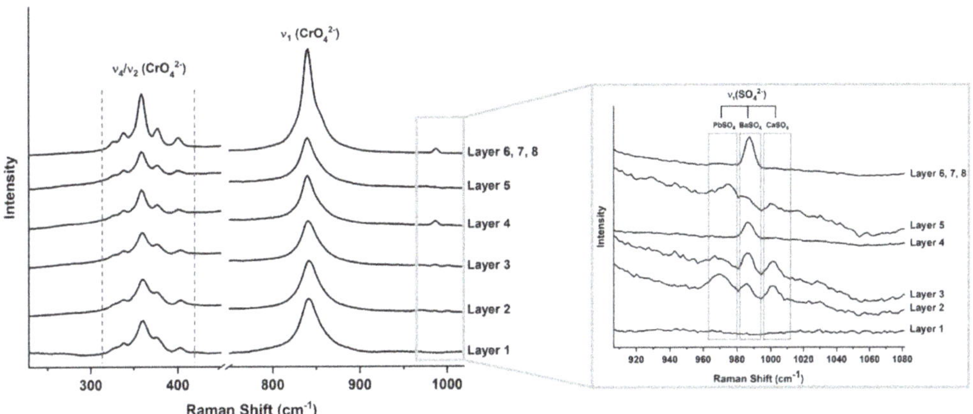

Figure 4. Raman spectra of the first 8 layers in the cross section of a sample taken from the yellow pennant in *Man-Eater*. The slight shifts to higher wavenumbers in Layers 1, 2, 3, and 5 in both the $\nu_1(CrO_4^{2-})$ stretching band and $\nu_4/\nu_2(CrO_4^{2-})$ bending multiplet suggests the presence of a solid solution of lead chromate and lead sulfate ($PbCr_{1-x}S_xO_4$). This is confirmed by the $\nu_1(SO_4^{2-})$ stretching mode observed between 973 and 978 cm^{-1} in the inset.

Curiously, the Raman spectrum of Layer 9 in the cross section showed no signs of chrome yellow, only that of the tetragonal rutile form of titanium white (TiO_2) with bands at 144 (B_{1g}), 445 (E_g), and 610 (A_{1g}) cm^{-1} [20]. The Raman spectrum exhibited a recently characterized luminescence emission pattern (1222, 1306, 1385, 1497, 1600, and 1686 cm^{-1}; see Figure 5 and Figure 7) attributed to neodymium (Nd^{3+}) ions substituting into the orthorhombic alkaline earth sulfates of titanium dioxide pigments, made through co-precipitation with barium sulfate ($BaSO_4$) or calcium sulfate ($CaSO_4$), where Nd^{3+} occurs naturally in ilmenite ($FeTiO_3$), the source ore for Ti [21]. Observing this pattern can help with dating issues, as it was only detected in works dating from 1945–1977. These co-precipitated pigments of lower tinting strength were more prevalent in industrial paints, in particular oils and alkyds. However, the presence of both $BaSO_4$ (988 cm^{-1}) and $CaSO_4$ (1017 cm^{-1}) complicates the identification of the type of co-precipitate, where either could have been added mechanically to the pigment mixture. In turn, this makes it more difficult to establish the date of the final repainting campaign since $BaSO_4$ and $CaSO_4$ coprecipitates were phased out in the late 1940s and 1970s, respectively. Internal records show that, after the *Salute to Calder* exhibition closed in 1970, a repaint was considered but never executed [22]. The Nd^{3+} luminesce pattern in the Raman spectrum places a last repainting within the accepted range of 1945–1977.

While no yellow pigment was detected via Raman spectroscopy in Layer 9, SEM-EDS (Figure 3) showed the top layer to contain cadmium unlike the remaining layers of paint, indicating the presence of the semiconductor pigment cadmium sulfide, known as cadmium yellow (P.Y. 37; C.I. 77199). Cd was also seen in the XRF spectra taken of the yellow pennant. SEM-EDS also showed this layer to be particularly rich in magnesium silicates and silicates that are used as fillers.

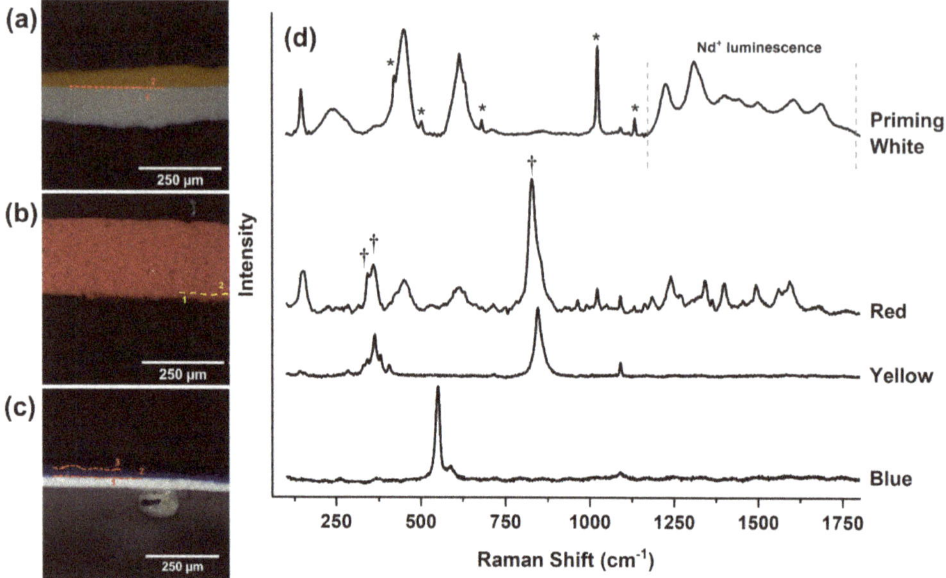

Figure 5. Microscopy images of three cross sections (**a**–**c**) taken from the Mobile with 14 Flags, or the maquette, as referred to in the text. The Raman spectrum of each of these colors (**d**) shows the presence of ultramarine in the blue, chrome yellow in the yellow, and P.R. 4 and molybdate orange (†) in the red. The white priming layers observed in (**a**,**c**) contain titanium white in rutile form deposited on $CaSO_4$ (*), which again shows a luminescence pattern (*) from a Nd^{3+} impurity. © 2021 MoMA, New York, NY, USA.

The presence of cadmium yellow could explain the pale appearance of the top yellow layer, as cadmium yellow is known to blanch with exposure to light, humidity, and environmental acid—a given for any outdoor sculpture. This exposure leads to the formation of cadmium sulfate ($CdSO_4$), which can react further with carbon dioxide (CO_2) to form cadmium carbonate ($CdCO_3$) [23]. While these moieties were not identified directly, this drastic fading is characteristic of cadmium yellows, even under gallery conditions [24]. It was also found that CdS degrades most when illuminated with blue light, which is fully absorbed and generates the highest photocurrent electrochemically [25]. Consequently, the abundance of energetic ultraviolet (UV) and blue light from solar radiation can promote more rapid decay of CdS. Oddly, the reverse is also true: cadmium yellow has been shown to darken considerably when embedded in an alkyd resin [26,27]. While that was not seen here, it points to the photoactivity of cadmium sulfide. Additionally, TiO_2 has a photocatalytic effect on some pigments when exposed to UV radiation [28], as when *Man-Eater* was installed outdoors. TiO_2 also exhibits photocatalytic chalking, or degradation of the paint film that exposes pigment particles, and could have further hastened the blanching of Layer 9 [29]. Conversely, the dark color observed visually in Layer 1 can perhaps be attributed to the photo-induced reduction in chromate ions to Cr (III) compounds, which is driven by both visible and UV light [23] and can markedly affect those chromate yellows of the rhombic varieties [18]. Sulfur-rich orthorhombic yellows are more prone to browning, possibly due to the increased solubility of $PbCrO_4$ and $PbCr_{1-x}S_xO_4$ in this phase, making more chromate ions available for redox reactions [13].

In contrast, the paint of the yellow pennant (Figure 5) in the maquette is still vibrant yellow. A cross section from the maquette showed only two layers, a white priming layer followed by a yellow one. The yellow was similarly identified as a chrome yellow, one that is probably a solid solution of lead chromate and lead sulfate ($PbCr_{1-x}S_xO_4$). As in Layers 1 through 3 in the cross section from *Man-Eater*, the symmetric $\nu_1(CrO_4{}^{2-})$ stretch is broadened and shifted to 845 cm^{-1}, and the $\nu_4(CrO_4{}^{2-})$ bending modes to 404 and 364 cm^{-1}, all of which are results of crystal compression. This paint is also characterized by a strong ν_1(C–O) symmetric stretch at 1090 cm^{-1} and a weak in-plane bend ν_4(C–O) at 717 cm^{-1}, both characteristic of calcium carbonate ($CaCO_3$). This confirms that the maquette and *Man-Eater* have two different yellow paints. Interestingly, the white priming layer also exhibited the same luminescence emission pattern attributed to Nd^{3+} ions in titanium-based whites. Both $CaSO_4$ (1020 cm^{-1}) and $CaCO_3$ (1090 cm^{-1}) are identified in this white, suggesting a different rutile-based paint pigment than in *Man-Eater*. The absence of $BaSO_4$ is more diagnostic for dating and places the painting sometime between 1959, when it left the museum and 1969, when it came back, confirming internal registrar records.

3.3. Blue Paint

As with the yellow pennant in *Man-Eater*, the blue pennant had bleached significantly, presenting as a brittle, light blue layer of paint. In chipped areas, darker blue layers of paint were visible below the light blue layer with the most vibrant blue layers closest to the rusted metal. Interestingly, one face of the blue pennant was slightly more vivid in color, which prompted the analysis of two different cross sections. From this point onward, the lighter side will be referred to as Side A, and the darker as Side B. This could be a result of the static nature of this mobile as discussed in internal correspondence, allowing Side A to be more exposed to the sun than Side B.

Analysis of the cross section from Side A (Figure 6) showed a dramatic rusting of the steel, infiltrating through the silver layer and into the first layer of paint. As in the yellow cross section, SEM-EDS confirmed the steel rust and the aluminum flake paint. In total, 6 layers could be delineated on Side A, with colors ranging from midnight to baby blue.

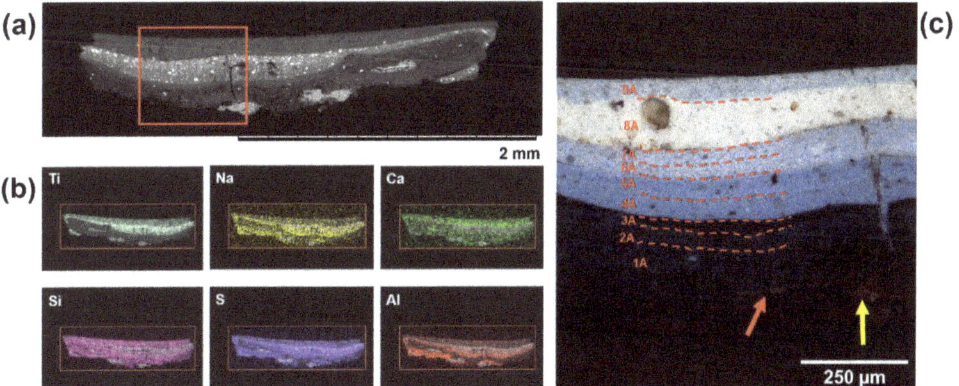

Figure 6. Cross section imaging of a sample taken from Side A of the blue pennant in *Man-Eater*. (**a**) BSE image of the cross section and (**b**) associated EDS mapping of relevant chemical elements, of particular note are the Ti and Ca in Layers 8A and 9A, the distribution of Al, S, Na, and Si associated with ultramarine. (**c**) Light microscope image at 20× magnification showing all 9 layers of paint, aluminum anticorrosive paint (red arrow), and the rusting steel (yellow arrow). Image copyright 2021 MoMA, New York, NY, USA.

The analysis of the first two paint layers of the lighter side, 1A and 2A, indicated the use of Prussian blue (P.B. 27; C.I. 77510) as the only blue pigment. $CaSO_4$ was also identified through Raman (1017 cm^{-1}). Prussian blue ($Fe^{III}[Fe^{II}(CN)_6]_{3-}$), was first synthesized accidentally by Dresbach in 1704 and became industrially produced by the nineteenth century [30]. It is the oldest synthetic coordination compound and has since found important use in the printing industry in addition to paints and artist materials. The dark blue color of Prussian blue is due to an intervalent transition between Fe^{II} and Fe^{III} through a coordinated cyano group (CN) where light in the orange-red region around 700 nm is absorbed [30]. The presence of Prussian blue was identified in the Raman spectrum (Figure 7) by the $1A_g$ ν(CN) stretching vibration at 2160 cm^{-1} and the E_g ν(CN) stretching vibration at 2090 cm^{-1} [31]. The spectrum also shows peaks for ν(Fe−C) stretching modes at 606 and 536 cm^{-1}, σ(Fe−CN−Fe) bending modes at 376 and 278, and a σ(Fe−C−Fe) deformation at 189 cm^{-1}. Perhaps most interesting is the shoulder at 2123 cm^{-1}, which corresponds to a CN^- stretch related to the $1A_g$ ν(CN) mode. This shoulder is most pronounced in the "soluble," or colloidal varieties of Prussian blue, where association with a cationic species such a K^+, $NH_4{}^+$, or Na^+ maintains charge balance, as opposed to the insoluble form that relies on a higher concentration of Fe^{III} [31,32]. Chemically, Prussian blue is prone to photoreduction and fading, the same quality that made the pigment valuable for cyanotypes, an early photographic method, and soluble varieties are far more sensitive to fading than the insoluble form [33]. The presence of extenders can also exacerbate photoinduced fading, and SEM-EDS showed this paint layer to be particularly rich in magnesium silicates and other silicates. This can explain the discoloration of Layers 1A and 2A and might have prompted repainting of the blue pennant with a darker blue Layer 3A, which was shown to contain ultramarine (P.B. 29; C.I. 77007) and a smaller amount of Prussian blue (Figure 7). The darker appearance of Layer 3A might correspond to the combination of two blues rather than a single color.

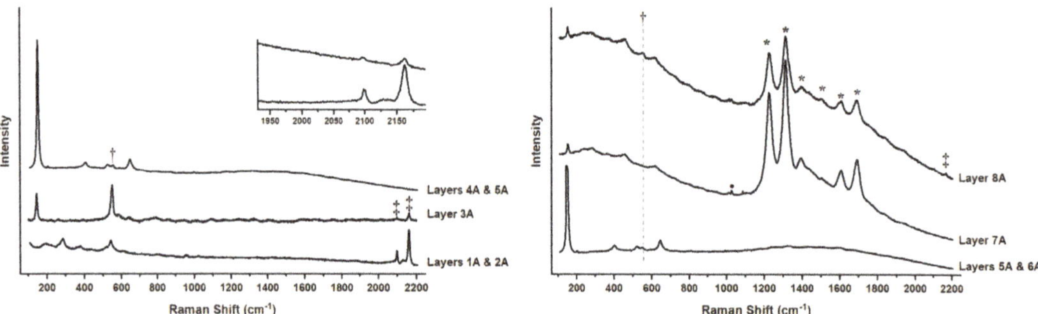

Figure 7. Raman spectra of the pigments in all 8 layers from Side A of the blue pennant in *Man-Eater*. Ultramarine (†) is observed at 550 cm^{-1} in Layers 3A, 4A, 5A, 6A, and 8 A; Prussian blue (‡) at 2090 and 2160 cm^{-1} in Layers 1A, 2A, 3A, and 8A. Titanium white was observed in anatase form in layers 3A, 4A, 5A, and 6A, whereas the rutile form was observed in Layers 7A and 8A, in addition to a luminescence pattern (*) from a Nd^{3+} impurity attributed to titanium white deposited on $CaSO_4$ (•).

By the 1940s, the use of synthetic ultramarine ($3Na_2O{\cdot}3Al_2O_3{\cdot}6SiO_2{\cdot}2Na_2S$) was commonplace after its laboratory preparation in 1828 by Guimet in France and Gmelin in Germany [34]. The incorporation of a sulfur radical ($S_3{}^-$) into the sodalite crystal lattice of sodium, aluminum, silicon, and oxygen acts as a chromophore. The broad absorption of green-yellow-orange visible light of this radical, centered around 600 nm, gives the pigment its signature blue color. This energy corresponds to an electronic transition between two singly occupied molecular orbitals [35]. Ultramarine was identified through Raman spectroscopy (Figure 7) by the $S_3{}^-$ radical symmetric stretch at 540 cm^{-1} [36]. Elemental analysis by XRF confirmed the presence of Al, Si, S, and K, whereas SEM-EDS showed the presence of the lighter Na. With a relatively low refractive index of 1.5, the opacity of the ultramarine is increased by the addition of white pigments, which is seen here with the addition of anatase (TiO_2), seen in the Raman spectrum at 144 cm^{-1}, and confirmed by the presence of Ti in SEM-EDS. The Raman spectrum of Layers 4A and 5A show a higher concentration of anatase resulting in a much lighter blue than Layer 3A; the two are similar in composition and could have been applied in two successive coats that contain ultramarine and anatase. Layers 6A and 7A are even lighter in color than the previous two, but they contain the same mixture of ultramarine and anatase and could also have been applied successively as part of one campaign. The white Layer 8A shows the Nd^{3+} luminescence pattern associated with rutile (TiO_2) precipitated on $CaSO_4$, which was also observed in Layer 9 of the yellow pennant. Finally, Layer 9A shows very small peaks for ultramarine and $BaSO_4$ in addition to the dominant Nd^{3+} luminescence pattern of TiO_2. Additionally, the detection of niobium (Nb) in the XRF spectra of the blue pennant further narrows down the type of rutile present. The presence of detectable amounts of this rare earth metal by XRF is indicative of the sulfate process for producing titanium white, where it remains after manufacture as an impurity from the ore [37]. As in the case of the pale-yellow Layer 9, TiO_2 has been shown to have a catalytic effect on the degradation and fading of Prussian blue, which could explain some of the lighter blues observed in those layers containing the white pigment [28]. Interestingly, the color of the blue pennant was named "light gray" in internal correspondence from 1970, far from the original deep Prussian Blue [22].

Similar to Prussian blue, ultramarine is prone to photoinduced fading; it is also highly sensitive to acids, which are prevalent in urban settings [38]. Furthermore, it has been shown that ultramarine pigment dispersed in an alkyd binder accelerates the degradation of the paint film and results in a bleached and brittle film [26,27]. Consequently, the combination of chemical sensitivity and catalytic effect can explain the pale color of the blue pennant.

Side B (Figure 8) is darker in comparison to Side A, but it also exhibits fading. The same penetration of steel rust up through the aluminum and the first paint layers is again observed. Layers 1B and 2B contains Prussian blue and $CaSO_4$; Layer 3B ultramarine, Prussian blue, and anatase; a thin Layer 4B followed by a thick Layer 5B contain ultramarine and anatase; a tan-colored Layer 6B with anatase; Layer 7B rutile precipitated on $CaSO_4$; and Layer 8B ultramarine with rutile precipitated on $CaSO_4$. After Layer 3B, there is a breakdown in the similarity between the two sides, as the lighter Layers 6A and 7A have no counterpart in the stratigraphy of Side B. The Raman spectrum of Layer 6B indicates it is composed of anatase, but the tan color can be attributed to the presence of Fe, possibly in the form of iron oxides, barites, and silicates, all detected by SEM-EDS. The purpose of this layer is unclear. Layers 7B and 8B have the same composition as Layers 7A and 8A were probably applied at the same time.

Figure 8. Cross section imaging of a sample taken from Side B of the blue pennant in *Man-Eater*, showing layer 6B, which is unique to this face of the pennant. Raman analysis indicates this layer to be anatase-based, and SEM-EDS shows it to be rich in iron, which gives the layer its tan color. © 2021 MoMA, New York, NY, USA.

In contrast, the paint of the blue pennant on the maquette is still vivid, and a cross section showed three layers, a white priming layer followed by two blue ones of the same compositions. The blue was identified as ultramarine, which is typical of Calder's mature palette [39]. This paint is also characterized by a ν_1(C–O) symmetric stretch at 1090 cm^{-1} characteristic of $CaCO_3$. This further confirms that the maquette and *Man-Eater* were painted at different times and with different blue paints. The white priming layer also exhibited the Nd^{3+} luminescence emission pattern attributed to Nd^{3+} ions in rutile, and both $CaSO_4$ (1020 cm^{-1}) and $CaCO_3$ (1090 cm^{-1}), all similar to the priming layer of the yellow pennant. These paints again do no match any of those found in the blue pennant of *Man-Eater*.

3.4. Red Paint

The extant red paint on the pennants in Man-Eater was the most vibrant of the three primary colors on the surface. Nevertheless, the paint was brittle and chipped; in some areas, the aluminum coating was visible, in others, the rusted metal was exposed. Multiple cross sections were taken from the five red pennants and all revealed an identical stratigraphy (Figure 9). Some of the red hues veered darker even to the naked eye, especially those closer to the iron surface save for Layer 1 (Figure 9). Similar to the yellow and blue, SEM-EDS confirmed the presence of a steel rust layer followed by an aluminum paint coating. Raman analysis (Figure 10) proved crucial in analyzing the red pigments, where every layer contained one or more β-naphthol organic red pigments. This class of pigments first became available at the turn of the 20th century by coupling a substituted aniline ring with β-naphthol [40]. It is worth noting that assigning a particular shade to β-naphthol organic red pigments can be challenging because the presence of fillers, particle size, and method of manufacture all affect their final color [13]. These pigments also range in photosensitivity, and for some, white reduction from a deep shade using diluents, such as titanium white, can make them prone to fading, possibly due to catalytic effects also observed on other organic reds [28]. These pigments are sensitive to solvents, acids, bases, and some even to water.

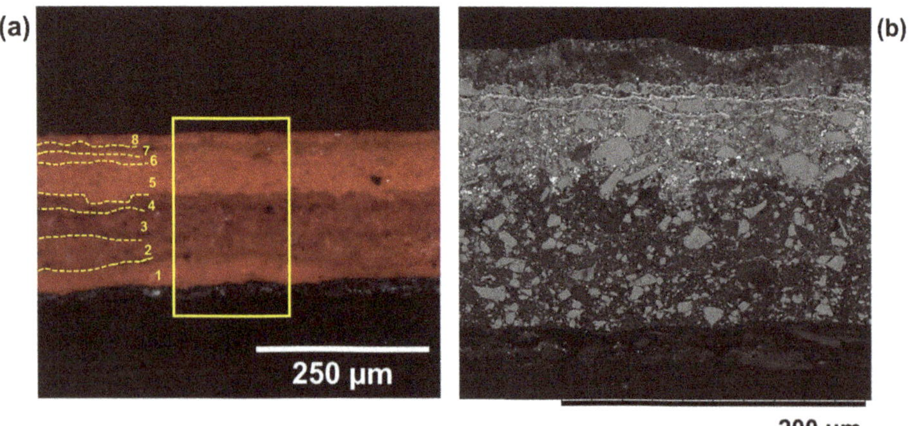

Figure 9. Cross section imaging of a sample taken from one of the red pennants in *Man-Eater*. (**a**) Light microscope image at 20× magnification showing all 8 layers of paint and aluminum anticorrosive paint, and (**b**) BSE image of the cross section that shows the division between layers 5, 6, and 7 in particular. © 2021 MoMA, New York, NY, USA.

In Layer 1, Parachlor Red (P.R. 6; C.I. 12090) was identified by Raman spectroscopy (163, 176, 219, 267, 331, 372, 394, 421, 465, 484, 533, 613, 635, 658, 710, 743, 763, 840, 861(sh), 985, 1091, 1103, 1132, 1140, 1158, 1180, 1218, 1243, 1265, 1291, 1326, 1392, 1447, 1474, 1485, 1555, 1565, 1584, 1605, 1623 (sh) cm^{-1}). P.R. 6 was first synthesized in 1906 and was used until the late 1980s when it rapidly fell out of favor due to high solubility in organic solvents and mineral spirits as well as poor lightfastness [13]. Nevertheless, this pigment was used in air-drying alkyd systems, because deep shades without much white reduction are relatively lightfast [40]. In the context of Calder's preference for matte paints, it is plausible that the original layer of paint was an alkyd one that contained P.R. 6. Chlorine $K\alpha_1$ and $K\alpha_2$ lines were obscured by Rh L-series lines and as such not discerned in XRF, and mapping with SEM-EDS only indicated that the paint was likely extended with silicates and/or magnesium silicates. Furthermore, the difficulty of detecting chlorine elementally is due to P.R. 6 having only one chlorine substituent. Layer 2 shows a mixture of P.R. 6 and Para Red (P.R. 1; C.I. 12070) in the Raman spectrum (185, 360, 410, 460, 1002, 1105, 1428, 1592 cm^{-1}). The latter was the first synthetic organic red discovered in 1895 but has since lost its industrial significance. P.R. 1 has a dull brown hue, and is not fast against organic solvents, acids, bases, and even water; it is also prone to darken on exposure to light [40].

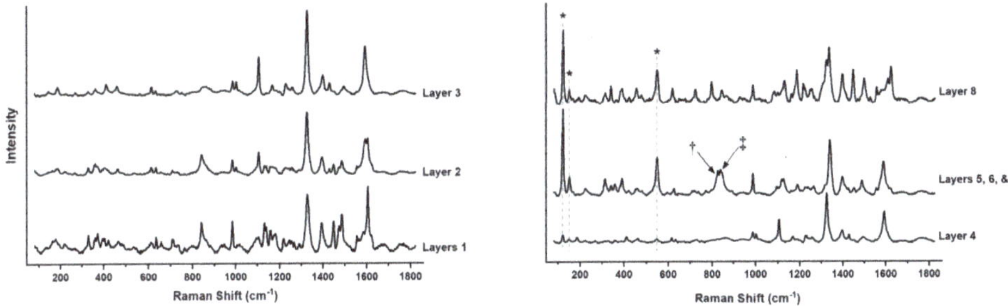

Figure 10. Raman spectra of all 8 Layers of a cross section from one of the red pennants in *Man-Eater*. All layers contain one or more β-naphthol red and all peak positions are listed in the text. In addition, Layers 4, 5, 6, 7, and 8 contain red lead (*); and layers 5, 6, and 7 molybdate orange (†) at 825 cm^{-1} and chrome yellow (‡) at 839 cm^{-1}.

Layers 3 and 4 are similar in composition and could have been applied in two coats succession. They both also contain P.R. 1 according to their Raman spectra (185, 326, 360, 410, 461, 614, 632, 729, 986, 1002, 1105, 1167, 1201, 1228, 1257, 1324, 1396, 1428, 1496, 1592 cm^{-1}), with SEM-EDS indicating the presence of a calcium-based filler in addition to silicates and/or magnesium silicates. However, Layer 4 also contains some red lead (Pb_3O_4) as evident by lattice bands at 121 (A_{1g}), 151(E_{2g}), and 547 (A_{1g}) cm^{-1} [41] in addition to lead $M\alpha$ and $M\beta$ lines in SEM-EDS. Red lead (P.R. 105; C.I. 77518), one of the earliest pigments to be artificially made, has been used as a red pigment since as early as the 5th B.C. century in China [42]. Today, it is manufactured by heating lead oxides and is mostly used as a protective pigment to passivate iron and steel and inhibit corrosion [13]. The choice of a paint containing red lead might have seemed appropriate at the time of repainting because of evidence of rusting on the red pennants.

Layers 5, 6, and 7 are rather complex paint layers, with at least 4 pigments present, one organic and three inorganics, and could be three coats applied in one campaign. Layer 5 seems to be painted over a physically degraded Layer 4 since it fills several voids to form a jagged layer. The organic pigment, chlorinated para red (P.R. 4; C.I. 12085) is a positional isomer of P.R. 6 and has a yellowish red hue (313, 340, 358, 594, 624, 708, 719. 736, 769, 890, 986, 1096, 1117, 1125, 1188, 1296, 1337, 1396, 1451, 1487, 1555, 1587, 1617 cm^{-1}). P.R. 4 has lost much of its commercial impact, as it is tinctorially weak and loses much of its light fastness in white reductions, while, in contrast, even full shades darken upon exposure to light [40]. The inorganics include red lead, chrome yellow—possibly the monoclinic form with few or no sulfates considering the position of the $\nu_1(CrO_4{}^{2-})$ stretch at 839 cm^{-1}—and molybdate orange (P.R. 104; C.I. 77605).

Molybdate orange is a solid solution of $PbCrO_4 \cdot PbMoO_4 \cdot PbSO_4$, used mainly in paints, coatings, and colored plastics. It was first industrialized in 1934-35, and commercial pigments contain ~10% lead molybdate ($PbMoO_4$) [38]. This pigment is often found in mixtures with chrome yellow to match the color of the orange basic lead chromate ($PbCrO_4 \cdot PbO$), which is no longer of commercial importance. Molybdate orange is also found in mixtures with organic reds to give an extended color range and impart lightfastness and weather resistance onto a paint film [13]. Molybdenum was first identified in the paint stratigraphy through non-invasive XRF and confirmed by Raman spectroscopy with a $\nu_1(CrO_4^{2-})$ symmetric stretch at 825 cm^{-1} presenting as a doublet with chrome yellow [43].

Layer 8 is a mixture of both red lead and toluidine red (P.R. 3; C.I. 12120). P.R. 3 is the most lightfast of the β-naphthol reds in deep shades and is manufactured and used on a large industrial scale. Primarily employed in air drying paints, it is the most important of the β-naphthol reds and was also used in printing inks, pastels, and watercolors [40]. Nevertheless, much as with the remaining β-naphthol reds, it suffered from a sensitivity to light and solvents and declined in popularity after the 1970s. SEM-EDS indicated Layer 8 to be extended with a calcium-based filler, silicates, and magnesium silicates. In comparison with *Man-Eater*, the red on the maquette appears to have been painted in two layers, a very thin preliminary one with a thicker coat of red on top. Unlike the blue and yellow pennants, the red paint was applied directly to the aluminum with no white layer. While Raman and SEM-EDS analysis indicated that they are similar in composition, the lower layer is visually more saturated in color. The pigments consisted of P.R. 4 and molybdate orange, the latter identified in the Raman spectrum by a $\nu_1(CrO_4^{2-})$ stretch at 827 cm^{-1} and a $\nu_1/\nu_2(CrO_4^{2-})$ bending multiplet at 341/359 cm^{-1}. Mo was observed in the XRF spectrum using the Tracer 5i and confirms the presence of molybdate orange. The paint also included TiO_2 in rutile form (449, 612 cm^{-1}) and $CaSO_4$ (1020 cm^{-1}). The B_{1g} mode at 144 is obscured due to a broad band at 155 ascribed to Pb–O lattice modes in lead chromate [44]. Similar to the rest of the primary colors on the maquette, the red pennant is different in shade and composition than in *Man-Eater*.

3.5. Black

The black paint from the lower pennants of *Man-Eater* showed two layers in the cross section, in addition to extensive rusting. The aluminum layer was not seen in the cross section but observed during treatment. This points to consistent surface preparation of the metal before applying any paint. Raman analysis of the layers proved difficult due to overwhelming fluorescence. Mapping with SEM-EDS was more fruitful here and confirmed XRF analysis, where Layer 1 was rich in Ca and P, which indicates the presence of bone black (P.Bk. 9; C.I. 77267). Bone black is defined as a carbonized product of collagen mixed with hydroxyapatite ($Ca_5(PO_4)_3$), and the latter is the source of Ca and P in the SEM-EDS and XRF [45]. Layer 2 did not show the presence of these secondary constituents and can indicate a carbonaceous black not derived from bone matter. The black-painted shapes on the maquette also had only two layers. Fluorescence was not problematic in this instance, and Raman analysis of Layers 1 and 2 indicated the presence of a carbon-based black in both layers, with D and G bands at 1324 and 1597 cm^{-1}, respectively [46]. The D band is indicative of disorder in the crystal structure of carbonaceous blacks, while the G band arises from C–C stretching [47]. While both layers are rich with large particles of Ca that were identified as $CaCO_3$ by Raman spectroscopy (156, 238, 714, 1090 cm^{-1}), only Layer 1 is rich in aluminosilicates based on SEM-EDS. Finally, Layer 2 interestingly showed the presence of Prussian blue by Raman spectroscopy with weak but readily identifiable 1Ag ν(CN) stretching vibration at 2160 cm^{-1} and the Eg ν(CN) stretching vibration at 2090 cm^{-1}. Curiously, XRF spectra indicate that the rods were painted with a bone black, where hydroxyapatite was detected by the presence of phosphorous, unlike the pennants.

3.6. Paint Binder

Obtaining a sample of the original paint layer from *Man-Eater* proved difficult considering the many layers of overpaint. In all samples, the presence of both pigments and fillers obscured most peaks in the fingerprint region of the μ-FTIR spectrum, generally in the mid-IR from 1800 to 500 cm^{-1} [48]. Diagnostic C–O and C–C skeletal vibrations in the fingerprint region were obscured by the broad peaks of pigments such as ultramarine (1096 cm^{-1}) and chrome yellow (851 and 819 cm^{-1}) and fillers such as calcium sulfate (1620, 1425 (sh), 1148 (sh), 1120, 670 cm^{-1}), calcium carbonate (1803, 1447, 1411, 879 cm^{-1}), and barium sulfate (1186, 1124, 1074, 982 cm^{-1}). All pigments and fillers seen in Raman analysis of the cross sections were confirmed by μ-FTIR but assigning them to a respective stratigraphy is not possible. However, samples from different representative areas were analyzed by transmission μ-FTIR in hopes of identifying any binding media from the now brittle paint.

A carbonyl (C=O) stretching band at 1734 cm^{-1} was detected in all μ-FTIR spectra from *Man-Eater*, with intensities varying from medium to very weak, exemplified here by a spectrum of a sample taken from the edge of the blue pennant (Figure 11). Detected across all spectra were C–H stretches at 2956 (sh), 2927, and 2855 cm^{-1}. Blue was also identified in this sample by peaks at 2098 [ν(CN)] and 1415 cm^{-1}, in addition to calcium sulfate (1620, 1425 (sh) 1148 (sh), 1120, 670 cm^{-1}). An unidentified peak at 794 cm^{-1} could belong to a silicate, perhaps quartz, considering the presence of silicon in the EDS mapping. This composition matches the Raman composition of the first paint layer in the blue cross sections, with only Prussian blue. While it is difficult to make a conclusive assessment without more chromatographic techniques, the identified bands in these spectra point towards an alkyd resin paint [49]. Alkyd paints were manufactured for commercial use, such as household or industrial paints, and only Winsor & Newton continues to have a line of artist-grade alkyd paints [49,50]. Calder's preference for these matte paints, especially those manufactured by Ronan, is well documented [8,9,14,39]. Most often, alkyd paints are polyester resin-based, made from combining polyhydric alcohol with a polybasic acid, to which monobasic drying oils are added, lending elasticity to an otherwise hard resin film. Here, the C=O band points to the polyester resin, as the fingerprint region is not particularly diagnostic for the drying oil; furthermore, oil C=O peaks are comparatively lower in absorption [49]. Conversely, the C–H bands probably belong to the drying oil component in alkyd paints, which can vary in concentration depending on the desired finish and drying time, among other factors [50]. Samples taken from the surface of the maquette are more representative of the binder in comparison with *Man-Eater* considering the absence of many layers of overpaint. The μ-FTIR spectra of the samples from the maquette showed the same C=O and C–H absorption bands as that of *Man-Eater*, indicating a possible alkyd binding medium as well.

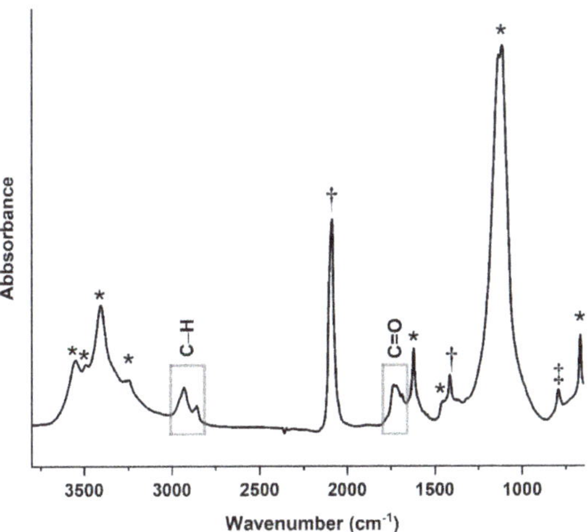

Figure 11. A μ-FTIR spectrum acquired from a sample taken from the edge of the blue pennant in *Man-Eater*. The presence of Prussian blue as was detected in Raman spectroscopy in Layers 1A and 1B (†), and not ultramarine, possibly indicates this sample as original. A carbonyl (C=O) stretching band at 1734 cm^{-1} and C–H stretches at 2956 (sh), 2927, and 2855 cm^{-1} point towards an alkyd binder, but any remaining diagnostic peaks are obscured by the presence of $CaSO_4$ (*) and a silicate (‡).

4. Conclusions

For MoMA's 2021 exhibition, *Alexander Calder: Modern from the Start*, the possibility arose for reviving *Man-Eater with Pennants*, a forgotten sculpture by the artist. The sculpture had been repainted several times early in its history; however, after its last exhibition in 1969–1970, extensive paint loss and rusting, due to outdoor exposure, were not addressed. As part of planning the restoration treatment for *Man-Eater*, cross sections and samples were taken for further analysis by Raman spectroscopy, SEM-EDS, and XRF to determine the pigments present and aid in the identification of the artist's original color choices. The original paint layers were identified as the ones closest to an aluminum anti-corrosive layer or those penetrated by drastic rusting of the metal. Through these analyses, we can establish an early primary palette for Calder: a darker blue based on Prussian blue, a brown-red based on P.R. 6, and a lemon yellow based on a sulfate-containing chrome yellow. Among the many layers of overpaint, the ones last applied can be dated to no later than the 1970's due to the presence of the Nd^{3+} rich titanium white, which seems plausible considering that the work has not been exhibited since 1970. Analysis of the associated maquette confirmed the later painting date as described in internal records based on the presence of the Nd^{3+} rich titanium white as a completely different palette consisting of ultramarine blue, a chrome yellow, and the yellowish-red P.R. 4 in combination with molybdate orange. Considering that the maquette was painted between 1959 and 1969 and does not appear to have been repainted since, the colors do not match those of the original paints in *Man-Eater*. To our knowledge, this is the first analysis of his early primary palette in the literature and presents a unique path forward for other treatments of works of Calder from this period held at other institutions and private collections worldwide.

As part of the treatment undertaken by Monumenta Art Conservation and Finishing LLC, the aged paint layers and surface rust were removed before applying an anti-corrosion primer, followed by epoxy paint base coats. The analysis of *Man-Eater* and the maquette helped determine the tonality of yellow, blue, and red used in the restoration campaign. Additionally, it was important to note the presence of toxic pigments, such as red lead, from a safety perspective. Most significantly, understanding the differences in pigment composition between the maquette and final sculpture prevented the use of inaccurate colors, as was done in other instances in the past [14]. The final color coats were applied by spray and brush. Now, half a century after its last exhibition, *Man-Eater with Pennants* is newly conserved and installed, ready to delight and charm visitors to MoMA's sculpture garden.

Author Contributions: A.H. carried out some sampling in addition to Raman Spectroscopy, μ-FTIR, SEM-EDS, microscopy, XRF data interpretation, and drafting this manuscript. M.R. carried out sampling for cross sections in addition to XRF. spectroscopy and digital microscopy. L.Z. carried out art historical research, mined MoMA's internal records, and drafted sections of the manuscript. A.M. carried out XRF interpretation and revisions of the manuscript. All authors have read and agreed to the published version of the manuscript.

Funding: This research received no external funding.

Data Availability Statement: Data available on request.

Acknowledgments: The authors are grateful to our partners in the restoration of *Man-Eater with Pennants*: Cara Manes, Associate Curator in the Department of Painting and Sculpture, The Museum of Modern Art, Abigail Mack and Ellen Rand at Monumenta Art Conservation and Finishing LLC, and Alexander S. C. Rower, President of the Calder Foundation, New York. We are also grateful for instrumental access by the Conservation Center at the Institute of Fine Arts, New York University. Finally, this project was possible with generous support from the Bank of America.

Conflicts of Interest: The opinions: findings, and conclusions or recommendations expressed in this publication are those of the authors and do not necessarily reflect those of the Museum of Modern Art. The authors declare no competing financial interest.

References and Notes

1. Toll, S.I. My Way: Calder in Paris. *Sewanee Rev.* **2010**, *118*, 589–602. [CrossRef]
2. Morris, G.L.K.; De Kooning, W.; Calder, A.; Glarner, F.; Motherwell, R.; Davis, S. What Abstract Art Means to Me. *Bull. Mus. Mod. Art.* **1951**, *18*, 15. [CrossRef]
3. Canning Miller, D. Letter to Sweeney, J.J., then curator at The Museum of Modern Art, New York. The Museum of Modern Art, Department of Painting and Sculpture files. 1945.
4. Goodwin, P. Letter to Sweeney, J.J., then curator at The Museum of Modern Art, New York. The Museum of Modern Art, Department of Painting and Sculpture files. 1945.
5. Sweeney, J.J. Letter to Goodwin, P., then curator at The Museum of Modern Art, New York. The Museum of Modern Art, Department of Painting and Sculpture files. 1945.
6. Sweeney, J.J. Letter to Canning Miller, D., then curator at The Museum of Modern Art, New York. The Museum of Modern Art, Department of Painting and Sculpture files. 1945.
7. Canning Miller, D. Letter to Calder, A., The Museum of Modern Art, Department of Painting and Sculpture files. 1945.
8. Calder, A. Internal museum correspondence with the artist. undated.
9. Pozzi, F.; Arslanoglu, J.; Nagy, E. Alexander Calder's Half-Circle, Quarter-Circle, and Sphere (1932): A complex history of re-painting unraveled. *Herit. Sci.* **2020**, *8*, 79. [CrossRef]
10. Price, B.A.; Pretzel, B.; Lomax, S. (Eds.) *Infrared and Raman Users Group Spectral Database*, 2007 ed.; IRUG: Philadelphia, PA, USA, 2009; Volume 1–2.
11. Fremout, W.; Saverwyns, S. Identification of synthetic organic pigments: The role of a comprehensive digital Raman spectral library. *J. Raman Spectrosc.* **2012**, *43*, 1536–1544. [CrossRef]
12. Bell, I.M.; Clark, R.J.; Gibbs, P.J. Raman spectroscopic library of natural and synthetic pigments (pre- $\approx$ 1850 AD). *Spectrochim. Acta Part A Mol. Biomol. Spectrosc.* **1997**, *53*, 2159–2179. [CrossRef]
13. Lewis, P.A. (Ed.) *Pigment Handbook*, 2nd ed.; Properties and Economics; Wiley-Interscience: New York, NY, USA, 1988; Volume 1.
14. Lodge, R.G.; Lodge, E.W. Notes for a History of "Calder Red" Color and Its Paints in the United States in Relation to the Re-coatings of Alexander Calder's Flamingo (1973) and La Grande Vitesse (1969) and Other Calder Stabiles. 32.
15. Zycherman, L. Correspondence with the Calder Foundation, New York, NY, USA. Personal communication, 2021.
16. Telephony. *Teleph. Am. Teleph. J.* **1911**, *60*, 12.
17. Liu, B.; Wei, J.; Yang, M.; Yin, F.; Xu, K. Effect of heat treatment on the mechanical properties of copper clad steel plates. *Vacuum* **2018**, *154*, 250–258. [CrossRef]
18. Kuhn, H.; Curran, M. Chrome Yellow and Other Chromate Pigments. In *Artists Pigments Handbook: Their History and Characteristics*; Feller, R.L., Ed.; Cambridge University Press: New York, NY, USA, 1987; Volume 1.
19. Monico, L.; Janssens, K.; Hendriks, E.; Brunetti, B.G.; Miliani, C. Raman study of different crystalline forms of $PbCrO_4$ and $PbCr1-xSxO_4$ solid solutions for the noninvasive identification of chrome yellows in paintings: A focus on works by Vin-cent van Gogh. *J. Raman Spectrosc.* **2014**, *45*, 1034–1045. [CrossRef]
20. Balachandran, U.; Eror, N. Raman spectra of titanium dioxide. *J. Solid State Chem.* **1982**, *42*, 276–282. [CrossRef]
21. Rogge, C.E.; Arslanoglu, J. Luminescence of coprecipitated titanium white pigments: Implications for dating modern art. *Sci. Adv.* **2019**, *5*, eaav0679. [CrossRef] [PubMed]
22. Memo, N.J. to Houlihan, P., Conservator. The Museum of Modern Art, Department of Painting and Sculpture files. 1970.
23. Miliani, C.; Monico, L.; Melo, M.J.; Fantacci, S.; Angelin, E.M.; Romani, A.; Janssens, K. Photochemistry of Artists' Dyes and Pigments: To-wards Better Understanding and Prevention of Colour Change in Works of Art. *Angew. Chem. Int. Ed.* **2018**, *57*, 7324–7334. [CrossRef] [PubMed]
24. Mass, J.L.; Opila, R.; Buckley, B.; Cotte, M.; Church, J.; Mehta, A. The photodegradation of cadmium yellow paints in Henri Ma-tisse's Le Bonheur de vivre (1905–1906). *Appl. Phys. A* **2013**, *111*, 59–68. [CrossRef]
25. Anaf, W.; Trashin, S.; Schalm, O.; van Dorp, D.; Janssens, K.; De Wael, K. Electrochemical Photodegradation Study of Semiconductor Pigments: Influence of Environmental Parameters. Anal Chem. *Am. Chem. Soc.* **2014**, *86*, 9742–9748.
26. Pintus, V.; Wei, S.; Schreiner, M. Accelerated UV ageing studies of acrylic, alkyd, and polyvinyl acetate paints: Influence of inorganic pigments. *Microchem. J.* **2016**, *124*, 949–961. [CrossRef]
27. Pagnin, L.; Calvini, R.; Wiesinger, R.; Weber, J.; Schreiner, M. Photodegradation Kinetics of Alkyd Paints: The Influence of Vary-ing Amounts of Inorganic Pigments on the Stability of the Synthetic Binder. *Front. Mater.* **2020**, *7*, 423. [CrossRef]
28. Van Driel, B.A. Titanium White, Friend or Foe?: Understanding and Predicting Photocatalytic Degradation of Modern Oil Paintings. Ph.D. Thesis, Delft University of Technology, Delft, The Netherlands, 2018.
29. van Driel, B.A.; Wezendonk, T.A.; van den Berg, K.J.; Kooyman, P.J.; Gascon, J.; Dik, J. Determination of early warning signs for photocatalytic degradation of titanium white oil paints by means of surface analysis. *Spectrochim. Acta A Mol. Biomol. Spectrosc.* **2017**, *172*, 100–108. [CrossRef] [PubMed]
30. Berrie, B. Prussian Blue. In *Artists Pigments Handbook: Their History and Characteristics*; West FitzHugh, E., Ed.; National Gallery of Art: Washington, DC, USA, 1997; Volume 3.
31. Moretti, G.; Gervais, C. Raman spectroscopy of the photosensitive pigment Prussian blue. *J. Raman Spectrosc.* **2018**, *49*, 1198–1204. [CrossRef]

32. Amat, A.; Rosi, F.; Miliani, C.; Sassi, P.; Paolantoni, M.; Fantacci, S. A combined theoretical and experimental investigation of the electronic and vibrational properties of red lead pigment. *J. Cult. Herit.* **2020**, *46*, 374–381. [CrossRef]
33. Kirby, K.; Saunders, D. Fading and Colour Change of Prussian Blue: Methods and Manufacture and the Influence of Extenders. *Natl. Gallery Tech. Bull.* **2004**, *25*, 73–91.
34. Plesters, J. Ultramarine Blue, Natural and Artificial. In *Artists Pigments Handbook: Their History and Characteristics*; Roy, A., Ed.; Cambridge University Press: New York, NY, USA, 1993; Volume 2.
35. Rejmak, P. Structural, Optical, and Magnetic Properties of Ultramarine Pigments: A DFT Insight. *J. Phys. Chem. C* **2018**, *12*, 29338–29349. [CrossRef]
36. Osticioli, I.; Mendes, N.; Nevin, A.; Gil, F.; Becucci, M.; Castellucci, E. Analysis of natural and artificial ultramarine blue pigments using laser induced breakdown and pulsed Raman spectroscopy, statistical analysis and light microscopy. *Spectrochim. Acta Part A Mol. Biomol. Spectrosc.* **2009**, *73*, 525–531. [CrossRef] [PubMed]
37. Laver, M. Titanium Dioxide Whites. In *Artists Pigments Handbook: Their History and Characteristics*; FitzHugh, E.W., Ed.; National Gallery of Art: Washington, DC, USA, 1997; Volume 3.
38. Buxbaum, G.; Pfaff, G. (Eds.) *Industrial Inorganic Pigments*; John Wiley & Sons: Hoboken, NJ, USA, 2006.
39. Schmelzer, P. The Problem of Paint. Sightlines. 2011. Available online: https://walkerart.org/magazine/the-problem-of-paint (accessed on 19 May 2021).
40. Herbst, W.; Hunger, K.; Wilker, G. *Industrial Organic Pigments: Production, Properties, Applications*, 3rd ed.; Wiley-VCH: Weinheim, Germany, 2004.
41. Burgio, L.; Clark, R.J.H.; Firth, S. Raman spectroscopy as a means for the identification of plattnerite (PbO_2), of lead pigments and of their degradation products. *Analysts* **2001**, *126*, 222–227. [CrossRef] [PubMed]
42. FitzHugh, E.W. Red Lead and Minium. In *Artists Pigments Handbook: Their History and Characteristics*; Feller, R.L., Ed.; Cambridge University Press: New York, NY, USA, 1987; Volume 1.
43. Chua, L.; Hoevel, C.; Smith, G.D. Characterization of Haku Maki prints from the "Poem" series using light-based techniques. *Herit. Sci.* **2016**, *4*, 25. [CrossRef]
44. Wilkins, R.W.T. The Raman spectrum of crocoite. *Miner. Mag.* **1971**, *38*, 249–250. [CrossRef]
45. van Loon, A.; Boon, J.J. Characterization of the deterioration of bone black in the 17th century Oranjezaal paintings using electron-microscopic and micro-spectroscopic imaging techniques. *Spectrochim. Acta Part B At. Spectrosc.* **2004**, *59*, 1601–1609. [CrossRef]
46. Daly, N.S.; Sullivan, M.; Lee, L.; Trentelman, K. Multivariate analysis of Raman spectra of carbonaceous black drawing media for the in situ identification of historic artist materials. *J. Raman Spectrosc.* **2018**, *49*, 1497–1506. [CrossRef]
47. Tomasini, E.P.; Halac, E.B.; Reinoso, M.; Di Liscia, E.J.; Maier, M.S. Micro-Raman spectroscopy of carbon-based black pigments. *J. Raman Spectrosc.* **2012**, *43*, 1671–1675. [CrossRef]
48. Derrick, M.R.; Stulik, D.; Landry, J.M. *Infrared Spectroscopy in Conservation Science*; Getty Conservation Institute: Los Angeles, CA, USA, 1999.
49. Learner, T.; Institute, G.C. *Analysis of Modern Paints*; Getty Conservation Institute: Los Angeles, CA, USA, 2004.
50. Standeven, H.A.L. *House Paints, 1900–1960, History and Use*; Getty Conservation Institute: Los Angeles, CA, USA, 2011.

 heritage

MDPI

Communication

Preliminary Investigations into the Alteration of Cadmium Orange Restoration Paint on an Ancient Greek Terracotta Krater

Georgina Rayner [1,*], Susan D. Costello [1], Arthur McClelland [2], Austin Akey [2] and Katherine Eremin [1]

1 Straus Center for Conservation and Technical Studies, Harvard Art Museums, 32 Quincy Street, Cambridge, MA 02138, USA; susan_costello@harvard.edu (S.D.C.); katherine_eremin@harvard.edu (K.E.)
2 Center for Nanoscale Systems, Harvard University, 11 Oxford Street, Cambridge, MA 02138, USA; amcclelland@cns.fas.harvard.edu (A.M.); aakey@fas.harvard.edu (A.A.)
* Correspondence: georgina_rayner@harvard.edu

Abstract: In preparation for exhibition, an ancient Greek terracotta krater received treatment which included selective in-painting with cadmium orange (CdSSe). After one year on display the object displayed disfiguring alteration in select areas of restoration. Cross-section analysis of samples taken from the object revealed that alteration only occurred in areas where the paint was in direct contact with darkened and abraded areas of the terracotta surface, in which analysis found the presence of chlorine. The alteration was recreated in mock-ups for more in-depth analysis. Using Raman spectroscopy and scanning electron microscopy with energy-dispersive X-rays (SEM-EDS) it was discovered that selenium-rich structures were forming throughout the paint films. The observed alteration is the result of degradation of the CdSSe pigment which occurs in the presence of chlorine and light. This research highlights the need for careful selection of restoration materials when dealing with objects suspected to contain residual chloride ions if desalination cannot be undertaken.

Keywords: alteration; cadmium orange; chlorine; selenium; terracotta

Citation: Rayner, G.; Costello, S.D.; McClelland, A.; Akey, A.; Eremin, K. Preliminary Investigations into the Alteration of Cadmium Orange Restoration Paint on an Ancient Greek Terracotta Krater. *Heritage* **2021**, 4, 1497–1510. https://doi.org/10.3390/heritage4030082

Academic Editor: João Pedro Veiga

Received: 28 June 2021
Accepted: 28 July 2021
Published: 29 July 2021

1. Introduction

The Harvard Art Museums Bell Krater: Torch Race (1960.344, dated c. 430-420 BCE, Figure 1a) was one of several Greek terracotta artefacts which received treatment in preparation for display in 2014. After one year of display the cadmium orange restoration paint on the krater had altered in color in some areas from orange to grey (Figure 1b,c). In comparison, other objects displayed in the same case also in-painted with the same paint showed no sign of alteration. The appearance was deemed unacceptable and re-treatment was required. To appropriately re-treat the krater an understanding of the cause of the alteration was essential and began with understanding the treatment history of the object.

There is no record of any conservation work performed on the krater before its bequest to the museum, however, the object had been re-assembled in the past and this treatment resulted in misplaced joins, abraded surfaces and noticeable over-paint. During a 2004 loan, an efflorescence was observed on the object. The efflorescence was analyzed by SEM-EDS and found to contain calcium and chlorine. FTIR analysis was also performed but the mineral could not be identified with the available spectral databases at that time.

When the spectrum was revisited in the early stages of this investigation, the expansion of the Infrared and Raman Users Group (IRUG) spectral database [1] led to the identification of calclacite ($Ca(CH_3COO)Cl.5H_2O$), a calcium chlorine acetate salt (see Figure 5 later). The formation of calclacite has previously been attributed to archaeological ceramics that have residual chlorides from burial or treatment with hydrochloric acid and have been stored in wooden cases [2]. Hydrochloric acid, along with other acids such as acetic or nitric, were used in the past to remove burial accretions [3]. For hydrochloric acid, if not completely removed from the object by rinsing, residual chlorine salts can remain and can react with the acid vapors from the wood resulting in the efflorescence. Based on the identification of the efflorescence, the assumption was made that the object underwent incomplete treatment with hydrochloric acid to remove burial accretions at some point in the past.

Figure 1. (**a**) Bell Krater: Torch Race, in the Manner of the Peleus Painter (c. 430-420 BCE, 1960.344) after conservation treatment in 2014. Harvard Art Museums/Straus Center for Conservation and Technical studies, bequest of David Moore Robinson. (**b**) Inset of male figure after conservation treatment in 2014. (**c**) Inset of male figure showing alteration of the restoration paint (highlighted by the white arrow) after one year of display.

Desalination was considered during the 2014 treatment of the object but was eventually deemed an unnecessarily risky treatment as restoration joins were considered stable and storage in a controlled environment had prevented the formation of any further efflorescence [4]. Instead, treatment of the object focused on aesthetic work. Disfiguring paint was removed, restoration fills levelled, and break lines were filled and sealed with acrylic materials, followed by in-painting. Cadmium orange acrylic paint (Golden Heavy Body Artist Acrylics in C.P. Cadmium Orange), containing CdSSe, was chosen for in-painting due to its ability to lighten the darkened, abraded areas of the terracotta. Titan Buff acrylic paint (Golden Heavy Body Artist Acrylics in Titan Buff), containing titanium white (TiO_2), was added as required to adjust the tone. Full details of the materials used for the treatment may be found in a separate publication [4].

Cadmium pigments offer a broad range of colors from light yellow to deep red. Cadmium yellow is composed of cadmium sulfide (CdS) and the incorporation of Se in increasing amounts changes the color from yellow through orange to red. The alteration of cadmium yellow has been reported in paintings by Henri Matisse [5–9], Edvard Munch [9–12], Vincent Van Gogh [13,14], Pablo Picasso [13,15] and others [13,16–18], all dating from the late 19th-early 20th century, before the stability of pigment was improved [16]. Alteration in areas identified as cadmium sulfide presents itself as either lightening or darkening of the once vibrant pigment.

Researchers have utilized a multitude of analytical tools, and most recently, synchrotron-based radiation techniques, to probe the degradation of the pigment. The observed alteration has been determined to be the result of degradation of the cadmium sulfide pigment through a photo-oxidation process. In the initial stages of the process cadmium sulfate is formed ($CdSO_4 \cdot xH_2O$) which then converts to cadmium carbonate ($CdCO_3$) and cadmium oxalate (formed from the breakdown of organic components such as binding media [14], CdC_2O_4). The formation of these white compounds contributes to the observed lightening of the pigment. The formation of cadmium oxide (CdO), a brown compound, has been associated with darkening [13].

The alteration of cadmium orange (CdSSe) in an artist's work has not previously been identified, however, in the field of semiconductor chemistry, the photo-oxidation of cadmium selenide (CdSe), has been observed. Studies determined that during irradiation of CdSe in the presence of oxygen, photo-oxidation occurs producing cadmium oxide (CdO) and selenium dioxide (SeO_2) [19,20].

Whilst the degradation of artist materials is not unexpected, in any museum, restoration work performed by conservators is expected to last many decades. Increasing access to technical analysis within museum communities has made it easier for appropriate materials to be chosen and those found not to be suitable, such as fugitive colors, can be avoided. In the unlikely event that restoration materials fail quickly, such as in the case presented here, unusual circumstances are likely involved. The theory developed that the krater had received incomplete treatment with hydrochloric acid in the past warranted further investigation in order to determine if this was the cause of the paint alteration and to inform how to proceed with re-treatment.

2. Materials and Methods

2.1. Samples from the Krater

Small samples were taken from the krater from areas of restoration with and without alteration and prepared as cross-sections allowing for the stratigraphy of the layers to be observed and analyzed by SEM-EDS. Additional small fragments of the paint film from degraded areas were collected for analysis by Fourier transform infrared (FTIR) spectroscopy, Raman spectroscopy, pyrolysis-Gas Chromatography Mass Spectrometry (pyrolysis-GCMS) and Scanning Electron Microscopy with Energy-Dispersive X-rays (SEM-EDS).

2.2. Samples of Paint Used for the Conservation Treatment

Paint used for the conservation treatment, Golden Heavy Body Artist Acrylics in C.P. Cadmium Orange and Titan Buff, were prepared as paint outs on glass microscope slides. The paint films were analyzed by X-ray Fluorescence (XRF) spectroscopy, FTIR, Raman and pyrolysis-GCMS.

2.3. *Mock-Ups*

Pieces of modern terracotta were acidified by covering with 6N hydrochloric acid (HCl) and allowed to dry at least overnight before painting. Each piece was painted with a swatch of cadmium orange and cadmium orange mixed with Titan Buff, mimicking the mixing of the paint in the conservation treatment. Non-acidified pieces of terracotta were prepared for direct comparison to the acidified pieces. The mock-up pairs were used for experiments to recreate the formation of the calclacite efflorescence and to investigate the role of light in the alteration of the cadmium orange paint layer. A final mock-up treated with 1N HCl was also produced to compare with the mock-ups treated with 6N HCl. The mock-ups were sampled as needed for preparation as cross-sections, as well as analysis by FTIR, pyrolysis-GCMS and SEM-EDS. Raman analysis was performed directly on the mock-ups.

2.4. *Experimental Methods*

2.4.1. Cross-Section Preparation

Samples for cross-section analysis were embedded in Bio-Plastic liquid polyester casting resin (Ward's Natural Science). Sections were ground to exposure using a Buehler Handimet 2 roll grinder with Carbimet abrasive paper rolls ranging in grit from 240–600. Samples were then polished using a Buehler Metaserv 2000 polisher with 6 μm and 1 μm Buehler MetaDi Monocrystalline Diamond Suspension.

2.4.2. Optical Microscopy

Cross-sections and paint swatches on the mock-ups were observed using a Zeiss Axio Imager.M2m upright microscope equipped with four objectives (5x, 10x, 20x and 50x) and a Zeiss Axiocam 512 Color digital camera. Images were captured using the Zeiss Zen 2.6 (blue edition) software. Visible light conditions utilized a halogen lamp and an EPI-polarization filter cube. Ultraviolet (UV) conditions utilized a mercury vapor lamp and a FITC filter cube (excitation BP450-490, beam splitter FT 510 and emission LP515).

2.4.3. Fourier Transform Infrared (FTIR) Spectroscopy

FTIR in transmission mode was performed using a Bruker Vertex 70 infrared bench spectrometer coupled to a Bruker Hyperion 3000 infrared microscope. Samples were compressed on to a diamond cell with a stainless-steel roller prior to analysis. Spectra were recorded between 4000 and 600 cm^{-1} at 4 cm^{-1} spectral resolution and 32 scans per spectrum. The collected spectra were compared to in-house and published databases.

2.4.4. Raman Spectroscopy

Two different systems were used during this study. Analysis of the fresh and degraded paint films were conducted using a Bruker Optics Senterra dispersive Raman microscope with an Olympus BX51M microscope. The 633 nm laser was used at 5 mW power with an integration time of either 5 or 10 and 10 co-additions.

Raman measurements of the dark needle-like structures formed in the altered paints were conducted on a Horiba XploRA One confocal Raman microscope with an Olympus LMPLFLN-BD 50x, NA 0.5, long working distance objective and 0.05 mW of 785 nm excitation light. A 1200 blaze grating was used. The detector was a 1024 × 256 scientific CCD that was thermoelectrically cooled to −70 °C. Two 90 s acquisitions were averaged together. Horiba's "Denoiser" smoothing algorithm was applied to the data. A polynomial baseline was fit to the data and subtracted. Spectra were compared with in-house and published databases.

2.4.5. Pyrolysis-Gas Chromatography Mass Spectrometry (Pyrolysis-GCMS)

Pyrolysis-GCMS analyses were performed using a CDS pyroprobe 5000 (platinum coil) interfaced to an Agilent Technologies 7890B GC system coupled with an Agilent Technologies 5977A MSD. The samples were placed in the center of a quartz tube. Pyrolysis was conducted at 600 °C for ten seconds. The samples were directed onto the column (HP-5ms Ultra Inert column, 30 m × 25 µm × 0.25 µm) by helium gas in a 20:1 split ratio. The GC oven was set with an initial temperature of 40 °C for 1 min prior to increasing in 10 °C/min intervals to 300 °C over a 26 min period then maintained for 20 min for a total acquisition time of 47 min.

2.4.6. X-ray Fluorescence (XRF) Spectrometry

A Bruker Artax XRF spectrometer with a Silicon Drift Detector (SDD) and a rhodium anode X-ray tube was used for analysis. The primary X-ray beam is collimated to give a spot size of 0.65 mm. Spectra were acquired for 100 s live time at 50 kV and 600 µA. A helium flux was used to increase the detection efficiency for light elements (atomic number of potassium and lower).

2.4.7. Scanning Electron Microscopy with Energy-Dispersive X-ray Spectrometry (SEM-EDS)

Two different systems were used during this study. Initial analysis of the cross-sections was conducted using a JEOL JSM-6460LV SEM with an Oxford Instruments 80 mm^2 X-MaxN X-ray spectrometer running the Oxford INCA software. The SEM was operated in low vacuum mode at a chamber pressure of 35 Pa, with an operating voltage of 20 kV, beam current circa 1 nA and working distance of 10 mm. The cross-sections were not coated prior to analysis.

More detailed analysis including mapping of the cross-sections and paint films were collected using an FEI Helios 660 Dual-Beam SEM/FIB equipped with an EDAX Octane Plus EDS system, with EDAX TEAM brand acquisition and mapping software. Samples were prepared for imaging by carbon-coating with a layer of amorphous carbon of between 50–100 nm thickness, for electrical discharge purposes. Images were collected using an accelerating voltage of 3 kV, beam current of 3.2 nA. EDS maps were acquired with an accelerating voltage of 15 kV and a beam current of 13 nA.

3. Results and Discussion

3.1. *Analysis of Cross-Sections from the Krater*

The discoloration of the paint was exclusively observed to occur in areas where the restoration paint was in contact with abraded areas around break lines in the ceramic. No alteration was observed when the paint was applied over the acrylic fills from the 2014 treatment. In Figure 2a, the terracotta of the krater is present in a cross-section prepared from an area where the paint altered, observable as a very uniform substrate layer. In Figure 2b, a cross-section prepared from an area where the paint did not alter, the paint is applied to a mixture of restoration materials instead of terracotta. Clearly different from the original terracotta (Figure 2a), this substrate is consistent with the presence of fill materials applied during the 2014 treatment and from the earlier, undated treatment.

The paint layer, measuring between 3–5 µm, appears almost completely white in the altered cross-section with only a few remaining colored particles. The paint layer is difficult to see in the unaltered cross-section, however with ultraviolet (UV) illumination a distinct green fluorescence from the paint layer is observable.

Analysis using SEM-EDS (Figure 3) confirmed the presence of chlorine (Cl) in the altered paint layer and in the terracotta (not shown in Figure 3). No Cl was observed in any of the layers in cross-sections prepared from the unaltered paint. This observation, combined with the identification of calclacite from the 2004 efflorescence, suggested that the alteration may be caused by an interaction between the cadmium orange restoration paint and residual chloride ions in the terracotta.

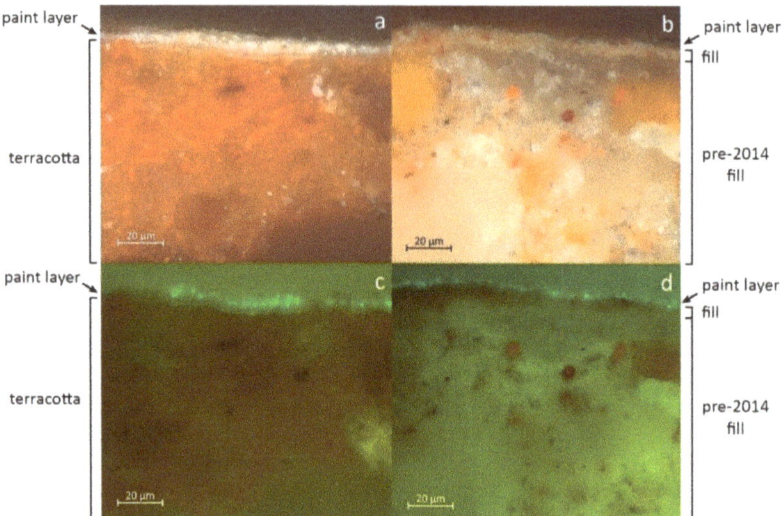

Figure 2. Microscope images of cross-section samples taken from the krater. (**a**) altered paint sampled from the male figure (pictured in Figure 1c) in which the restoration paint sits directly on top of the terracotta surface. (**b**) unaltered paint sampled from an area in which the paint was not in contact with the terracotta due to the presence of fill materials. (**c**,**d**) represent the same areas depicted in (**a**,**b**) taken with UV illumination (FITC filter cube).

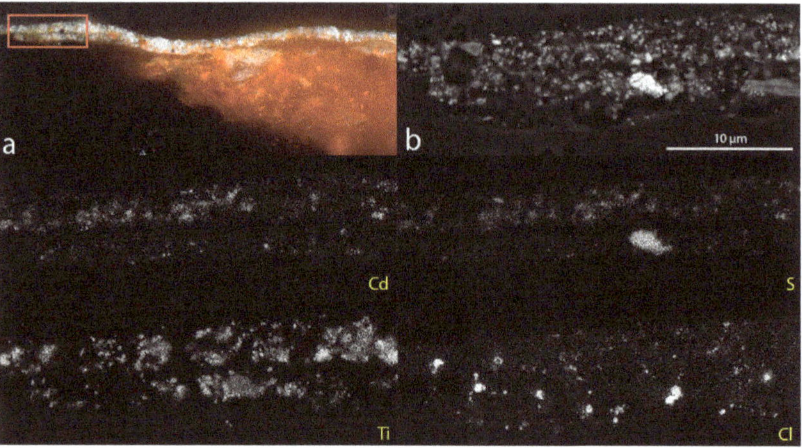

Figure 3. (**a**) Microscope image of cross-section prepared from a sample taken from an area of alteration on the krater. (**b**) SEM backscattered electron image of the altered paint film which has been highlighted in (**a**) by the red rectangle with EDS maps below showing Cd, S, Ti and Cl channels.

3.2. Characterization of the Fresh Paint

The paint film prepared from Golden Heavy Body Artist Acrylics in C.P. Cadmium Orange is uniformly orange in color with no obvious inclusions. Elemental analysis by XRF of the cadmium orange paint confirmed the presence of cadmium (Cd), sulfur (S) and selenium (Se), consistent with the cadmium orange pigment. Barium (Ba) and zinc (Zn) were additionally identified in the paint.

The Raman spectrum of the cadmium orange paint (Figure 4a) contains bands assigned to the longitudinal optical (LO) phonon and overtone (2LO) of CdS at 296 and 595 cm^{-1}, respectively. The LO phonon for CdSe is typically observed at around 200 cm^{-1} but may shift in wavenumbers based on the pigment composition, and as such may be accounted for by either or both of the peaks at 192 and 214 cm^{-1} [21]. The presence of barium sulfate, identifiable in Figure 4a by the peak at 988 cm^{-1} representing the symmetric stretching of the SO_4 group, indicates that the paint contains the lithopone version of cadmium orange [16]. The mineral form of zinc could not be identified from analysis, however, the identification of lithopone may suggest the presence of zinc sulfide (ZnS) and the UV fluorescence observed in the paint, discussed below, may be attributed to the presence of zinc oxide (ZnO).

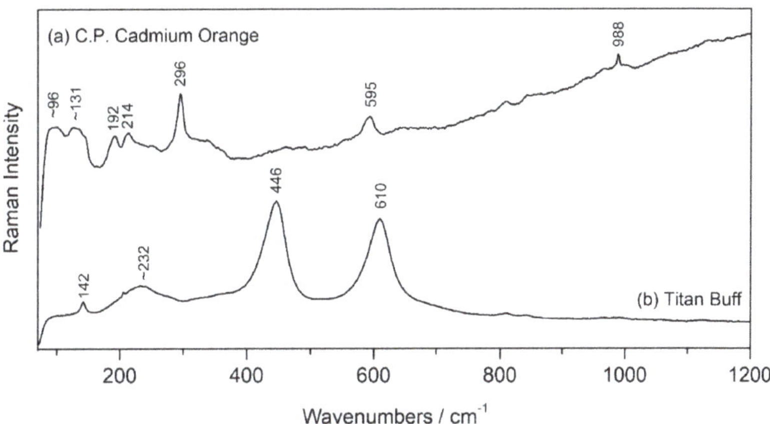

Figure 4. Raman spectra obtained from (**a**) fresh C.P. Cadmium Orange acrylic paint (Golden Heavy Body Artist Acrylics) and (**b**) fresh Titan Buff acrylic paint (Golden Heavy Body Artist Acrylics).

The cadmium orange paint has the same distinct green UV fluorescence visible in the cross-section images shown in Figure 2c,d. The fluorescence occurs from very fine particles distributed evenly throughout the paint matrix. Multiple cadmium orange reference samples and pure barium sulfate displayed no fluorescence under the same UV conditions. A reference sample for lithopone contained a scattering of particles which fluoresced pale yellow, typical of zinc oxide (ZnO). Zinc oxide may be present in lithopone in trace amounts, and whilst the fluorescence of zinc oxide is not a match for the green fluorescence observed in the cadmium orange paint film, it is the only pigment identified to have a fluorescence that may be present. It is speculated that the fluorescence color has altered as a result of the paint preparation, however, further study is required.

The paint film prepared from Golden Heavy Body Artist Acrylics in Titan Buff is off-white in color and includes both red and black inclusions. XRF analysis of the paint film showed the presence of titanium (Ti) and Raman analysis identified the presence of rutile (titanium dioxide, TiO_2), based on the presence of B_{1g}, E_g and A_{1g} Raman active modes of the of the rutile single crystal at 142, 446 and 610 cm^{-1}, respectively (Figure 4b) [22]. The red and black inclusions were identified as hematite (Fe_2O_3) and rutile, respectively, with Raman. The identification of rutile in the black inclusions suggests the use of the natural form of the mineral in addition to the white synthetic form used for the bulk of the paint [23].

Importantly, no Cl was identified in either of the paints, supporting the theory that residual chloride ions in the terracotta are involved in the observed alteration of the paint. FTIR analysis confirmed the presence of an acrylic binder in both the cadmium orange and Titan Buff paints, which was identified to contain a combination of methyl methacrylate, n-butyl acrylate and n-butyl methacrylate by pyrolysis-GCMS.

3.3. Mock-Up Study

Mock-ups were created to allow for more extensive study of the paint alteration as the amount of altered material from the krater was limited. The mock-ups were created in pairs with one terracotta piece acidified using 6N hydrochloric acid and the other without. Each mock-up was painted with a swatch of the pure cadmium orange paint and a swatch of the cadmium orange paint mixed with the Titan Buff, mimicking the conservation treatment.

Initial mock-ups were designed to recreate the conditions which are assumed to have contributed to the formation of the efflorescence observed in 2004. One mock-up pair was sealed in a bag with a small volume of acetic acid, approximately 2 mL, to recreate the proposed scenario which caused the formation of the calclacite efflorescence on the krater. White efflorescence formed on the acidified mock-up after three days. The efflorescence was sampled, analyzed by FTIR and identified as calclacite (Figure 5). No efflorescence was observed on the non-acidified mock-up.

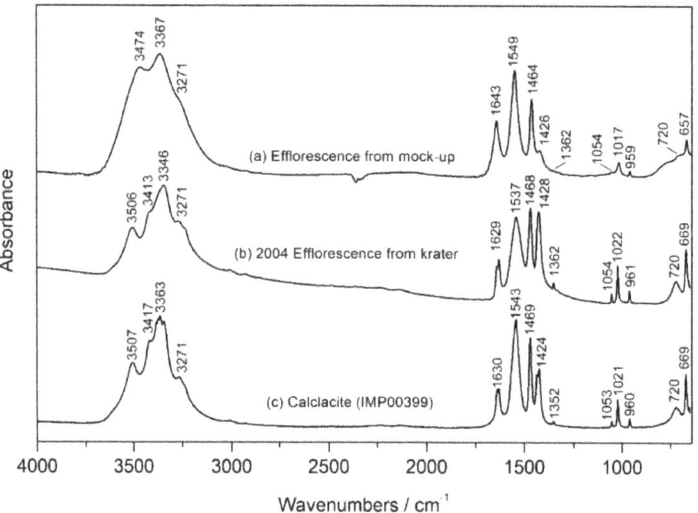

Figure 5. FTIR results for efflorescence formed on (**a**) a mock-up treated with hydrochloric acid and exposed to acetic acid, compared with (**b**) the efflorescence that formed on the krater whilst on loan in 2004 and (**c**) a reference spectrum for calclacite from the IRUG database.

Next, a series of mock-ups were created to study the effects of both chloride ions and light on the alteration of the pigment, aiming to investigate whether display conditions had contributed to the alteration. Mock-up pairs (acidified and non-acidified), were placed in bright natural light (1 and 2), non-direct light (3 and 4) and in the dark (5 and 6). The mock-ups in direct light received direct natural light through a window which filters out the majority of UV radiation. Those placed in indirect light were sheltered from direct natural light, mimicking the closest scenario to the display conditions of the object. The mock-ups which received no light were kept in a closed draw.

During the first three months of exposure, mock-ups in direct light received approximately 9 h of natural daylight per day. The amount of light these mock-ups received during daylight hours varied between 207–2624 footcandles, the equivalent of approximately 2140–28,400 Lux, with the measured UV content of the light varying between 5.7–14.4 mW/Lumen (determined using an ELSEC 765 Environmental Monitor). The lower values are associated with overcast, cloudy days and the higher values with bright, sunny days. The amount of light received by the mock-ups in indirect light during the same time frame remained below 10 footcandles (108 Lux) and UV measurements were consistently zero.

The acidified mock-ups in direct and indirect light (mock-ups 2 and 4, respectively) began to noticeably alter within one month with the cadmium orange darkening to brown and the cadmium orange/Titan Buff mixture changing to grey, matching the observed alteration on the krater. The acidified mock-up placed in the dark (mock-up 6) displayed no visible change after one year and no change was seen on the non-acidified mock-ups at any light level. Figure 6 shows the mock-ups after one year.

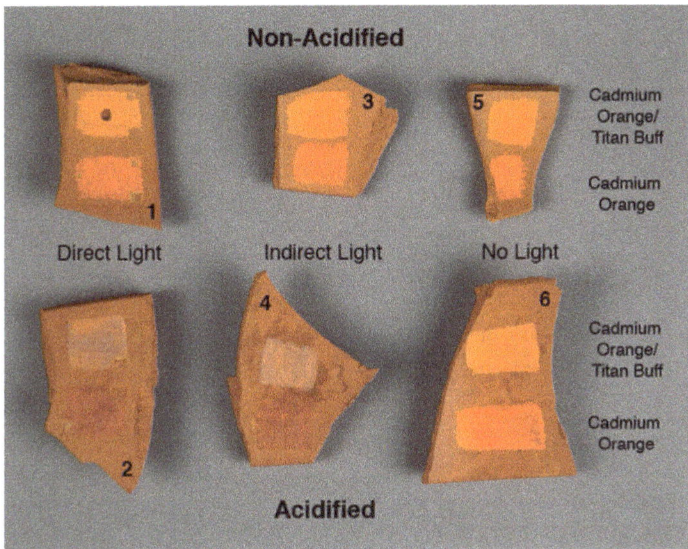

Figure 6. Mock-ups created to investigate the alteration of cadmium orange pictured after one year of exposure to direct light (mock-ups 1 and 2, left), indirect light (mock-ups 3 and 4, center) and no light (mock-ups 5 and 6, right). The pieces in the top row were not acidified prior to painting. Each piece of terracotta is painted with a swatch of cadmium orange/Titan Buff on the top half and cadmium orange on the bottom half.

The mock-ups used for this study were treated with 6N hydrochloric acid to produce a quick reaction. A mock-up treated with 1N hydrochloric acid was created for comparison with the more harshly acidified samples. The same alteration was observed on the mock-up acidified with 1N hydrochloric acid, albeit occurring at a reduced rate in comparison to the 6N counterpart taking at least three months for any noticeable alteration to occur.

3.4. Characterization of the Altered Paint from the Mock-Ups and the Krater

A greyed appearance on terracotta ceramics in-painted with acrylic paints has been observed before and was attributed to residual solvent retained in the terracotta substrate migrating to the surface carrying lighter pigments with it [24]. This was considered as a possibility, however, other terracotta ceramics treated in the same manner as the krater did not show the same alteration and not all of the mock-ups altered as might be expected if this was the case. The migration of pigments could also not be confirmed from SEM-EDS mapping of the paint films in cross-section, with the elements associated with the paint layers remaining relatively evenly distributed across the paint layer (see Figure 3).

A key observation in the alteration of the paint films on mock-ups 2 and 4 was the formation of dark needle-like structures which are visible only with high magnification in both the cadmium orange paint swatch and the cadmium orange/Titan Buff swatch, the latter shown in Figure 7a. Whilst these needle-like structures formed on both mock-ups, the formation was more extensive in the mock-up exposed to direct light (mock-up 2), suggesting that high light exposure acts as a catalyst for formation but is not required. The analysis described here is focused on the cadmium orange/Titan Buff altered paint film on mock-up 4 as it most closely resembles the paint mixture and conditions under which alteration occurred on the krater.

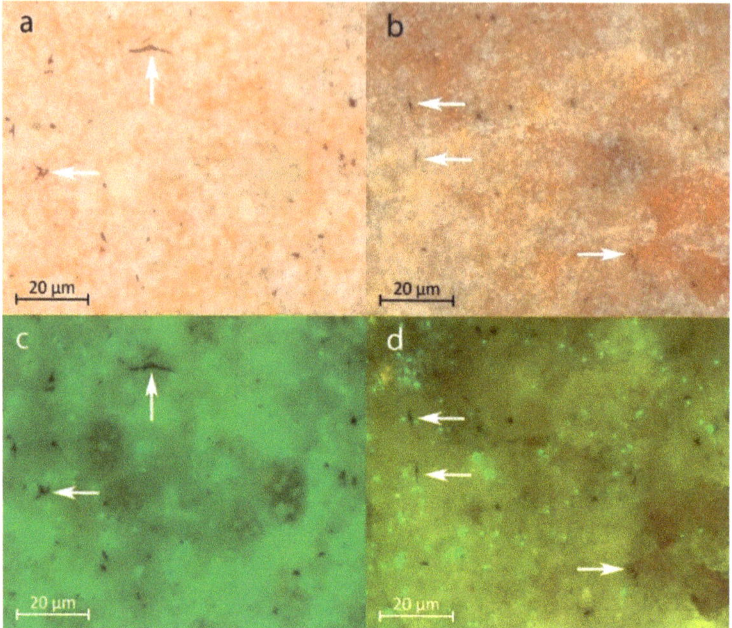

Figure 7. Dark needle-like structures formed in (**a**) the cadmium orange/Titan Buff paint mixture on mock-up 4 (exposed to indirect light) and (**b**) the altered paint layer on the krater. (**c**,**d**) represent the same areas depicted in (**a**,**b**) taken with UV illumination (FITC filter cube). White arrows have been used to highlight some of the larger needle-like structures.

A sample of the altered paint from the krater was revisited and the same dark needle-like structures were observed (Figure 7b). These structures are a little more difficult to see in the visible and UV image (Figure 7b and d, with some structures highlighted by white arrows), due to their small size which is comparable to the dark inclusions present in the Titan Buff paint.

SEM-EDS imaging of the altered paint from mock-up 4 shows that the majority of the dark needle-like structures, which are in the order of 1/2 µm long, can be seen just underneath the surface of the paint film whilst a few protrude slightly from the surface. SEM-EDS mapping revealed that the structures are rich in Se (Figure 8a). None of the other elements associated with the pigments (Cd, S and Ti) or Cl appear to be associated with the structures. Raman spectra collected from the structures have a unique sharp Raman peak around 234 cm^{-1} which is a match to a reference spectrum of Se available through the RRUFF^{TM} project database (R050656, Figure 9) [25]. Comparison with the literature indicates that this Raman band corresponds to the trigonal, polymeric form of Se (t-Se_n) [26]. Clusters of dark needle-like structures protruding from the paint surface, similar in size to the Se-rich structures observed in the mock-up paint, were identified in the altered paint taken from the krater with SEM-EDS mapping again revealing that the structures are rich in Se (Figure 8b) and the Raman spectrum matches the reference for Se already discussed (Figure 9).

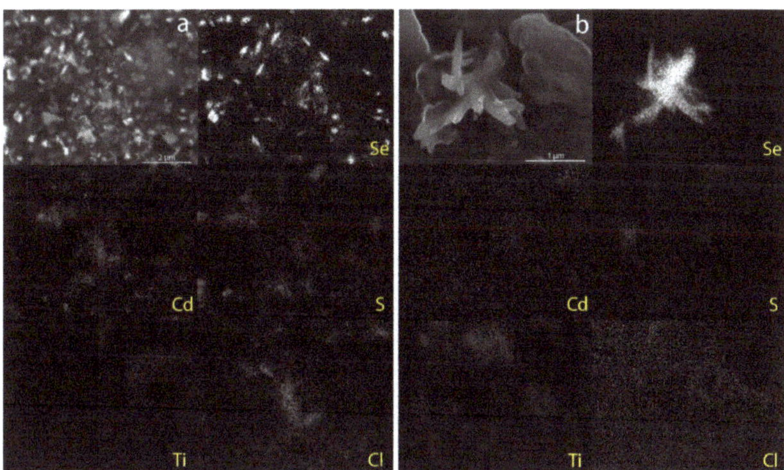

Figure 8. (**a**) SEM backscattered electron image of surface of mock-up 4 with EDS maps showing Se, Cd, S, Ti and Cl channels. (**b**) SEM backscattered electron image of paint taken from altered region of the Krater, showing a branched structure located at the surface with EDS maps showing Se, Cd, S, Ti and Cl channels.

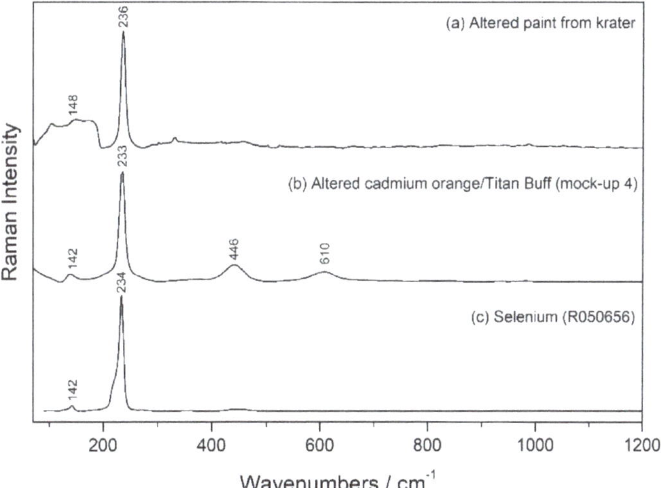

Figure 9. Raman spectra obtained from (**a**) a dark needle-like structure from the altered paint taken from the krater and (**b**) a dark needle-like structure within the altered cadmium orange/Titan Buff paint from mock-up 4 compared with (**c**) a reference spectrum of selenium available through the RRUFF™ database. The broader peaks visible in spectrum (**b**) at 446 and 610 cm^{-1} are associated with the rutile in the Titan Buff paint in the paint matrix surrounding the needle-like structures.

The FTIR spectrum of the altered paint from both the mock-ups and krater remained relatively unchanged in comparison to the spectra of the fresh paint. Whilst the formation of different cadmium compounds (for example CdC_2O_4, $CdCO_3$) may be identified in the FTIR spectrum [12,15], the strong signal from the acrylic binder prevented their identification if present. No apparent degradation of the acrylic binder was observed in the pyrolysis-GCMS chromatogram.

Raman spectra of the paint surrounding the Se-rich structures are shown for the altered paint from the krater, the altered cadmium orange/Titan Buff swatch from mock-up 4 and the altered cadmium orange swatch from mock-up 2 in Figure 10. The only recognizable feature from the fresh paint is the presence of rutile in the spectra from the krater and mock-up 4 (Figure 10a,b, respectively). The peaks associated with the cadmium orange pigment (shown in Figure 4a, with peak locations shaded in orange in Figure 10) are absent in all three spectra. Instead, the presence of Se remains visible (highlighted by the red shaded area in Figure 10) and a new peak has formed in the mock-ups at 251 cm^{-1}, noted in Figure 10 by the area shaded blue. This peak remains unidentified, seemingly not associated with any of the previously identified degradation products of cadmium pigments ($CdSO_4 \cdot xH_2O$, $CdCO_3$, CdC_2O_4 or CdO).

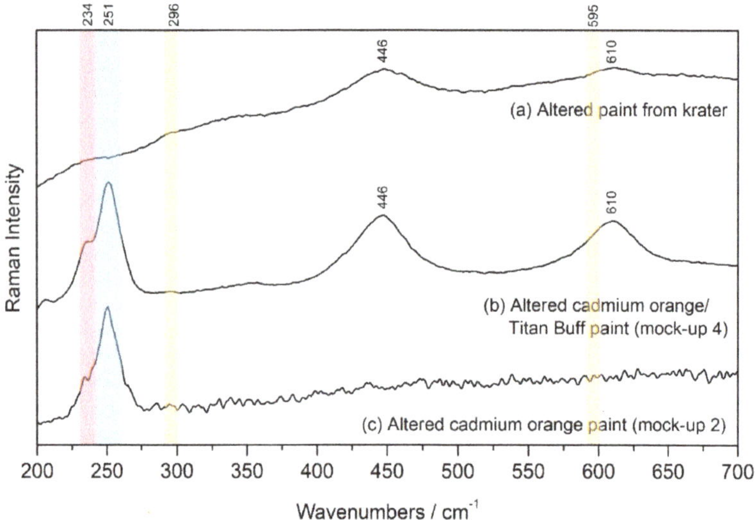

Figure 10. Detailed region from 200–700 cm^{-1} of the Raman spectra obtained from the altered paint surrounding the Se-rich structures from (**a**) the krater, (**b**) the cadmium orange/Titan Buff paint from mock-up 4 and (**c**) the cadmium orange paint from mock-up 2. The orange shaded areas indicate where the peaks associated with the cadmium orange pigment were located in this region (296 and 595 cm^{-1}, now absent). The red shaded area indicates the peak associated with selenium at 234 cm^{-1} and the blue shaded area highlights the new peak observed at 251 cm^{-1}. The broad peaks centered at 446 and 610 cm^{-1} are associated with rutile from the Titan Buff paint.

4. Conclusions

Using complementary analytical techniques it was determined that the alteration observed for the restoration paint applied to the krater is the result of degradation of the cadmium orange (CdSSe) pigment. This conclusion is supported by the identification of selenium-rich needle-like structures of the trigonal Se_n form within the paint film after alteration occurs, as observed in both the altered paint from the krater and in the mock-ups created to study the cause of the alteration. Based on the behavior of the restoration paint on the ancient Greek krater and on the mock-ups, alteration occurs only in areas of acidified ceramic and in the presence of light. It is hypothesized that the residual chloride ions in the terracotta, resulting from either burial or incomplete treatment with HCl to remove burial accretions, and light exposure act as catalysts for the observed alteration. This is consistent with observations in the literature in which mobile chloride ions have been associated with promoting the oxidation of cadmium sulfide [12] as well as making the pigment more sensitive to photo-oxidation [5,27].

Based on previous research it was anticipated that cadmium would form white (e.g., $CdSO_4$, $CdCO_3$) or brown (CdO) compounds, however, no cadmium-containing compounds were identifiable during this preliminary study. This research will hopefully be extended in the future to fully deduce the cadmium–containing degradation products and further investigate the role of acidity/pH in the alteration. Together, this will allow for a better understanding of the full mechanistic process in which the degradation of cadmium orange occurs.

For the re-treatment of the krater, the altered paint film was removed and non-cadmium containing pigments (Golden Fluid Acrylics cadmium red and yellow) used for in-painting after mock-ups determined them to be stable under the conditions tested during this study [4]. This same treatment will be used in the future for other ceramics known, or suspected, to have been treated with hydrochloric acid.

Author Contributions: Conceptualization, S.D.C., K.E. and G.R.; methodology, S.D.C., K.E. and G.R.; formal analysis, G.R., K.E., A.M. and A.A.; investigation, G.R., K.E., A.M. and A.A.; resources, S.D.C.; writing—original draft preparation, G.R.; writing—review and editing, G.R., S.D.C., K.E., A.M. and A.A.; visualization, G.R.; supervision, K.E. All authors have read and agreed to the published version of the manuscript.

Funding: This work was performed in part at the Center for Nanoscale Systems (CNS), a member of the National Nanotechnology Coordinated Infrastructure Network (NNCI), which is supported by the National Science Foundation under NSF award no. ECCS-2025158. CNS is part of Harvard University.

Institutional Review Board Statement: Not applicable.

Informed Consent Statement: Not applicable.

Data Availability Statement: The data presented in this study is available on request from the corresponding author. Publicly available datasets were used for reference material in this study. This data can be found here: http://www.irug.org/search-spectral-database, accessed on 9 June 2021; https://rruff.info, accessed on 9 June 2021.

Conflicts of Interest: The authors declare no conflict of interest. The funders had no role in the design of the study; in the collection, analyses, or interpretation of data; in the writing of the manuscript, or in the decision to publish the results.

References

1. Price, B.A.; Pretzel, B.; Quillen Lomax, S. (Eds.) Infrared and Raman Users Group Spectral Database. Available online: www.irug.org (accessed on 9 June 2021).
2. Fitzhugh, E.W.; Gettens, R.J. Calclacite and Other Efflorescent Salts on Objects Stored in Wooden Museums Cases. In *Science and Archaeology*; Brill, R.H., Ed.; MIT Press: Cambridge, MA, USA; London, UK, 1971; pp. 91–101.
3. Johnson, J.S.; Erickson, H.M.; Iceland, H. Identification of Chemical and Physical Change during acid Cleaning of Ceramics. *MRS Proc.* **1995**, *352*, 831–837. [CrossRef]
4. Costello, S.D.; Rayner, G.; Eremin, K. The degradation of cadmium orange restoration paint on an ancient Greek terracotta vase. In Proceedings of the ICOM-CC 18th Triennial Conference Preprints, Copenhagen, Denmark, 4–8 September 2017.
5. Mass, J.L.; Opila, R.; Buckley, B.; Cotte, M.; Church, J.; Mehta, A. The photodegradation of cadmium yellow paints in Henri Matisse's Le Bonheur de vivre (1905–1906). *Appl. Phys. A* **2012**, *111*, 59–68. [CrossRef]
6. Mass, J.; Sedlmair, J.; Patterson, C.; Carson, D.; Buckley, B.; Hirschmugl, C. SR-FTIR imaging of the altered cadmium sulfide yellow paints in Henri Matisse's Le Bonheur de vivre (1905–6)—Examination of visually distinct degradation regions. *Analyst* **2013**, *138*, 6032–6043. [CrossRef] [PubMed]
7. Pouyet, E.; Cotte, M.; Fayard, B.; Salome, M.; Meirer, F.; Mehta, A.; Uffelman, E.S.; Hull, A.; Vanmeert, F.; Kieffer, J.; et al. 2D X-ray and FTIR micro-analysis of the degradation of cadmium yellow pigment in paintings of Henri Matisse. *Appl. Phys. A* **2015**, *121*, 967–980. [CrossRef]
8. Voras, Z.E.; Deghetaldi, K.; Wiggins, M.B.; Buckley, B.; Baade, B.; Mass, J.L.; Beebe, T.P. ToF–SIMS imaging of molecular-level alteration mechanisms in Le Bonheur de vivre by Henri Matisse. *Appl. Phys. A* **2015**, *121*, 1015–1030. [CrossRef] [PubMed]
9. Mass, J.L.; Uffelman, E.; Buckley, B.; Grimstad, I.; Vila, A.; Delaney, J.; Wadum, J.; Andrews, V.; Burns, L.; Florescu, S.; et al. Portable X-ray Fluorescence and Infrared Fluorescence Imaging Studies of Cadmium Yellow Alteration in Paintings by Edvard Munch and Henri Matisse in Oslo, Copenhagen, and San Francisco. In *Smithsonian Contributions to Museum Conservation No. 5, The Noninvasive Analysis of Pianted Surfaces: Scientific Impact and Conservation Practice*; Doherty, T., Nevin, A., Eds.; Smithsonian Institution Scholarly Press: Washington, DC, USA, 2016; pp. 53–64.

10. Plahter, U.; Topalova-Casadiego, B. The Scream by Edvard Munch: Painting techniques and colouring materials. In *The National Gallery Technical Bulletin 30th Anniversary Conference Postprints*; Spring: London, UK, 2011; pp. 244–252.
11. Levin, B.D.; Nguyen, K.X.; Holtz, M.E.; Wiggins, M.B.; Thomas, M.G.; Tveit, E.S.; Mass, J.L.; Opila, R.; Beebe, T.; Muller, D.A. Detection of CdS Nanoparticles and Implications for Cadmium Yellow Paint Degradation in Edvard Munch's The Scream (c. 1910, Munch Museum). *Microsc. Microanal.* **2017**, *23*, 1910–1911. [CrossRef]
12. Monico, L.; Cartechini, L.; Rosi, F.; Chieli, A.; Grazia, C.; De Meyer, S.; Nuyts, G.; Vanmeert, F.; Janssens, K.; Cotte, M.; et al. Probing the chemistry of CdS paints in The Scream by in situ noninvasive spectroscopies and synchrotron radiation X-ray techniques. *Sci. Adv.* **2020**, *6*, eaay3514. [CrossRef] [PubMed]
13. Leone, B.; Burnstock, A.; Jones, C.; Hallebeek, P.; Boon, J.; Keune, K. The deterioration of cadmium sulphide yellow artists' pigments. In Proceedings of the ICOM-CC 14th Triennial Conference Preprints, The Hague, The Netherlands, 12–16 September 2005; Volume 2.
14. Van Der Snickt, G.; Janssens, K.; Dik, J.; De Nolf, W.; Vanmeert, F.; Jaroszewicz, J.; Cotte, M.; Falkenberg, G.; Van Der Loeff, L. Combined use of Synchrotron Radiation Based Micro-X-ray Fluorescence, Micro-X-ray Diffraction, Micro-X-ray Absorption Near-Edge, and Micro-Fourier Transform Infrared Spectroscopies for Revealing an Alternative Degradation Pathway of the Pigment Cadmium Yellow in a Painting by Van Gogh. *Anal. Chem.* **2012**, *84*, 10221–10228. [CrossRef] [PubMed]
15. Comelli, D.; MacLennan, D.; Ghirardello, M.; Phenix, A.; Patterson, C.S.; Khanjian, H.; Gross, M.; Valentini, G.; Trentelman, K.; Nevin, A. Degradation of Cadmium Yellow Paint: New Evidence from Photoluminescence Studies of Trap States in Picasso's Femme (Époque des "Demoiselles d'Avignon"). *Anal. Chem.* **2019**, *91*, 3421–3428. [CrossRef] [PubMed]
16. Fielder, I.; Bayard, M. Cadmium Yellows, Oranges and Reds. In *Artists' Pigments: A Handbook of Their History and Characteristics*; Feller, R.L., Ed.; Cambridge University Press: Cambridge, UK, 1986; pp. 65–108.
17. Kolkena, L.; Blok, V.; van der Berg, K.J.; Megens, L.; Keune, K.; van Loon, A. A phenomenological atlas of degraded cadmi-um yellow oil paint in paintings by Piet Mondrian and some of his contemporaries. In Proceedings of the ICOM-CC 17th Triennial Conference Preprints, Melbourne, Australia, 15–19 September 2014.
18. Van der Snickt, G.; Dik, J.; Cotte, M.; Janssens, K.; Jaroszewicz, J.; De Nolf, W.; Groenewegen, J.; Van der Loeff, L. Characterization of a Degraded Cadmium Yellow (CdS) Pigment in an Oil Painting by Means of Synchrotron Radiation Based X-ray Techniques. *Anal. Chem.* **2009**, *81*, 2600–2610. [CrossRef] [PubMed]
19. Hines, D.A.; Becker, M.A.; Kamat, P.V. Photoinduced Surface Oxidation and Its Effect on the Exciton Dynamics of CdSe Quantum Dots. *J. Phys. Chem. C* **2012**, *116*, 13452–13457. [CrossRef]
20. Katari, J.E.B.; Colvin, V.L.; Alivisatos, A.P. X-ray Photoelectron Spectroscopy of CdSe Nanocrystals with Applications to Studies of the Nanocrystal Surface. *J. Phys. Chem.* **1994**, *98*, 4109–4117. [CrossRef]
21. Bischof, T.; Ivanda, M.; Lermann, G.; Materny, A.; Kiefer, W.; Kalus, J. Linear and Nonlinear Raman Studies on CdSxSe1-x Doped Glasses. *J. Raman Spectrosc.* **1996**, *27*, 297–302. [CrossRef]
22. Krishnamurti, D. The Raman spectrum of rutile. *Proc. Math. Sci.* **1962**, *55*, 290–299. [CrossRef]
23. Laver, M. Titanium Dioxide Whites. In *Artists' Pigments: A Handbook of Their History and Characteristics*; West Fitz-Hugh, E., Ed.; Oxford University Press: Oxford, UK, 1997; pp. 295–355.
24. Koob, S. The Conservation and Restoration of Greek Vases: Loss Compensation for Publication and Museum Display. In *Konservieren Oder Restaurieren: Die Restaurierung Griechischer Vasen von der Antike bis Heute; Beihefte zum Corpus Vasorum Antiquorum Band III*; Verlag C. H. Beck: Munich, Germany, 2007; pp. 113–117.
25. Lafuenete, B.; Downs, R.T.; Yang, Y.; Stone, N. The power of databases: The RRUFF project. In *Highlights in Mineralogical Crystallography*; Armbruster, T., Danisi, R.M., Eds.; W. De Gruyter: Berlin, Germany, 2015; pp. 1–30. Available online: https://rruff.info (accessed on 9 June 2021).
26. Goldan, A.H.; Li, C.; Pennycook, S.J.; Schneider, J.; Blom, A.; Zhao, W. Molecular structure of vapor-deposited amorphous selenium. *J. Appl. Phys.* **2016**, *120*, 135101. [CrossRef]
27. Bube, R.H.; Thomsen, S.M. Photoconductivity and Crystal Imperfections in Cadmium Sulfide Crystals. Part I. Effect of Impurities. *J. Chem. Phys.* **1955**, *23*, 15–17. [CrossRef]

 heritage

MDPI

Article

Dyes of a Shadow Theatre: Investigating *Tholu Bommalu* Indian Puppets through a Highly Sensitive Multi-Spectroscopic Approach

Alessandro Ciccola [1], Ilaria Serafini [1,*], Giulia D'Agostino [2], Belinda Giambra [3], Adele Bosi [1,4], Francesca Ripanti [5], Alessandro Nucara [5], Paolo Postorino [5], Roberta Curini [1] and Maurizio Bruno [2]

1 Department of Chemistry, Sapienza Università di Roma, Piazzale Aldo Moro 5, 00185 Rome, Italy; alessandro.ciccola@uniroma1.it (A.C.); adele.bosi@uniroma1.it (A.B.); roberta.curini@uniroma1.it (R.C.)
2 Department of Biological, Chemical and Pharmaceutical Sciences and Technologies (STEBICEF), Università Degli Studi di Palermo, Viale Delle Scienze, Ed. 17, 90128 Palermo, Italy; giuliadagostino@outlook.com (G.D.); maurizio.bruno@unipa.it (M.B.)
3 Cultural Heritage Restorer, Department of Biological, Chemical and Pharmaceutical Sciences and Technologies (STEBICEF), Università Degli Studi di Palermo, Viale Delle Scienze, Ed. 17, 90128 Palermo, Italy; info@belindagiambra.it
4 Department of Earth Sciences, Sapienza Università di Roma, Piazzale Aldo Moro 5, 00185 Rome, Italy
5 Department of Physics, Sapienza Università di Roma, Piazzale Aldo Moro 5, 00185 Rome, Italy; francesca.ripanti@uniroma1.it (F.R.); alessandro.nucara@roma1.infn.it (A.N.); paolo.postorino@roma1.infn.it (P.P.)
* Correspondence: ilaria.serafini@uniroma1.it

Citation: Ciccola, A.; Serafini, I.; D'Agostino, G.; Giambra, B.; Bosi, A.; Ripanti, F.; Nucara, A.; Postorino, P.; Curini, R.; Bruno, M. Dyes of a Shadow Theatre: Investigating *Tholu Bommalu* Indian Puppets through a Highly Sensitive Multi-Spectroscopic Approach. *Heritage* **2021**, *4*, 1807–1820. https://doi.org/10.3390/heritage4030101

Academic Editor: Diego Tamburini

Received: 30 June 2021
Accepted: 3 August 2021
Published: 16 August 2021

Abstract: *Tholu Bommalu* are typical leather puppets of the traditional Indian shadow theatre. Two of these objects are part of a collection in the International Puppets Museum "Antonio Pasqualino" (Palermo, Sicily, Italy), which can count on one hundred-seventy-three of artifacts. These Indian puppets were investigated to obtain information related to the use of dyes for their manufacturing through a multi-technical approach exploiting the combination of highly sensitive spectroscopic techniques. Wet cotton stubbons were used to entrap small particles of dyes on the fibers from the art objects for the consequent analyses. Visible Light Micro-Reflectance spectroscopy was employed for the preliminary identification of the molecular class of dyes directly on the swabs, while Surface Enhanced Raman Scattering allowed the identification of the specific dye. Several synthetic dyes belonging to different typologies of coloring compounds were identified. The study resulted in an interesting overview of dyes used in recent *Tholu Bommalata* manufacturing through the combination of micro-invasive techniques directly on the sampling substrate.

Keywords: reflectance spectroscopy; SERS; synthetic dyes; *Tholu Bommalata*; puppets

1. Introduction

Tholu Bommalata is the traditional shadow theatre of the Telugu Indian states of Andhra Pradesh, Telangana, and Karnataka. The expression literally means "dance of leather puppets", from "atta", meaning dance, and "*Tholu Bommalu*", meaning leather puppets [1]. According to the literature, the use of these puppets dates back to 200 B.C. under the dynasty of Satavahana [2]. To realize a shadow theatre, puppeteers press the dolls behind a backlit screen, so the audience only see dancing shadows [3]. The main areas of storytelling are the Indian epics of *Ramāyana* and *Mahabarata*, and the sacred Hindū texts *Purāṇa*, even though nowadays, the epics are no longer narrated, replaced by contemporary themes such as reforestation or family life scenes. Dolls can be moved from anyone, but only a skillful *sutradhar* (literally "wire mover") can give them life. He is the leader of a familiar-run troupe, where everyone has a specific role: dancer, singer, narrator, and actor [4]. In all the Indian traditional puppet theatres, *Tholu Bommalu* are the biggest ones

(120–180 cm height), and those with the highest mobility have joints along the neck, arms, and legs [5]. Every puppet is made of leather, the origin of which was not arbitrary: at the beginning, deer was used to represent Gods, goat for saints or common people, and buffalo for demons, while today goat skin is the most used. Dolls are charged with an enormous spiritual value. Indeed, they are made of leather, an impure material for Indians, and, thus, they need a purification rite to represent sacred characters: several weeks before the performance, plenty of rituals are made to offer the puppets to Gods. At the end of each performance, people let them go to the banks of Gange River, as a real funeral [1]. According to sources [3], the first puppeteers used to paint *Tholu Bommalu* with natural colors mixed with water, later replaced by synthetic dyes. Today, *Tholu Bommalata* is a kind of dying art: fifty years ago, more than 180 troupes were active in 30 different Indian districts; today there are only 9.

Here, we studied the two *Tholu Bommalu* (Figure 1), made in 1978 by the Ramana Murthy theatre, which are part of a 173-artifact collection stored at the International Puppets Museum "Antonio Pasqualino" in Palermo, Sicily, Italy. They represent Prince Rāma and Princess Sītā, the main characters of *Ramāyana.*

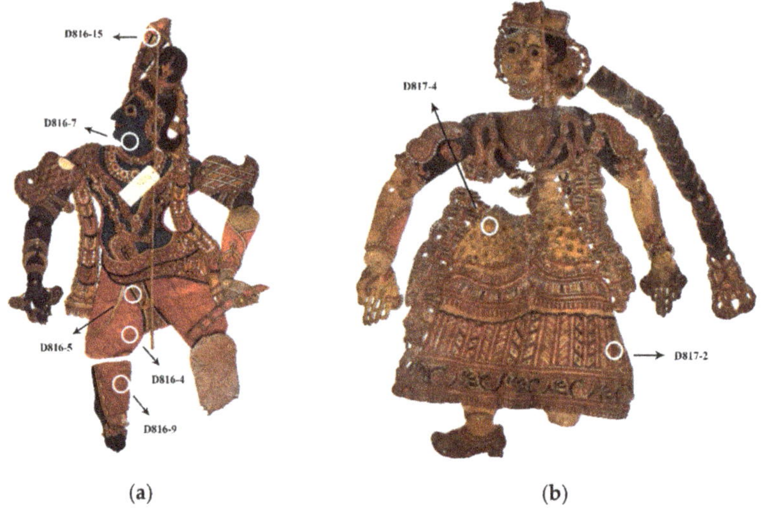

Figure 1. Sampling areas from *Tholu Bommalu* of Rāma (**a**) and Sītā (**b**).

The Rāma puppet (Figure 1a), as seventh reincarnation of Visnu, is pictured in profile and with blue skin: both aspects represent divine characters. It also has almond-shaped eyes, a big nose, and thin lips; it is entirely decorated with rich jewels that underline his social status, everything made with pink, red, blue, violet, and black dyes. Both arms and legs were probably added to the puppet in a second stage, replacing missing parts: they indeed show different decorations and colors—blue and violet—than the rest of the puppet.

The Princess Sītā puppet (Figure 1b), as a not divine character, is pictured frontally, and entirely made with black, red, and yellow dyes. As with the other puppet, it has almond-shaped eyes, a big nose, but thicker lips. She wears a dress full of flower decorations and rings in her hands and nose. They both wear *tilaka* on their forehead, a sacred sign, the mark of God and symbol that purifies the body.

The Sītā and Rāma *Tholu Bommalu* are made of goat skin, identified by the Globe Institute of Copenhagen using ZooMS, short for Zoo Archaeology by Mass Spectrometry. It is a technique that uses the slow evolution of collagen as a molecular barcode to read the identity of leather or bones. The method uses a well-established approach, peptide mass fingerprinting, allied to high throughput Time of Flight Mass Spectrometry. Samples are identified by differences in the mass of the peptides, which arise as a result of sequence differences between species [6]. Each part of the body is a separated piece of leather, one linked to the other by a thin string, observed and identified as cotton by a light microscope [7]. Between the junctions, there is a little piece of leather that make this part more resistant in the points of greatest fragility. To be held by *suthradar*, they are assembled on a wooden stick, recognized as cane wood by light microscope observation [8]. They are full of holes of different shapes and sizes marking every decoration line: light can pass through them during the performance, producing an incredible effect of lights and shadows. Preparatory drawing is carved on one side of the leather, while both sides are colored with a translucent effect. There is a similar varnish layer upon the surface on both the sides: according to the literature [3], it could be neem or coconut oil, used by puppeteers to preserve dolls from time damages.

As mentioned above, several colors were used to achieve the brilliant shades present in the puppets. During the conservation study of these objects, it was noticed that the related colorants were sensitive to water in terms of solubility. Consequently, it was evaluated that organic dyes could be present and their identification could be very interesting: with reference to the recent origin of the dolls, synthetic dyes could be used and their characterization could confirm the transition from natural dyes to synthetic ones in *Tholu Bommalata* production. For this aim, a multi-technical approach was dedicated to their identification and characterization. Visible light Fiber Optic Reflectance Spectroscopy [9–12], coupled to an optical microscope (micro-FORS), was employed to obtain preliminary information about the class of dyes: reflectance spectra were acquired directly on stubbons used to sample to dyes during the conservation treatments. The same samples were then analyzed with Surface Enhanced Raman Scattering (SERS) [13–18], which is based on the great enhancement of the Raman signal of an analyte in close proximity of a metal nanostructured substrate, allowing it to obtain vibrational spectra that, otherwise, would be affected by a dramatic fluorescence background. In the last decades, SERS spectroscopy has showed its great analytical potentiality [19], and it is nowadays increasingly used for the characterization of colorants in art objects. Starting from its first application on ancient textiles [20], this technique has been used for the identification of natural and synthetic dyes in paintings and dyed objects [13,21,22]. The evolution from simple metal colloids to nanostructured SERS substrates [23–28] and metal-nanoparticles loaded sampling devices [29–34] provided several analytical strategies, which were exploited for the detection of both natural and synthetic dyes [14,16,18,23,35–38]. The main advantage of SERS is highly represented by high sensitivity, which is in charge for lower detection limit than High Performance Liquid Chromatography, while the main drawback is the low reproducibility which could derive from local interaction between the nanostructure and mixtures of analytes through different functional groups.

Here, the combination of SERS and FORS data allowed to individuate several synthetic dyes used for the manufacturing of these objects and provided new data for consequent further analyses.

2. Materials and Methods

2.1. Samples

Two leather puppets from the *Tholu Bommalu* collection stored at the International Puppets Museum "Antonio Pasqualino" were studied. Seven samples were collected by rubbing mildly sterile stubbons on the surface of the puppets, wet with distilled water, five from Rāma (D816-4, D816-5, D-816-7, D816-9, D816-15) (Figure 1a) and two from Sītā (D817-2, D817-4) (Figure 1b).

2.2. Visible Light Micro-Reflectance Spectroscopy Analysis

Visible Light Reflectance spectra were acquired through a BWTEK fiber optic spectrophotometer Exemplar LS, coupled to a microscope through a fiber optic cable and a specific adaptor. Sampling stubbons were observed at the microscope in order to individuate the most colored areas and, under illumination with an external light source (Tungsten lamp), visible light reflectance spectra were acquired. Three spectra were recorded for every stubbon (integration time: 4 s; scans: 5; integration step: 1 nm) and an average spectrum was obtained. For the spectral assignment, first derivative and apparent absorption (Log(1/R) conversion, where R is the Reflectance) analyses of the spectra were performed to confirm information deriving from simple reflectance spectrum [9–11,39]. In order to minimize interference of the cotton substrate, spectra of the stubbon were acquired where no dye particles or spots were present, and after processing, the deriving apparent absorption spectrum was subtracted, as a blank, from the sample ones. All the spectra were attributed through comparison to reference materials spectra, literature, and databases [9–11,40].

2.3. Raman and SERS Analysis

Conventional Raman and SERS experiments were performed with a Horiba Jobin-Yvon HR-Evolution spectrometer equipped with a microscope and a 632 nm laser. A motorized mapping stage was used for inspecting the sample and collecting the Raman signal from specific locations on the sample.

For conventional Raman measurements, a preliminary collection of spectra was carried out. Set conditions for the spectra acquisition were: 100× objective magnification and the laser intensity varied between 0.15 and 15 mW in order to maximize the Raman signal and to observe clear features of the investigated compounds. The Ag-reduced colloid was prepared according to the protocol developed by Leopold and Lendl [41]. Briefly, a solution of $AgNO_3$ 1×10^{-3} M in MilliQ Water was prepared. Separately, the same volumes of a solution of $NH_2OH{\cdot}HCl$ 6×10^{-2} M and a solution of NaOH 1×10^{-1} M were mixed together. Ten mL of $NH_2OH{\cdot}HCl$ solution was added to 100 mL of $AgNO_3$ solution under stirring, with direct formation of a colloid. The colloid was left under stirring for 20 min, and it was used after its production. Aggregation of the colloid was induced by dropping 20 μL of 0.01 M $MgSO_4$ solution: 200 μL of Ag colloid was inserted in an Eppendorf tube and 20 μL of 0.01 M $MgSO_4$ solution was added and stirred to induce nanoparticle aggregation. The aggregated colloid was poured on some colored fibers sampled from the stubbons, and it was left to dry. SERS spectra were acquired in correspondence of Ag nanoclusters close to the stubbon fibers or on them. In order to evaluate eventual spectral interferences deriving from the colloid, spectra of the blank (200 μL of Ag colloid and 20 μL of $MgSO_4$) were also collected in the dried form. Set conditions for the SERS spectra acquisition were: 50× objective magnification, laser intensity varied between 0.15 and 0.38 mW according to the sample, maximum accumulation time of 5 s per scan, 60 scans maximum. Generally, six spectra for every typology of analyzed sample were acquired. For the spectral band assignment and compound identification, experimental spectra were compared to databases and literature [42–54].

3. Results

3.1. Visible Light Micro-Reflectance Spectroscopy Analysis

Sample D816-4, which derives from a pink-reddish area, presents a characteristic reflectance spectrum (Figure S1a, Supplementary Materials) where a first weak maximum is visible at 466 nm, followed by an increase of reflectance starting at 540 nm; the sigmoid curve presents a marked inflection point at 581 nm, while another one is observable at 690 to 695 nm. A less-defined inflection point should be present at 432 nm, but the individuation is not easy due to the higher noise at around 400 nm (lower intensity of the lamp emission). The related apparent absorption spectrum (Figure S1b, Supplementary Materials) presents a maximum at ~550 nm, with a shoulder at ~510 nm. A lower intensity band is observable at ~670 nm. The sample D816-9 presents similar features (Figure S2a, Supplementary Materials): the reflectance is the same in the 400–500 nm range, but it starts increasing around 510 nm with a complex trend. Indeed, several inflection points are observable, as highlighted from the first derivative spectrum. In particular, an inflection point is still visible at 560–580 nm, while, from the first derivative, the other ones should be present at 425, 617, and 648 nm, confirming the complexity of the spectrum. In the apparent absorption spectrum (Figure S2b, Supplementary Materials), there is a variation in intensity for the bands at 510 and 550 nm: the first one is more intense than the second, while a new low intensity band is observable ~600 nm. The band at 670 nm is not visible, whereas a new band is visible at ~690 nm.

In addition, for the sample D816-7 (Figure 2a), corresponding to a green-blue area of Rāma's skin, the reflectance spectrum obtained complex results: two weak reflectance maxima are observable around 480 and 600 nm, while an increase of reflectance is observable over 615 nm, with an inflection point at 669 nm. With reference to the derivative spectrum, another inflection point is present at 651 nm, while less defined ones could be present at 431 and at 500 to 525 nm (the second results in a very flat maximum in the derivative spectrum). The corresponding apparent absorption spectrum presents a main maximum at ~643 nm, with a minor flat band around 500 nm (Figure 2b).

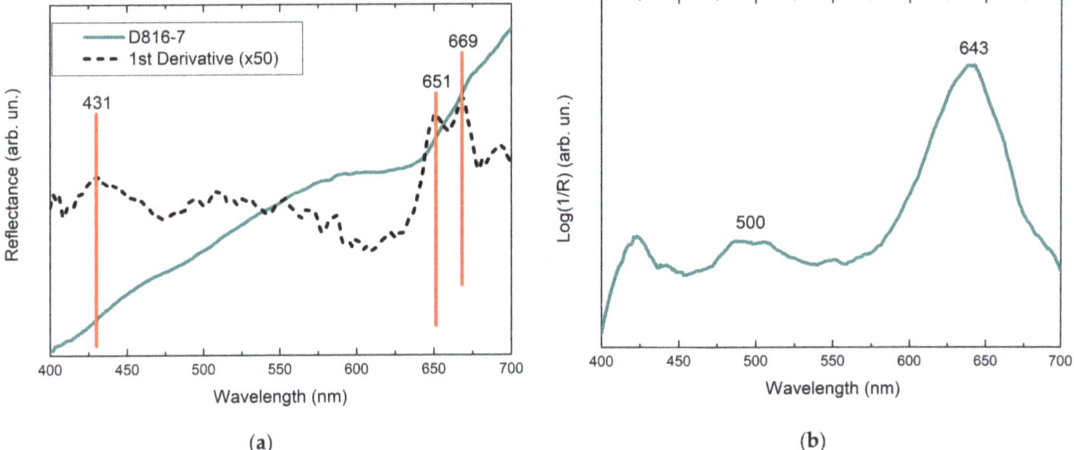

Figure 2. (**a**) Visible light reflectance spectrum obtained for D816-7 sample with the corresponding first derivative (red lines highlight the inflection points) and (**b**) related apparent absorption spectrum (background subtracted).

The reflectance spectra obtained for the samples D816-5 and D816-15 do not present informative features (Figures S3 and S4, Supplementary Materials): the reflectance is very low, increasing without any evident spectral feature, and the spectral noise is remarkable. The related first derivative spectra are characterized by continuous oscillations and high noise, not allowing for a clear identification of the inflection points. The apparent absorption results in a spectrum where the presence of characteristic signals is not evident. Only a very low intensity band at around 690 nm is observable after background subtraction, but it cannot be excluded that the processing could be responsible for the artifacts in this case. The absence of clear absorption signals could be in agreement with the black color of the extracted dye.

For the Sītā puppet, the FORS spectrum acquired for the D817-2 sample presents a first reflectance increase over 410 nm, reaching a maximum at around 470 nm and followed by a great increase from 525 nm. The reflectance slope decreases over 600 nm, while, in the first derivative spectra, the inflection points are observable at 424, 570, and 655 nm (Figure 3a). In the apparent absorption spectrum, a defined band is centered at 530 nm, with a broadening at around 490 nm, while a low intensity broad band is visible at 680 nm (Figure 3b).

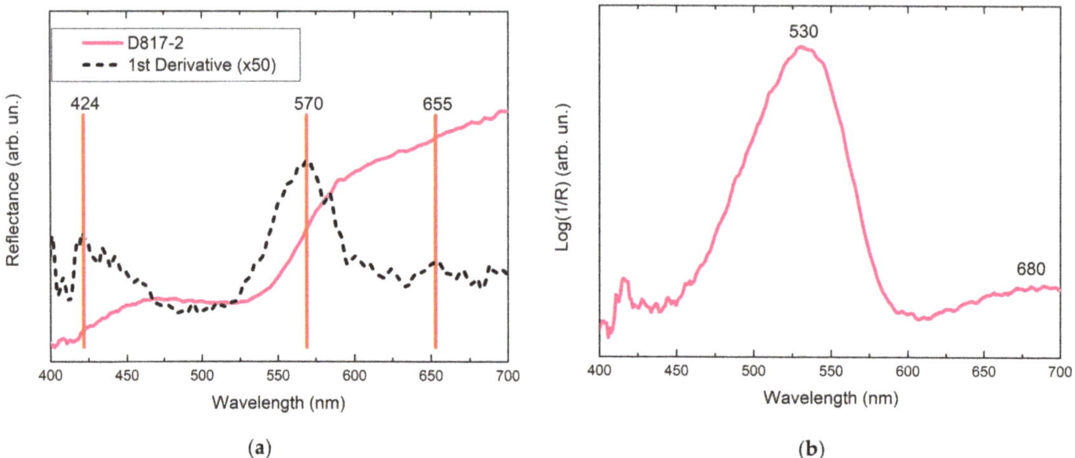

Figure 3. (**a**) Visible light reflectance spectrum obtained for the D817-2 sample with the corresponding first derivative (red lines highlight the inflection points) and (**b**) related apparent absorption spectrum (background subtracted).

The D817-4 FORS spectrum presents a more complex behavior: the reflectivity (Figure S5a, Supplementary Materials) increases over 403 nm, the slope seems to decrease around 470 nm only to grow again over 480 nm; a second decrease of the slop is observable at 640 nm, but the reflectivity increases again to over 650 nm. The first derivative spectrum highlights an inflection point at 440 nm and a second one at 550–560 nm, but the features of the obtained derivative spectrum are less clear in comparison to the previous case. In the apparent absorption spectrum, two broad bands at 506 nm (with shoulder ~540 nm) and at 674 nm are observable (Figure S5b, Supplementary Materials).

3.2. Raman and SERS Analysis

Preliminary Raman analyses were performed directly on the samples focusing on dye spots or particles. However, the fluorescence background did not allow acquiring meaningful spectra. The collection of spectra at the highest intensity of the laser was attempted, but this only resulted in degradation of the sample with the formation of amorphous carbon or in the individuation of signals of the cellulose from the stubbons (Figure S6, Supplementary Materials).

With respect to the SERS spectra obtained for sample D816-4, it was possible to distinguish some common characteristic peaks (Figure 4, top). In particular, in most spectra, two broad bands around 461 and 477 cm^{-1} are observable, while more defined peaks appear at 622, 680, and 735 cm^{-1}. At higher wavenumbers, two peaks are observable at 1004 and 1049 cm^{-1}, and other bands are distinguishable at 1213, 1255, 1301, 1329, 1364, 1436, 1461, 1484, 1509, 1565, 1577, 1582, 1601, 1623, and 1646 cm^{-1}. These signals change in intensity and width in the different spectra.

All the spectra of the D816-9 sample (Figure S7, Supplementary Materials) present peaks at 452 (shoulder), 479, 680, 1002, 1049, 1207, 1447, 1485, 1578, and 1623 cm^{-1}. It is interesting to notice that several bands observed for the D816-9 sample are also present in the spectra of the D816-4 sample.

In the case of sample D816-7, instead, the acquired spectra (Figure 5, top) present high reproducibility: there are peaks at 436, 457, 479, 530, 620, 644, 677, 693, 732, 761, 804, 827, and 918 cm^{-1}, and defined signals at around 1030, 1176, 1221, 1298, 1364, 1399, 1429, 1476 (shoulder), 1530, 1586, and 1621 cm^{-1} are clearly observable. In two spectra, further sharp bands are distinguishable at 546, 601, 1488, and 1576 cm^{-1}.

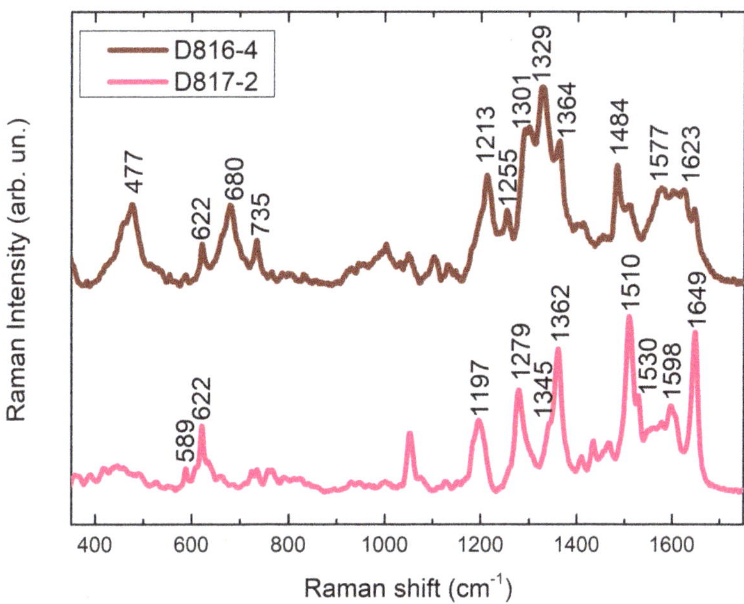

Figure 4. Comparison of SERS spectra obtained for the D816-4 (top) and D817-2 (bottom) samples. Some peak wavenumbers are evidenced for clarity. The spectrum of sample D817-2 presents a clear correspondence with Rhodamine B SERS spectrum [50] (literature spectrum not reported in the image).

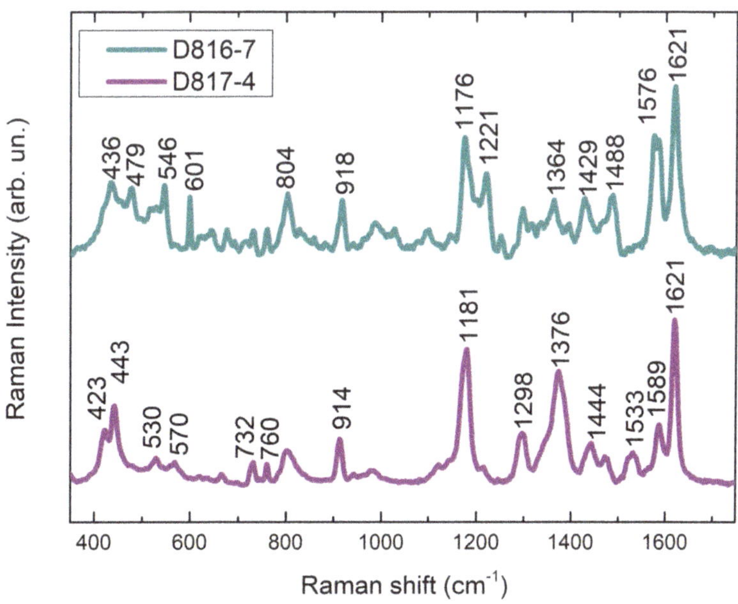

Figure 5. Comparison of SERS spectra obtained for the D816-7 (top) and D817-4 (bottom) samples. Some peak wavenumbers are evidenced for clarity. The spectra of samples presents a clear correspondence with the SERS spectra of Malachite Green [44,46] (D816-7) and Crystal Violet [44,46,54] (present in both the samples, see Discussion; literature spectra not reported in the image).

The SERS spectra obtained for D816-5 sample present a good reproducibility, even if their relative intensities can change in the different spectra. In most of them, peaks at 479, 656, 691, 733, 1139, 1225, 1300, 1340, 1443, 1458, 1585, and 1628 cm^{-1} are present. In some spectra, further peaks can be identified: one spectrum presents bands at 434, 518, 570, 716, 810, 1001, 1207, and 1315 cm^{-1}, which result generally sharper than the above-mentioned ones (Figure 6, top), while in another spectrum, intense signals at 798, 1404, and 1539 cm^{-1} are observable. For the D816-15 sample, the spectra (Figure 6, bottom) present a general reproducibility, even if, also in this case, change in relative intensities are observed; the main spectral features are at 422, 440, 479, 519, 562, 593, 679, 717, 813, 857, 914, 1003, 1033, 1047, 1130, 1183, 1209, 1252, 1323, 1335, 1445, 1458 (shoulder), 1531, 1586, and 1627 cm^{-1}. Some similarities with the spectra of sample D816-5 are evident.

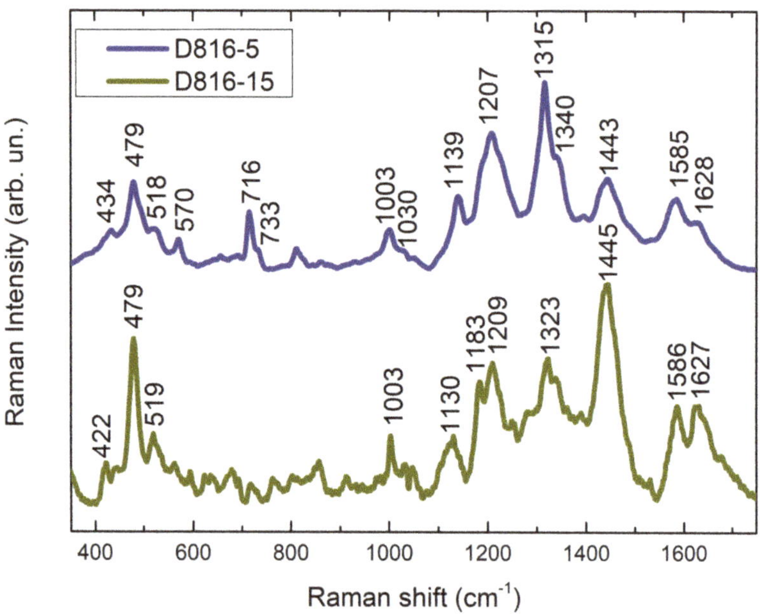

Figure 6. Comparison of SERS spectra obtained for the D816-5 (top) and D816-15 (bottom) samples.

About the samples from Sītā puppet, the SERS spectra acquired show a great reproducibility. For the D817-2 sample, all the spectra present distinguishable peaks at 589, 622, 634 (shoulder), 661, 726, 737, 768, 1050, 1129, 1185 (shoulder), 1197, 1279, 1345 (shoulder), 1362, 1437, 1471, 1510, 1530, 1578, 1598, and 1649 cm^{-1} (Figure 4, bottom). The SERS spectra of the D817-4 sample (Figure 5, bottom), instead, present reproducible bands at 423, 443, 530, 570, 622, 636, 665, 732, 745 (shoulder), 760, 802, 834 (shoulder), 914, 944, 977, 1124, 1181, 1217, 1298, 1345 (shoulder), 1376, 1444, 1473, 1524, 1533, 1565, 1589, and 1621 cm^{-1}. The general pattern of this spectra presents high similarity with the spectra obtained for the D816-7 sample.

4. Discussion

The main results obtained from data analysis are recapped in Table 1. From the FORS and SERS results, different typologies of synthetic dyes were undoubtedly used for the puppets' manufacturing. In some cases, the identification of the present dyes is straightforward: for the D817-2 sample, the apparent absorption shoulder at 490 nm and the maximum at 530 nm are attributable to a xanthene pink dye [11,40]. Among the several colorants belonging to this family, which includes Eosin, Rhodamine, and Phloxine, the SERS spectrum (Figure 5, bottom) allows a clear identification of Rhodamine-based dyes [42,49,50].

Moreover, among the two main dyes of this typology (Rhodamine B and Rhodamine 6G), the acquired spectra present higher similarities with the Rhodamine B spectra reported in the literature [50]. Analogously, the maximum at 640 nm observable in the apparent absorption spectra of the D816-7 sample suggests the presence of a triphenylmethane dye, such as Malachite Green and Patent Blue V [11]. The SERS spectra obtained for this sample confirm these preliminary data, because both their general patterns and the Raman bands are characteristic of this family of dyes. However, it is difficult to identify the specific dye: the experimental spectra, indeed, show great similarities with both Crystal Violet and Malachite Green dyes, both belonging to the same class [44]. In addition, Crystal Violet's SERS spectrum is very similar to the Methyl Violet one; these three dyes should be thus considered for the attribution [54,55]. With reference to the acquired spectra (Figure 6, top), some bands (732, 760 cm^{-1}) suggest the presence of Crystal Violet, while others (1176, 1221, 1364, 1399 cm^{-1}) suggest the presence of Malachite Green. Taking into account the FORS spectra, a possible hypothesis consists in the identification of Malachite Green as the principal dye, with minor amounts of Crystal Violet (which, although in a minor quantity, is reported to be more easily detectable through SERS spectroscopy in comparison to Malachite Green) [44].

Table 1. Assignment of the dyes for the different samples, with reference to their appearance and their spectral features.

Sample	Color of the Sampled Area	Appearance of Sample on the Stubbon	Apparent Absorption Bands (nm)	SERS Peaks (cm^{-1})	Dye Assignment
D816-4	Pinkish red	Red stain	501, 550	622, 680, 735, 1364, 1436, 1509, 1565, 1582, 1601, 1646 (Rhodamine B); 637, 714, 1329, 1452, 1577, 1623 (Eosin Y)	Rhodamine B, Eosin Y
D816-5	Black	Black spot	No main feature	See Discussion	Mixture of dyes (?)
D816-7	Blue-green	Black-greenish particles	640	1221, 1364, 1399, 1429 (Malachite Green); 732, 761, 1476, 1530 (Crystal or Methyl Violet); further intense peaks are present but common to both the dyes (see Discussion)	Malachite Green, Crystal (or Methyl) Violet
D816-9	Pinkish red	Red stain	510, 550	479, 1184, 1623	Eosin Y (?)
D816-15	Black	Black spot	No main feature	See Discussion	Mixture of dyes (?)
D817-2	Red	Red stain	490 (shoulder), 530	622, 768, 1129, 1185 1197, 1279, 1345, 1362, 1437, 1510, 1530, 1578, 1598, 1649	Rhodamine B
D817-4	Yellow with black decorations	Pale orange with some black particles	506, 540 (Mono-azo dye)	423, 443, 530, 570, 732, 760, 802, 914, 944, 1181, 1298, 1376, 1444, 1473, 1533, 1589, 1621 (Crystal or Methyl Violet)	Mono-azo dye (?), Crystel (or Mehyl) Violet

The presence of a triarylmethane dye is also suggested for the D817-4 sample, but, in this case, the identification of the molecule is clear: the SERS spectrum (Figure 5, bottom) presents a fulfilling matching with the SERS spectrum of Crystal Violet (or Methyl Violet), and there is no ambiguity with other compounds of the same class (e.g., fuchsine based dyes) [42,54], whose characteristic peaks are not evident in the spectra. Nevertheless, there is an actual contradiction with the FORS results: the main apparent absorption band of Crystal Violet should be at 594 nm, but it is not visible in the experimental spectrum. According to the literature, the main matching of the apparent absorption spectrum bands at 506 and 540 nm occurs with the spectra of orange mono-azo dyes, for instance, Lithol Fast Scarlet RNP or Orange II [11]. A possible hypothesis to explain this unusual behavior could be formulated with reference to the sampling area. The stubbon was used to sample the dye from a part of the puppet with an orange-yellow background, and this is in agreement with the bands observed in the FORS spectra. However, this area is decorated with a dark motif. Considering the high SERS detectability of Crystal and Methyl Violet, the minor amount of this dye present in the decoration represents the molecules effectively detected with this technique, while the orange dye (likely a mono-azo) could be not visible for a lower affinity to the silver nanoparticles or for the lower Raman cross-section.

For the remaining samples, the identification of the present dyes is less easy, but some hypotheses can be formulated. The FORS spectra obtained for the D816-4 sample suggest the presence of a xanthene dye: in particular, the apparent absorption bands at 501 and 550 nm present a good matching with the Visible Light absorption of Rhodamine B. Moreover, one of the SERS spectra (Figure 4, top) presents some peaks at 622, 735, 1051, 1364, 1510, and 1645 cm^{-1} (observable also in the spectrum of D817-2, with eventual reduced shifts in wavenumbers), which can confirm the presence of this colorant [50]. Other bands at 637, 714, 1329, 1452, 1577, and 1623 cm^{-1} could be indicative of Eosin Y [14,43,47,48]. For the D816-9 sample, even if some signals confirm the presence of Eosin Y (479, 1184, 1623 cm^{-1}), the lower quality of the SERS spectra, along with the change in intensity in the apparent absorption spectra, does not allow for inferring any plausible hypothesis on the composition of the area. However, two samples derived from close areas of the same color and, thus, a similar composition is expected.

The two last samples, D816-5 and D816-15, represent the most complex ones for the interpretation. The FORS spectra are not really informative because of the low and noisy reflectivity spectra and consequent artifacts in the apparent absorption spectra. However, the SERS spectra (Figure 6) present several affinities, thus allowing to hypothesize a similar composition. With reference to the literature, it is hard to individuate a dye whose reported spectrum completely matches the experimental ones. Moreover, the variation in intensity of some peaks in different spectra suggests that several dyes could be present in mixture. For instance, peaks at 658, 1003, 1030, 1130, 1183, 1209, 1585, and 1627 cm^{-1} could be indicative of the presence of Sudan Black B [51,52], even if the relative intensity of bands does not match the literature spectrum of this dye [51,52]. On the other side, peaks observable in some spectra at 570, 733, 1323, 1443, and 1585 cm^{-1} could be indicative of the use of carminic acid dye [15,53], even if some characteristic signals are missing. Moreover, if Sudan Black B could represent the main colorant of the sampled black area, the carminic acid could not explain its color. In order to verify and confirm these hypotheses, a separative analytical technique should be used to discriminate the different colorants (for instance, High Performance Liquid Chromatography coupled to Photo-Diode-Array HPLC or High Resolution Mass Spectrometry).

From the analytical point of view, the combination of two laboratory spectroscopic techniques resulted useful for the identification of the present dyes. However, even if the micro-FORS provided preliminary hints, we remark that the acquisition of spectra on sampling stubbons resulted critical. If the acquired spectra resulted highly significant for the identification in some samples (e.g., D816-7, D817-2), the individuation of marker bands was not easy for other ones, especially if the concentration of the sampled dye was not high and the interference from the background—likely due to the fibrous stubbon matrix—was remarkable. For instance, in the case of the D816-5 and D816-15 samples, no main spectral feature is clearly observable. Nonetheless, the general performances obtained for the micro-FORS approach applied to the sampling stubbons suggest that an on-site FORS analysis would have provided remarkable results: the higher concentration of dyes on the objects and the absence of background interference from the stubbon would represent improvement factors for the acquisition of informative Vis-Light reflectance spectra. In this study, it was not possible to acquire FORS spectra in situ, however, its application presents a high potential not only for its non-invasiveness but also for the quality of obtained data. On the other side, the stubbon sampling was fundamental for the vibrational spectra collection. Indeed, the acquisition of on-site Raman spectra would likely not be useful for similar matrices. The portable Raman spectrometers are, in general, less performative than benchtop instruments, especially when highly fluorescent analytes are present, as in this case. Moreover, as highlighted by the preliminary measurement, the conventional Raman spectra resulted not useful for the identification of the synthetic dyes, and the micro-invasive addition of SERS colloid was fundamental in obtaining high quality vibrational spectra. Even in the case of samples containing Malachite Green and Crystal Violet, whose standard Raman spectra are reported in the literature, it was not possible to detect any clear spectral feature attributable to the dye. Probably, the high fluorescence of these dyes could be increased by the presence of degradation products and impurities from the art object. Moreover, focusing directly on the dye particles on the stubbons was more difficult in comparison to focusing on Ag nanoclusters close to them, because the fibers are more susceptible to move under the lens, with consequent loss of focus. Actually, for similar historical objects where the presence of synthetic dyes is highly probable due to the manufacturing period, a sampling could be necessary, but the high sensitivity of SERS spectroscopy allows for minimizing the number of sampled analytes (or the amount of analyzed sample). In order to maximize the results achievable through an integrated spectroscopic approach, the utilized protocol, enriched by a preliminary on-site FORS analysis, is suggested for similar matrices with synthetic dyes to address the following chromatographic analyses and when it is not possible to sample fragments for the analysis with low sensitivity methodologies.

5. Conclusions

In this study, a whole spectroscopic approach involving FORS and SERS spectroscopies was employed in order to identify different synthetic dyes used for the manufacturing of *Tholu Bommalu* leather puppets from the 1970s. The two techniques were directly applied on sampling stubbons, used for the extraction of the dyes from the objects of interest. This approach resulted successful for the identification of coloring molecules in some parts of the puppets: indeed, the data separately obtained from the two techniques reciprocally confirmed the attributions (Malachite Green and Crystal/Methyl Violet for the D816-7 sample, Rhodamine B for D817-2). In other cases (D816-4, D817-4) the use of SERS after FORS analyses provided further data and allowed to draw a richer picture on the composition of the materials: for the D816-4 sample, Rhodamine B was found in probable mixture with Eosin Y, while for the D817-4 sample, a combination of an azo dye for the yellow part and Crystal/Methyl Violet for the dark decorations was hypothesized. Finally, in the case of some areas (the D816-5, D816-9, and D816-15 samples), the analyses provided preliminary hypotheses about the colorants, which are likely present in mixture. The achieved results are of paramount importance in the perspective of selecting the proper methodology for a consequent separative analysis, which could confirm the results obtained from the spectroscopic approach and solve the doubts regarding the whole composition of complex mixtures of dyes. Moreover, the reported experiments confirmed the transition from natural dyes to synthetic ones in the manufacturing of *Tholu Bommalu* puppets.

Supplementary Materials: The following are available online at https://www.mdpi.com/article/10.3390/heritage4030101/s1, Figure S1: (a) Visible light reflectance spectrum obtained for the sample D816-4 with the corresponding first derivative (red lines highlight the inflection points) and (b) related apparent absorption spectrum (background subtracted); Figure S2: (a) Visible light reflectance spectrum obtained for the sample D816-4 with the corresponding first derivative (red lines highlight the inflection points) and (b) related apparent absorption spectrum (background subtracted); Figure S3: (a) Visible light reflectance spectrum obtained for the sample D816-5 with the corresponding first derivative (red lines highlight the inflection points) and (b) related apparent absorption spectrum (background subtracted); Figure S4: (a) Visible light reflectance spectrum obtained for the sample D816-15 with the corresponding first derivative (red lines highlight the inflection points) and (b) related apparent absorption spectrum (background subtracted); Figure S5: (a) Visible light reflectance spectrum obtained for the sample D817-4 with the corresponding first derivative (red lines highlight the inflection points) and (b) related apparent absorption spectrum (background subtracted);

Figure S6: Conventional Raman spectra obtained for the samples in the range 350–1050 cm^{-1} (top) and in the range 1050-1750 cm^{-1} (bottom); Figure S7: SERS spectrum obtained for D816-9 sample.

Author Contributions: Conceptualization: A.C., I.S., A.N., B.G. and M.B.; investigation: A.C. (Raman and SERS spectra, interpretation of FORS spectra), A.B. (FORS spectra), A.N. (interpretation of FORS spectra); resources, P.P. and M.B.; main writing—preparation of original draft, A.C., I.S. and G.D.; writing—review and editing, A.C., I.S., G.D. and F.R.; supervision: R.C., P.P. and M.B. All authors have read and agreed to the published version of the manuscript.

Funding: This research received no external funding.

Institutional Review Board Statement: Not applicable.

Informed Consent Statement: Not applicable.

Data Availability Statement: Not applicable.

Acknowledgments: We are grateful to the International Puppets Museum "Antonio Pasqualino" in Palermo, Sicily, Italy, for the granted access to the historical objects and the collaboration on this research.

Conflicts of Interest: The authors declare no conflict of interest.

References

1. Sahapedia. Available online: https://www.sahapedia.org/ (accessed on 29 June 2021).
2. Autiero, S. Danzando Nella Luce. Il Teatro Delle Ombre Nell'India Meridionale. In *Il Principe e la sua Ombra*; National Museum of Oriental Art "G. Tucci": Roma, Italy, 2013; p. 41.
3. Chandra Sekhar, A. A Selected Crafts of Andhra Pradesh. In *Census of India 1961*; The Government Press: New Delhi, India, 1961.
4. Sarma, D.; Homen, A. *Storytelling and Puppet Traditions of India*; Indira Gandhi National Centre for the Arts: New Delhi, India, 2010.
5. Sorensen, N. Tolu Bommalu Kattu: Shadow Theater. *J. South Asian Nat. His.* **1975**, *10*, 1–19.
6. Collins, M.; Buckely, M.; Grundy, H.; Thomas Oates, J.; Wilson, J.; van Doorn, N. ZooMS: The Collagen Barcode and Fingerprints. *Spectrosc. Eur.* **2010**, *22*, 6–10.
7. Campo, G.; Bagan, R.; Oriols, N. *Identificaciò de Fibres: Suports Tèxtils de Pintures*; Genelaitat de Catalunya, Departament de Cultura i Mitjans de Comunicació: Barcelona, Spain, 2009.
8. Ollendorf, A.L.; Mulholland, S.C.; Rapp, G.J. Phytolith Analysis as a Means of Plant Identification: Arundo Donax and Phragmites Communis. *Ann. Bot.* **1988**, *61*, 209–214. [CrossRef]
9. Cosentino, A. FORS Spectral Database of Historical Pigments in Different Binders. *E Conserv. J.* **2015**, 54–65. [CrossRef]
10. Aceto, M.; Agostino, A.; Fenoglio, G.; Idone, A.; Gulmini, M.; Picollo, M.; Ricciardi, P.; Delaney, J.K. Characterisation of Colourants on Illuminated Manuscripts by Portable Fibre Optic UV-Visible-NIR Reflectance Spectrophotometry. *Anal. Methods* **2014**, *6*, 1488–1500. [CrossRef]
11. Montagner, C.; Bacci, M.; Bracci, S.; Freeman, R.; Picollo, M. Library of UV-Vis-NIR Reflectance Spectra of Modern Organic Dyes from Historic Pattern-Card Coloured Papers. *Spectrochim. Acta Part A Mol. Biomol. Spectrosc.* **2011**, *79*, 1669–1680. [CrossRef]
12. Bacci, M.; Picollo, M.; Trumpy, G.; Tsukada, M.; Kunzelman, D. Non-Invasive Identification of White Pigments on 20Th-Century Oil Paintings by Using Fiber Optic Reflectance Spectroscopy. *J. Am. Inst. Conserv.* **2007**, *46*, 27–37. [CrossRef]
13. Pozzi, F.; Leona, M. Surface-Enhanced Raman Spectroscopy in Art and Archaeology. *J. Raman Spectrosc.* **2016**, *47*, 67–77. [CrossRef]
14. Cesaratto, A.; Leona, M.; Pozzi, F. Recent Advances on the Analysis of Polychrome Works of Art: SERS of Synthetic Colorants and Their Mixtures with Natural Dyes. *Front. Chem.* **2019**, *7*, 1–12. [CrossRef]
15. Chen, K.; Leona, M.; Vo-Dinh, K.C.; Yan, F.; Wabuyele, M.B.; Vo-Dinh, T. Application of Surface-Enhanced Raman Scattering (SERS) for the Identification of Anthraquinone Dyes Used in Works of Art. *J. Raman Spectrosc.* **2006**, *37*, 520–527. [CrossRef]
16. Candela, R.G.; Lombardi, L.; Ciccola, A.; Serafini, I.; Bianco, A.; Postorino, P.; Pellegrino, L.; Bruno, M. Deepening inside the Pictorial Layers of Etruscan Sarcophagus of Hasti Afunei: An Innovative Micro-Sampling Technique for Raman/SERS Analyses. *Molecules* **2019**, *24*, 3403. [CrossRef]
17. Ciccola, A.; Serafini, I.; Ripanti, F.; Vincenti, F.; Coletti, F.; Bianco, A.; Fasolato, C.; Montesano, C.; Galli, M.; Curini, R.; et al. Dyes from the Ashes: Discovering and Characterizing Natural Dyes from Mineralized Textiles. *Molecules* **2020**, *25*, 1417. [CrossRef]
18. Germinario, G.; Ciccola, A.; Serafini, I.; Ruggiero, L.; Sbroscia, M.; Vincenti, F.; Fasolato, C.; Curini, R.; Ioele, M.; Postorino, P.; et al. Gel Substrates and Ammonia-EDTA Extraction Solution: A New Non-Destructive Combined Appraoch Useful for the Identification of Anthraquinone Dyes from Wool Textiles. *Microchem. J.* **2020**, *155*, 104780. [CrossRef]
19. Stiles, P.L.; Dieringer, J.A.; Shah, N.C.; Van Duyne, R.P. Surface-Enhanced Raman Spectroscopy. *Annu. Rev. Anal. Chem.* **2008**, *1*, 601–626. [CrossRef] [PubMed]
20. Guineau, B.; Guichard, V. Identification des colorants organiques naturels par microspectrometrie Raman de resonance et par effet Raman exalte de surface (SERS). In *ICOM Committee for Conservation, 8th Triennial Meeting, Sydney, Australia, 6–11 September 1987: Preprints*; ICOM: Paris, France, 1987; pp. 659–666.

21. Leona, M.; Londero, P.S.; Lombardi, J.R. 10 Years of Surface-Enhanced Raman Spectroscopy in Art and Archaeology. *Microsc. Microanal.* **2014**, *20* (Suppl. 3), 2006–2007. [CrossRef]
22. Serafini, I.; Ciccola, A. Nanotechnologies and Nanomaterials: An Overview for Cultural Heritage. In *Nanotechnologies and Nanomaterials for Diagnostic, Conservation and Restoration of Cultural Heritage*; Lazzara, G., Fakhrulli, R., Eds.; Elsevier: Amsterdam, The Netherlands, 2018; pp. 325–380.
23. Bruni, S.; De Luca, E.; Guglielmi, V.; Pozzi, F. Identification of Natural Dyes on Laboratory-Dyed Wool and Ancient Wool, Silk, and Cotton Fibers Using Attenuated Total Reflection (ATR) Fourier Transform Infrared (FT-IR) Spectroscopy and Fourier Transform Raman Spectroscopy. *Appl. Spectrosc.* **2011**, *65*, 1017–1023. [CrossRef]
24. Bruni, S.; Guglielmi, V.; Pozzi, F.; Mercuri, A.M. Surface-Enhanced Raman Spectroscopy (SERS) on Silver Colloids for the Identification of Ancient Textile Dyes. Part II: Pomegranate and Sumac. *J. Raman Spectrosc.* **2011**, *42*, 465–473. [CrossRef]
25. Bruni, S.; Guglielmi, V.; Pozzi, F. Historical Organic Dyes: A Surface-Enhanced Raman Scattering (SERS) Spectral Database on Ag Lee-Meisel Colloids Aggregated by NaClO4. *J. Raman Spectrosc.* **2011**, *42*, 1267–1281. [CrossRef]
26. Bruni, S.; Guglielmi, V.; Pozzi, F. Surface-Enhanced Raman Spectroscopy (SERS) on Silver Colloids for the Identification of Ancient Textile Dyes: Tyrian Purple and Madder. *J. Raman Spectrosc.* **2010**, *41*, 175–180. [CrossRef]
27. Leona, M. Microanalysis of Organic Pigments and Glazes in Polychrome Works of Art by Surface-Enhanced Resonance Raman Scattering. *Proc. Natl. Acad. Sci. USA* **2009**, *106*, 14757–14762. [CrossRef] [PubMed]
28. Nie, C.S.; Feng, Z. Simple Preparation Method for Silver SERS Substrate by Reduction of AgNO3 on Copper Foil. *Appl. Spectrosc.* **2002**, *56*, 300–305. [CrossRef]
29. Platania, E.; Lofrumento, C.; Lottini, E.; Azzaro, E.; Ricci, M.; Becucci, M. Tailored Micro-Extraction Method for Raman/SERS Detection of Indigoids in Ancient Textiles. *Anal. Bioanal. Chem.* **2015**, *407*, 6505–6514. [CrossRef]
30. Platania, E.; Lombardi, J.R.; Leona, M.; Shibayama, N.; Lofrumento, C.; Ricci, M.; Becucci, M.; Castellucci, E. Suitability of Ag-Agar Gel for the Microextraction of Organic Dyes on Different Substrates: The Case Study of Wool, Silk, Printed Cotton and a Panel Painting Mock-Up. *J. Raman Spectrosc.* **2014**, *45*, 1133–1139. [CrossRef]
31. Ricci, M.; Lofrumento, C.; Castellucci, E.; Becucci, M. Microanalysis of Organic Pigments in Ancient Textiles by Surface-Enhanced Raman Scattering on Agar Gel Matrices. *J. Spectrosc.* **2016**, *2016*. [CrossRef]
32. Lofrumento, C.; Ricci, M.; Platania, E.; Becucci, M.; Castellucci, E. SERS Detection of Red Organic Dyes in Ag-Agar Gel. *J. Raman Spectrosc.* **2013**, *44*, 47–54. [CrossRef]
33. Leona, M.; Decuzzi, P.; Kubic, T.A.; Gates, G.; Lombardi, J.R. Nondestructive Identification of Natural and Synthetic Organic Colorants in Works of Art by Surface Enhanced Raman Scattering. *Anal. Chem.* **2011**, *83*, 3990–3993. [CrossRef]
34. Doherty, B.; Brunetti, B.G.; Sgamellotti, A.; Miliani, C.A. Detachable SERS Active Cellulose Film: A Minimally Invasive Approach to the Study of Painting Lakes. *J. Raman Spectrosc.* **2011**, *42*, 1932–1938. [CrossRef]
35. Pozzi, F.; Van Den Berg, K.J.; Fiedler, I.; Casadio, F.A. Systematic Analysis of Red Lake Pigments in French Impressionist and Post-Impressionist Paintings by Surface-Enhanced Raman Spectroscopy (SERS). *J. Raman Spectrosc.* **2014**, *45*, 1119–1126. [CrossRef]
36. Calà, E.; Benzi, M.; Gosetti, F.; Zanin, A.; Gulmini, M.; Idone, A.; Serafini, I.; Ciccola, A.; Curini, R.; Whitworth, I.; et al. Towards the Identification of the Lichen Species in Historical Orchil Dyes by HPLC-MS/MS. *Microchem. J.* **2019**, *150*, 104140. [CrossRef]
37. Castro, R.; Pozzi, F.; Leona, M.; Melo, M.J. Combining SERS and Microspectrofluorimetry with Historically Accurate Reconstructions for the Characterization of Lac Dye Paints in Medieval Manuscript Illuminations. *J. Raman Spectrosc.* **2014**, *45*, 1172–1179. [CrossRef]
38. Cañamares, M.V.; Reagan, D.A.; Lombardi, J.R.; Leona, M. TLC-SERS of Mauve, the First Synthetic Dye. *J. Raman Spectrosc.* **2014**, *45*, 1147–1152. [CrossRef]
39. Ciccola, A.; Tozzi, L.; Romani, M.; Serafini, I.; Ripanti, F.; Curini, R.; Vitucci, F.; Cestelli Guidi, M.; Postorino, P. Lucio Fontana and the Light: Spectroscopic Analysis of the Artist's Collection at the National Gallery of Modern and Contemporary Art. *Spectrochim. Acta Part A Mol. Biomol. Spectrosc.* **2020**, *236*, 1–8. [CrossRef]
40. Pérez-Arantegui, J.; Rupérez, D.; Almazán, D.; Díez-de-Pinos, N. Colours and Pigments in Late Ukiyo-e Art Works: A Preliminary Non-Invasive Study of Japanese Woodblock Prints to Interpret Hyperspectral Images Using in-Situ Point-by-Point Diffuse Reflectance Spectroscopy. *Microchem. J.* **2018**, *139*, 94–109. [CrossRef]
41. Leopold, N.; Lendl, B. A New Method for Fast Preparation of Highly Surface-Enhanced Raman Scattering (SERS) Active Silver Colloids at Room Temperature by Reduction of Silver Nitrate with Hydroxylamine Hydrochloride. *J. Phys. Chem. B* **2003**, *107*, 5723–5727. [CrossRef]
42. Sessa, C.; Weiss, R.; Niessner, R.; Ivleva, N.P.; Stege, H. Towards a Surface Enhanced Raman Scattering (SERS) Spectra Database for Synthetic Organic Colourants in Cultural Heritage. The Effect of Using Different Metal Substrates on the Spectra. *Microchem. J.* **2018**, *138*, 209–225. [CrossRef]
43. Centeno, S.A.; Hale, C.; Carò, F.; Cesaratto, A.; Shibayama, N.; Delaney, J.; Dooley, K.; van der Snickt, G.; Janssens, K.; Stein, S.A. Van Gogh's Irises and Roses: The Contribution of Chemical Analyses and Imaging to the Assessment of Color Changes in the Red Lake Pigments. *Herit. Sci.* **2017**, *5*, 1–11. [CrossRef]
44. Schneider, S.; Brehm, G.; Freunscht, P. Comparison of Surface-Enhanced Raman and Hyper-Raman Spectra of the Triphenylmethane Dyes Crystal Violet and Malachite Green. *Phys. Status Solidi* **1995**, *189*, 37–42. [CrossRef]

45. Cañamares, M.V.; Chenal, C.; Birke, R.L.; Lombardi, J.R. DFT, SERS, and Single-Molecule SERS of Crystal Violet. *J. Phys. Chem. C* **2008**, *112*, 20295–20300. [CrossRef]
46. Gühlke, M.; Heiner, Z.; Kneipp, J. Surface-Enhanced Hyper-Raman and Raman Hyperspectral Mapping. *Phys. Chem. Chem. Phys.* **2016**, *18*, 14228–14233. [CrossRef]
47. Greeneltch, N.G.; Davis, A.S.; Valley, N.A.; Casadio, F.; Schatz, G.C.; Van Duyne, R.P.; Shah, N.C. Near-Infrared Surface-Enhanced Raman Spectroscopy (NIR-SERS) for the Identification of Eosin Y: Theoretical Calculations and Evaluation of Two Different Nanoplasmonic Substrates. *J. Phys. Chem. A* **2012**, *116*, 11863–11869. [CrossRef]
48. Narayanan, V.A.; Stokes, D.L.; Vo-Dinh, T. Vibrational Sprectral Analysis of Eosin Y and Erytheosin B—Intensity Studies for Quantiative Detection by Dyes. *J. Raman Spectrosc.* **1994**, *24*, 415–422. [CrossRef]
49. Michaels, A.M.; Nirmal, M.; Brus, L.E. Surface Enhanced Raman Spectroscopy of Individual Rhodamine 6G Molecules on Large Ag Nanocrystals. *J. Am. Chem. Soc.* **1999**, *121*, 9932–9939. [CrossRef]
50. Sun, C.H.; Wang, M.L.; Feng, Q.; Liu, W.; Xu, C.X. Surface-Enhanced Raman Scattering (SERS) Study on Rhodamine B Adsorbed on Different Substrates. *Russ. J. Phys. Chem. A* **2015**, *89*, 291–296. [CrossRef]
51. Zhao, Y.; Yamaguchi, Y.; Liu, C.; Li, M.; Dou, X. Rapid and Quantitative Detection of Trace Sudan Black B in Dyed Black Rice by Surface-Enhanced Raman Spectroscopy (SERS). *Spectrochim. Acta Part A Mol. Biomol. Spectrosc.* **2019**, *216*, 202–206. [CrossRef]
52. Muehlethaler, C.; Ng, K.; Gueissaz, L.; Leona, M.; Lombardi, J.R. Raman and SERS Characterization of Solvent Dyes: An Example of Shoe Polish Analysis. *Dye. Pigment.* **2017**, *137*, 539–552. [CrossRef]
53. Whitney, A.V.; Van Duyne, R.P.; Casadio, F. An Innovative Surface-Enhanced Raman Spectroscopy (SERS) Method for the Identification of Six Historical Red Lakes and Dyestuffs. *J. Raman Spectrosc.* **2006**, *37*, 993–1002. [CrossRef]
54. Cesaratto, A.; Lombardi, J.R.; Leona, M. Tracking Photo-Degradation of Triarylmethane Dyes with Surface-Enhanced Raman Spectroscopy. *J. Raman Spectrosc.* **2017**, *48*, 418–424. [CrossRef]
55. Geiman, I.; Leona, M.; Lombardi, J.R. Application of Raman Spectroscopy and Surface-Enhanced Raman Scattering to the Analysis of Synthetic Dyes Found in Ballpoint Pen Inks. *J. Forensic Sci.* **2009**, *54*, 947–952. [CrossRef]

heritage

MDPI

Article

Identifying Brazilwood's Marker Component, Urolithin C, in Historical Textiles by Surface-Enhanced Raman Spectroscopy

Brenda Doherty [1,*], Ilaria Degano [2], Aldo Romani [3], Catherine Higgitt [4], David Peggie [4], Maria Perla Colombini [2] and Costanza Miliani [5]

1 CNR Istituto di Scienze e Tecnologie Chimiche (SCITEC), Dipartimento di Chimica, Biologia e Biotechnologie, Università degli Studi di Perugia, Via Elce di Sotto 8, 06123 Perugia, Italy
2 Dipartimento di Chimica e Chimica Industriale, Universitò di Pisa, Via Giuseppe Moruzzi 13, 56124 Pisa, Italy; ilaria.degano@unipi.it (I.D.); maria.perla.colombini@unipi.it (M.P.C.)
3 Centro di Eccellenza SMAArt, Dipartimento di Chimica, Biologia e Biotechnologie, Università degli Studi di Perugia, Via Elce di Sotto 8, 06123 Perugia, Italy; aldo.romani@unipg.it
4 Scientific Department, National Gallery London, Trafalgar Square, London WC2N 5DN, UK; Catherine.Higgitt@ng-london.org.uk (C.H.); David.Peggie@ng-london.org.uk (D.P.)
5 Istituto per la Scienza del Patrimonio Culturali (ISPC-CNR), Via Guglielmo San Felice 8, 80134 Napoli, Italy; costanza.miliani@cnr.it
* Correspondence: brenda.doherty@cnr.it

Abstract: The fugitive nature of the colorants obtained from sappanwood (*Caesalpinia sappan* L.) or the South American species commonly known as 'brazilwoods' (including other *Caesalpinia* species and *Paubrasilia echinata* (Lam.)) makes the identification of brazilwood dyes and pigments in historic artefacts analytically challenging. This difficulty has been somewhat alleviated recently by the recognition and structural elucidation of a relatively stable marker component found in certain brazilwood dyes and pigments—the benzochromenone metabolite urolithin C. This new understanding creates an ideal opportunity to explore the possibilities for urolithin C's localization and identification in historical artefacts using a variety of analytical approaches. Specifically, in this work, micro-destructive surface-enhanced Raman spectroscopic methods following a one-sample two-step (direct application of the colloid and then subsequent exposure of the same sample to HF before reapplication of the colloid) approach are utilized for the examination of four historical brazilwood dyed textiles with the results confirmed via HPLC-DAD analysis. It is shown that characterization of reference urolithin C is possible, and diagnostic features of this molecule can also be traced in faded historical linen, silk and wool textiles, even in the presence of minor quantities of flavonoid, indigoid and tannin components. The exploitation of the same micro-sample through a series of SERS analyses affords a fuller potential for confirming the characterization of this species.

Keywords: urolithin C; brazilein; brazilwood marker component; historical textile

Citation: Doherty, B.; Degano, I.; Romani, A.; Higgitt, C.; Peggie, D.; Colombini, M.P.; Miliani, C. Identifying Brazilwood's Marker Component, Urolithin C, in Historical Textiles by Surface-Enhanced Raman Spectroscopy. *Heritage* **2021**, *4*, 1415–1428. https://doi.org/10.3390/heritage4030078

Academic Editor: Lucia Burgio

Received: 24 June 2021
Accepted: 22 July 2021
Published: 25 July 2021

Publisher's Note: MDPI stays neutral with regard to jurisdictional claims in published maps and institutional affiliations.

1. Introduction

Brazilwood dyestuffs and pigments are derived from a number of closely related and often historically confused species of trees of the Leguminosae family. Heartwood from the so-called 'soluble redwoods', sappanwood (*Caesalpinia sappan* L.) from Southeast Asia and several South American species commonly called brazilwoods (including *Caesalpinia brasiliensis* L. and *Paubrasilia echinata* (Lam.)), contain brazilin, the main colorant in brazilwood dyes. Brazilin (Figure 1a) is a tetracyclic homoisoflavonoid which converts to a more deeply red colored molecule, brazilein, through autoxidation of a hydroxyl group into a carbonyl group (Figure 1b). Other phenolic components isolated from brazilwood include xanthone, coumarin and various chalcones, flavones and other homoisoflavonoids [1].

The fugitive nature of brazilwood has been documented since the Middle Ages, and its use as both a dyestuff and a pigment has been regulated or proscribed as a result, particularly for use as a sole colorant [2]. It was, however, less expensive than other natural red dyestuffs and gave an attractive color when fresh, so found great use as a textile dye, an organic lake pigment in manuscripts (where light exposure was less of an issue) and in paintings. Evidence of the use of brazilwood-derived lake pigments has been found in paintings by a diverse range of artists, from Raphael to Rembrandt and Van Gogh [3–5]. For textiles, the dyestuff was generally extracted in neutral or basic aqueous solutions to which the mordanted textile was added. To prepare brazilwood lake pigments, various recipes are known. In some, the dyestuff is extracted from the raw material under alkaline conditions, then added with alum to precipitate the pigment from the solution. Alternatively, the dyestuff could be extracted in a slightly acidic solution (e.g., alum solution) and the pigment precipitated on addition of an alkaline solution or of a source of calcium carbonate or sulphate [1,6,7].

The analytical examination of brazilwood's main colorants is generally straightforward for the characterization of unaltered homoisoflavonoid constituents by non-destructive FT-Raman and Infrared spectroscopies as well as micro-destructive chromatographic methods alike [8–10]. However, accurately identifying aged and faded brazilwood-based dyes and pigments in historic artworks is challenging. Brazilwood's renowned lack of permanence has limited the identification of severely faded brazilwood in artefacts, in the heritage science domain, to only the most sensitive and specific micro-destructive techniques, such as HPLC with spectrophotometric or mass spectrometric detection [11]. Such studies have indicated the presence of a minor non-dyestuff component that appears to be a brazilwood marker, highlighted by Nowik [10]. This marker component has been recently identified as 3,8,9-trihydroxy-6H-benzo[c]chromen-6-one, urolithin C (Figure 1c), and unequivocally linked to brazilwood by Peggie et al. [12].

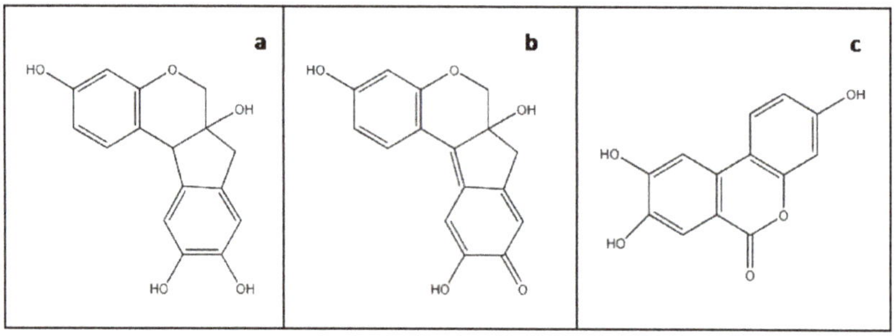

Figure 1. Molecular structures of (**a**) brazilin, (**b**) brazilein and (**c**) marker component urolithin C.

Urolithin C had initially been thought to be a degradation product of brazilein, yet its natural and unquantified presence in the heartwood has been highlighted, suggesting its unrelated presence [13]. Furthermore, observations by Peggie et al. have established a link between an alkaline extraction method and the increased presence of urolithin C in brazilwood dyes and lake pigments. Tamburini has subsequently noted that a sample dyed with sappanwood at alkaline pH showed a higher relative abundance of urolithin C than in a sample dyed at neutral pH, but also highlighted therein a much lower abundance of protosappanins [14]. Such deductions may provide information as to the origins of urolithin C within these dyes and pigments. In nature, tannin constituents play a major role in the formation of the urolithin metabolite. Studies have observed that in animal guts microbia can metabolize ellagitannin and ellagic acid into dibenzopyran-6-one derivatives with different hydroxyl substitutions, through the loss of one of the two lactones present in ellagic acid (lactonase/decarboxylase) followed by subsequent removal of hydroxyls through dehydroxylase activity [15,16]. In the context of the study of historical textiles and pigments, what is of particular interest is that urolithin C appears to be relatively stable, and still persists in degraded samples of brazilwood dyes or pigments even when other colorant components have been lost [13]. This provides an ideal opportunity for further analytical methods to be examined to assess their suitability in determining the presence of urolithin C.

Surface-enhanced Raman spectroscopy (SERS) is an ideal candidate for such studies, and is being increasingly adopted for the investigation of natural and synthetic organic dyes (and pigments derived from such colorants) associated with historical artefacts by combining Raman spectroscopy's fingerprinting ability with plasmonic-enabled enhanced sensitivity. The SERS effect is due to a combination of an electromagnetic (EM) enhancement associated with plasmon excitation in metal particles and a chemical (CHEM) enhancement associated with the transfer of electrons from the analyte molecule to/from the metal particles in both ground and excited states, often forming a metal–molecule bond. The application of SERS techniques has expanded in key application areas of heritage science in the last twenty years [17] and is nowadays competitive in terms of sample requirements to HPLC for ultra-sensitive and selective constituent identifications with current in situ potential. The construction of in-house databases and optimized analytical protocols help to improve the reproducibility of measurements for the characterization of the low quantities and/or unknown components that are typically encountered in the analysis of historic samples. Although still a micro-destructive and qualitative technique when used for the analysis of real, historical samples, the versatility of SERS methodologies can allow a single sample to be used in a series of sequential analyses. First, a single micro-sample can be mixed with an activated colloidal solution and investigated as such. Following drying of this same sample, it can undergo suitable acid-based pre-treatments such as treatment with nitric acid or controlled exposure to hydrofluoric acid vapors before examination [18]. SERS screening methods are therefore potentially of great value for the investigation of naturally aged artefacts suspected of containing low concentrations of brazilwood dyestuffs and pigments which are of historical relevance and which necessitate less destructive investigations or even in situ analyses when sampling is prohibited, or where chromatographic methods are unavailable. The use of conventional SERS methodologies is most likely to be successful in cases where the brazilwood colorant is potentially present in the most superficial layer. The use of this technique when analyzing paint samples is, however, much more complicated, as high degrees of heterogeneity can be expected, given that the brazilwood is present as a lake pigment that may have been mixed with other pigments. The embedding of such samples in resins or other materials as cross-sections further adds to the complexity. The use of SERS methodologies thus seems most viable, at this stage, as a first approach, to use during the examination of textiles dyed with brazilwood, even in a faded condition.

In this work a reference sample of urolithin C was examined by classical Raman and colloidal SERS to understand whether characteristic features of the relatively persistent urolithin C marker component can be successfully distinguished. Based on the results of this study and additional work on reference samples, colloidal SERS was used to examine four historical textiles (with pre-treatments as necessary). Samples of the four textiles had been previously analyzed by HPLC-DAD during restoration campaigns (data included in the restoration documentation stored by the corresponding institutions) and were all known to contain urolithin C together with varying proportions of surviving brazilwood colorant components, other dyestuffs and a range of other molecules typical of historical dyed textile samples.

2. Materials and Methods

Urolithin C was purchased from Dalton Research Molecules, 349 Wildcat Road, Toronto, Ontario, M3J 2S#11, Canada, purity greater than 95% by HPLC, and used as such. Brazilin was purchased from ICN Biomedical Inc. Cat #205613. The powdered brazilwood lake reference was prepared at the National Gallery in 2003 from Sappan Lignum (*Caesalpinia sappan*), no further details of the recipe are available. Details of the four historical textile samples, two from tonacelles (ecclesiastical garments) and two from tapestries examined, are given in Table 1.

Samples S#11 and S#5 belong to the series of the Valois Tapestries, woven in Brussels in the sixteenth century and belonging to the Gallerie degli Uffizi, Florence, directed by Eike Schmidt. The series consists of eight sixteenth-century tapestries, representing Catherine de Medici and her family observing courtly festivities, collectively known as The Valois Tapestries [19]. The samples were collected during the restoration performed by Restauro Tessile di Bayer e Perrone da Zara, Florence, Italy, directed by Alessandra Griffo (Gallerie degli Uffizi) and funded by the Friends of the Uffizi, chaired by Maria Vittoria Colonna Rimbotti, with the contribution of Mrs. Veronica Atkins. Samples S#27 and S#9 are part of a collection of eighteenth-century Italian ecclesiastical garments belonging to a public authority, which was subjected to a restoration campaign in 2011–2012 during which the private company 'Giordano Passarella' was appointed, who performed the sampling.

Table 1. Outline of samples investigated in this work.

Sample Abbreviation	Provenance	Description
S#5	Valois Tapestries, *Fontainebleau*, c. 1576, based on a design by Antoine Caron, woven under the direction of Master MPG, Brussels (Gallerie degli Uffizi, Florence)	Brown wool
S#11	Valois Tapestries, *Whale*, c. 1576, based on a design by Antoine Caron, woven under the direction of Master MPG, Brussels (Gallerie degli Uffizi, Florence)	Pale orange silk core fiber of a golden metal thread
S#27	Ecclesiastical garment	Red–brown linen tunic support
S#9	Ecclesiastical garment	Orange–brown silk chenille

2.1. *High Performance Liquid Chromatography (HPLC-DAD/HPLC-ESI-Q-ToF-MS)*

For HPLC-DAD analyses, a HPLC system 2089 quaternary gradient with a diode array spectrophotometric detector MD-2010 was used, equipped with an AS-950 autosampler (Jasco International Co., Tokyo, Japan). Injection volume was 20 μL. ChromNav software was used to carry out data acquisition and data analysis. The chromatographic separation was performed on an analytical reversed-phase column TC-C18 (2) (250 × 4.6 mm, 5 μm for samples S#27 and S#9 and 150 × 4.6 mm, 5 μm for samples S#11 and S#5) with a TC-C18 (2) pre-column (12.5 × 4.6 mm, 5 μm), both from Agilent Technologies (Palo Alto, Santa Clara, CA, USA). The eluents were A, trifluoroacetic acid (TFA 0.1% v/v) in bidistilled water, and B, trifluoroacetic acid (TFA 0.1% v/v) in HPLC-grade acetonitrile. The flow rate was 1.0 mL/min and the program was 15% B for 5 min, then to 50% B in 25 min, then to 70% B in 10 min, to 100% B in 1 min and then hold for 5 min; re-equilibration took 13 min. The separation took place at room temperature (25 °C). The detector operated with spectra acquisition in the range of 200–650 nm every 0.8 s with 4 nm resolution.

For HPLC-ESI-Q-ToF analyses, a HPLC 1200 Infinity coupled to a Jet Stream ESI-Q-ToF 6530 Infinity detector was used, equipped with an Agilent Infinity autosampler (Agilent Technologies, Palo Alto, Santa Clara, CA, USA). Injection volume was 4 μL for S#27 and 10 μL for S#9 and S#5. MassHunter® Workstation Software (B.04.00) was used to carry out mass spectrometer control, data acquisition and data analysis. The chromatographic separation was performed on an analytical reversed-phase column Poroshell 120 EC-C18 column (3.0 mm × 75 mm, 2.7 μm particle size) with a Zorbax Eclipse plus C-18 guard column (4.6 mm × 12.5 mm, 5 μm particle size), both from Agilent Technologies (Palo Alto, Santa Clara, CA, USA). The eluents were A, formic acid (FA 1% v/v) in LC-MS-grade water, and B, formic acid (FA 1% v/v) in LC-MS-grade acetonitrile. The flow rate was 0.4 mL/min and the program was 15% B for 2.6 min, then to 50% B in 13 min, to 70% B in 5.2 min, to 100% B in 0.5 min and then hold for 1 min; re-equilibration took 10 min. During the separation, the column was thermostated at 30 °C. The mass spectrometer operated in ESI ionization in negative mode and the working conditions were: drying gas (N_2, purity > 98%) temperature 350 °C and 10 L/min flow; capillary voltage 4.5 KV; nebulizer gas pressure 35 psig; sheath gas temperature 375 °C and 11 L/min flow; and fragmentor voltage 175 V. High-resolution MS and MS/MS spectra (CID voltage 30 V) were acquired in negative mode in the range 100–1000 m/z at a scan rate of 1.04 spectra/sec (N_2, purity 99.999%). Auto-calibration was performed daily using Agilent tuning mix HP0321 (Agilent Technologies) prepared in acetonitrile.

All samples were treated with 200 μL of 0.1% Na_2EDTAaq/DMF (1:1, v/v) and extracted at 60 °C for 60 min in an ultrasonic bath; the extract was filtered on PTFE (0.45 μm) prior to injection.

2.2. Surface-Enhanced Raman Spectroscopy

SERS measurements were carried out utilizing modified Lee and Meisel citrate-reduced silver colloids by the reduction of silver nitrate (Aldrich silver nitrate 99.9%) with sodium citrate (Aldrich sodium citrate dehydrate 99%) [20]. The colloid has a characteristic absorption maximum at 426 nm and FWHM of 110 nm, as measured with a Hewlett Packard 8453 photodiode array UV–Vis spectrometer (following a 1:9 dilution with ultrapure water to observe maximum absorbance within the instrumental range). SERS measurements were firstly carried out by adding a 5 μL drop of colloid aggregated with magnesium sulphate directly onto micro-samples. On drying, the same micro-samples were subsequently exposed to hydrofluoric acid vapors for 5 min in a closed Eppendorf, then dried and examined again with a 5 μL drop of aggregated colloid [18]. Spectra were recorded using a laboratory Jasco NRS#3100 spectrophotometer with an argon laser emitting at 514 nm (grating 1200 lines/mm), and spectra could be obtained between 2 and 10 min after addition of the colloid and remained constant in quality until the evaporation of the liquid. Data were collected over the range 150–1800 cm^{-1} with exposure times from 1–7 s and 3–10 accumulations. Laser power was maintained between 0.6 and 2 mW with an overall spectral resolution of ~4 cm^{-1}. Polystyrene was used for instrument calibration. Raw data are shown, free from any manipulation other than being overlaid in graphs for reasons of clarity.

3. Results and Discussion

3.1. Brazilin Reference and Brazilwood Pigment

SERS spectra could be readily obtained from the brazilin reference sample, the main constituent in brazilwood dye sources, utilizing the conventional colloidal SERS approach at pH quasi neutral (Figure 2i) and reported in Table 2. In order to obtain a higher quality SERS spectrum from a reference sample of a brazilwood lake pigment, it was instead necessary to routinely pre-treat the sample with HF acid vapors to liberate the main dye constituent from its inorganic substrate, an action that allows for a greater adsorption of dye molecules onto the SERS active substrate (Figure 2ii). On comparison of these spectra, and on account of the differing SERS experimental procedures, spectral behaviors such as frequency shifts and peak intensity fluctuations can be noticed. Such behavior is due to the varied chemisorption of the dye/hydrolyzed components on the silver colloidal surfaces on formation of a complex. This can lead to a modification of the polarizability and vibrational modes and may also enable charge transfer to occur. The change in intensity of bands in the SERS spectra can be attributed to the relative orientation of the analyte with respect to the silver. A broad background remains observable in both spectra due to residual intrinsic fluorescence, impurities or the SERS continuum of undefined physical origin [21].

Table 2. Bands observed in the Raman and SERS spectra (in cm^{-1}, excited at 532 nm) and tentative band assignments. Marker bands for the recognition of each material are given in bold.

Brazilin Reference	Brazilwood Lake Pigment	Urolithin C	Urolithin C	Tentative Assignments [9,20,22–24]
SERS	**SERS on Hydrolysis**	**SERS**	**Raman**	
	1619		**1616**	ν (C=C) + ν(C = O)
1562	**1564**	1589 1569		ν(C=C) + δ(OH)
		1530	**1530**	ν(C=C) and ν_{as}(O-C-O)
1515	1500			
1497				
1429				ν (C=C) + δ(ring) + δ(COH)
			1395	ν(C-C) and δ(HCC)
	1377			C=C(=O)=C=C
1352			1354	ν(C-O), δ(OCC) + δ(CH2)
		1335	1313	υ_{sy}(OCO)
1259		1255	1252	ν (C-O) + ν(C-C) + $\delta(CH_2)$
1219		1227	1208	ν (C-O) + ν(C-C)
1168	1176	**1167**	**1167**	δ(C-H) and ν(C-C)
1028				in-plane δ(CH)
		990	**987**	ν (C=C) + ν(C=O)
			958	ρ (CH_2)
925		935		
813				
715	728	704	719	γ (CO) + γ(CH)
676		**661**		γ (CH) + δ(CC-O)
623		611		γ (CH)
549	549	550		def.(ring)
466	**467**	469		def.(ring)
		390		
370				

ν: stretching; δ: bending; γ: out-of-plane bending; as: asymmetric; and sy: symmetric.

The intense broad band at 1562–4 cm^{-1} attributed to aromatic ring vibrations with δ(OH) and ν(C=C) contributions together with the ring deformations at 466–7 cm^{-1} are present in both spectra and represent the diagnostic SERS bands indicative of colorants in brazilwood. The ν(C=C(=O)-C=C) system located at 1377 cm^{-1} for the brazilwood pigment is attributed to its brazilein content, and it is unobservable for the brazilin reference which instead has a peak at 1352 cm^{-1} attributed to the ν(C-O), δ(OCC) and δ(CH2) modes. The peak at 1619 cm^{-1}, apparent in the hydrolyzed lake samples, is possibly again due to brazilein and attributed to ν(C=C) and ν(C=O) and can also be seen as a weak shoulder (marked with an asterisk) in the brazilin reference. This could in fact indicate a small amount of brazilein in the brazilin reference sample due to natural oxidation. It is likely that the brazilwood constituents are attached to the silver colloidal surface via the C=O and OH groups, similar to the flavonoid class of molecules [23].

Further weak peaks (marked by asterisks in Figure 2ii), which appear at 1515 and 1497 cm^{-1} in the brazilin reference and which possibly relate to the ν(C=C), are replaced by a small shoulder at 1518 cm^{-1} in the hydrolyzed lake. The band at 549 cm^{-1}, attributed to the δ(ring), appears less intense in the brazilwood pigment. Other spectral regions instead show pronounced frequency shifts possibly governed by changes in the polarizability of the analyte–metal complexes. The δ(CCH) and ν(C-C) bands shift from 1168 cm^{-1} in the brazilin reference to 1176 cm^{-1} for the brazilwood pigment. The weak 715 cm^{-1} band for the brazilin, attributed to the γ(CO) + γ(CH), is shifted to 728 cm^{-1} in the pigment. The bands at 1219 cm^{-1} (ν(C-O) + ν(C-C)), 1259 cm^{-1} (ν(C-O) + ν(C-C) + δ(CH2)), 1028 cm^{-1} (in-plane δ(CH)) and 676 cm^{-1} (γ(CH) + δ(CC-O)) are all instead only observed in the brazilin [13,25,26].

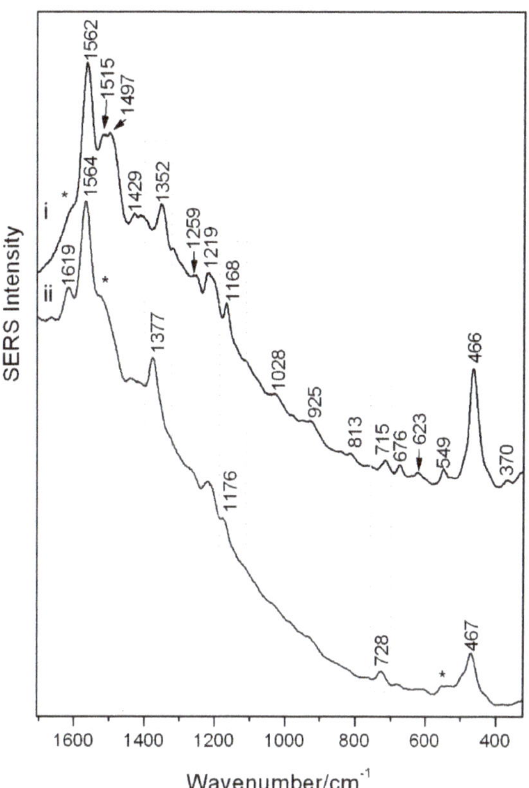

Figure 2. SERS spectra of (i) reference sample of brazilin reference (regular colloidal SERS) and (ii) a brazilwood pigment (pre-treated with HF). Reproducible peaks which vary widely in intensity are denoted with an asterisk.

3.2. Urolithin C Standard

Although fluorescence was relatively high, adequate classical Raman scattering of the brazilwood marker component, urolithin C, could be recorded using laser excitation at 514 nm as shown in Figure 3a with main bands highlighted in Table 2. Bands resulting from the phenanthrene skeleton can be tentatively assigned to the ν(CC) and ν(CO) at 1616 cm^{-1} and to the ν(CC) and ν_{ass}(O-C-O) at 1530 cm^{-1} [24]. Bands visible at ~1400 cm^{-1} may be due to the ν(C-C) and δ(HCC) and the out-of-plane CH modes at 719 cm^{-1}. SERS measurements of the same micro-sample of urolithin C reveals two predominant spectra as given in Figure 3b and reported in Table 2. Fluctuations for the bands at 1589 and 1569 cm^{-1}, related to ν(C=C) + δ(OH) modes, could be observed during spectra collection as there was a constant movement of colloidal/urolithin C particles within the examined aqueous droplet. While this can be related to the common blinking phenomenon, the reproducibility of the spectra could actually indicate a changeable interaction of the hydroxyl groups with the silver surface [27–29].

The bands at 1530, 1167 and very weak 990 cm^{-1} (ring breathing vibration) visible in the Raman spectrum are the only ones also featured in the SERS spectra without any great shift. Significant differences in the relative intensity and position of peaks can be observed when the spectra shown in Figure 3a,b are compared. These are the result of SERS surface selection rules and are due to a combination of the orientation of the analyte on the colloidal surface and specific Raman mode symmetry, and very likely in the formation of the surface complex where Raman modes in the molecule disappear on interaction with the surface and instead other modes or new modes may be activated [21].

No bands at higher wavenumber than 1600 cm^{-1} are visible under SERS conditions, although there is a large enhancement of the 1530 cm^{-1} band (ν(CC) and tentatively ν_{assym}(OCO)), accompanied by a band at 1335 cm^{-1}, possibly attributed to the ν_{symm}(OCO). The 1167 cm^{-1} SERS band results from ν(C-C) and δ(CH). Multiple strong bands at low wavenumbers are due to out-of-plane deformation modes of C=C bonds with a visible δ(C=O) in the range 600–660 cm^{-1}. All this data collectively provides information as to the orientation of urolithin C when chemisorbed or in close interaction with the silver. It is possible that the overall behavior of urolithin C, interacting with the silver colloid via both the carbonyl and hydroxyl functional groups, may be related directly to concentration effects. Possibly at lower concentrations urolithin C can initially orient somewhat flatter, while instead at higher concentrations (near to and above monolayer coverage) it is possibly forced to adsorb end-on due to crowding, resulting in the aforementioned signal fluctuations [30].

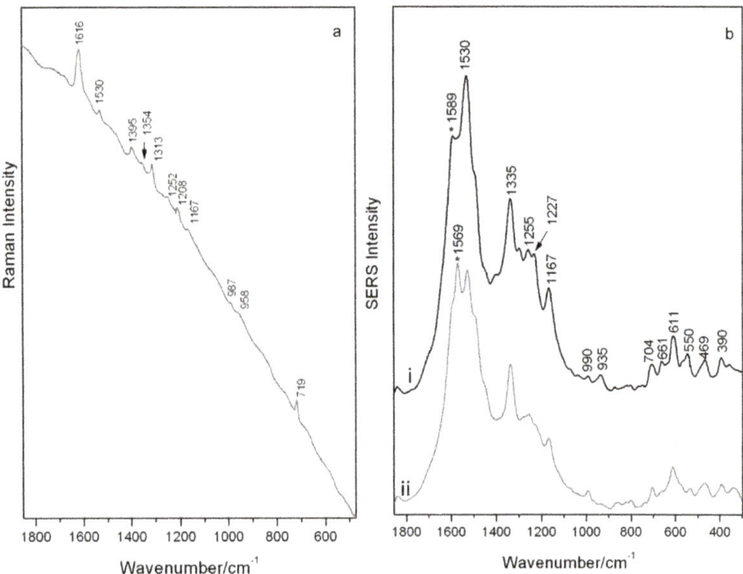

Figure 3. Spectra of urolithin C (λ 514 nm) by (**a**) classical Raman and (**b**) SERS (regular colloidal SERS) with reproducible yet fluctuating intensity of bands as highlighted by * in main traces i and ii.

3.3. Historical Textile Samples

The aforementioned spectral features and behaviors of both of brazilwood's main colorant components, and the marker component, urolithin C, were used to guide the interpretation of the SERS spectra obtained from micro-samples of historic brazilwood-dyed linen, silk and wool textiles.

In order to provide a validation of the results obtained through SERS, the historical textile samples had previously been investigated by chromatographic methods (HPLC-DAD) and were all known to contain urolithin C. In some samples additional colorants, namely flavonoids, indigoids and tannins, could be identified. All HPLC results are summarized in Table 3 and are described in more detail below in relation to the SERS results.

Table 3. Description of historical samples with main components identified by HPLC-DAD and SERS (* = minor species not detected by DAD but confirmed by tandem mass spectrometric detection).

Sample Abbreviation	Dyestuffs and Other Components Identified		
	HPLC-DAD and HPLC-ESI-Q-ToF	SERS	
		Colloidal	HF Pre-Treatment
S#5	Brazilwood (Urolithin C, sappanol *); 4-hydroxybenzoic acid	Urolithin C 4-hydroxybenzoic acid	Urolithin C 4-hydroxybenzoic acid
S#11	Brazilwood (Urolithin C); Ellagic acid; young fustic (sulfuretin); 4-hydroxybenzoic acid	Urolithin C	Urolithin C Ellagic acid
S#27	Brazilwood (Urolithin C, sappanol *, brazilein *); logwood (hematein *, hematoxylin *); ellagic acid; 4-hydroxybenzoic acid	Urolithin C Neoflavonoid-Brazilein/haematin?	Urolithin C, Logwood-Haematin?
S#9	Brazilwood (Urolithin C, sappanol *); young fustic (sulfuretin, sulfuretin-glucoside *, fisetin *); flavonoid dye (luteolin); indigo or woad (indigotin); 4-hydroxybenzoic acid	Urolithin C, Luteolin, Flavonoid-fustic?	Urolithin C, Luteolin, Flavonoid-fustic?

In the chromatogram (insert Figure 4a) of the extract of the brown wool sample S#5, shown at 275 nm, the peak for urolithin C could be detected together with very low quantities of sappanol, a molecular marker of redwood-type dyes [14,31]. A component of tannin dyes but also a known degradation marker of both flavonoid compounds and protein-based threads, 4-hydroxybenzoic acid, was also detected [32].

S#5 was investigated by colloidal SERS directly on the sample, which was followed with hydrofluoric acid pre-treatment, the resulting spectra shown in Figure 4a(i,ii), respectively. The well-resolved spectrum of S#5 with direct colloidal SERS can be seen in Figure 4 ai where the bands associated with urolithin C (1530, 1340, 1167, 987 and 654 cm^{-1}) can be observed. Following HF pre-treatment—and unlike the other samples described below—there is no improvement in the quality of the SERS response (Figure 4a(ii)), although weak bands associated with urolithin C can still be observed at 1530 and 660 cm^{-1}. It is likely that in the spectra for this sample the 4-hydroxybenzoic acid identified by HPLC analysis can be observed. Although there is a partial overlap with the marker component, the bands at 1618, 1390, 1017 and 830 cm^{-1} correspond to those reported in the literature for this molecule [33]. The results obtained from this sample confirm that urolithin C can be detected by SERS in samples where little or no brazilwood colorant components survive, even in samples containing high proportions of 4-hydroxybenzoic acid as is commonly observed in historical textile samples.

In the chromatogram of the extract of the orange silk sample S#11 (insert Figure 4b) shown at 275 nm, ellagic acid is the most intense peak, followed by urolithin C, fisetin and 4-hydroxybenzoic acid. This suggests that S#11 had been dyed with brazilwood and young fustic, consistent with previous studies [34] performed on silk core fibers of golden metal threads. Ellagic acid could have been added to the dye bath to improve the final color, used as a mordant or may have been co-extracted from the bark used when dyeing with young fustic.

In Figure 4b(i,ii), a noisy spectrum was obtained for the direct colloidal SERS where the bands related to the presence of the urolithin C marker component, namely 1530 and 660 cm^{-1}, can be observed. Following acid pre-treatment of this sample, a much more defined spectrum is obtained (Figure 4b(ii)) and the signals identifying the urolithin C constituent can be better appreciated (1530, ~1330 and ~660 cm^{-1}). The presence of strong bands in the lower wavenumber range attributed to the ring ν(CC) perhaps indicate a more planar orientation in relation to the colloidal surface. Further bands with reasonable intensity at 1605, 1470 and 1287 cm^{-1} can be ascribed to the co-presence of ellagic acid in this sample [35], as highlighted by the HPLC analysis. The bands at 1567, 1370, 885 and 548 cm^{-1} collectively provide information regarding the presence of flavonoid materials but do not permit determination of the specific dye source [25]. No bands exclusively pertaining to 4-hydroxybenzoic acid could be identified. This result highlights the selective nature of SERS when multi-components of varying solubilities and concentrations can interact with the Ag colloid, and where some components are likely preferentially adsorbed over others. In this case, implementing a further SERS extraction step with methanol/HCl could be of some benefit.

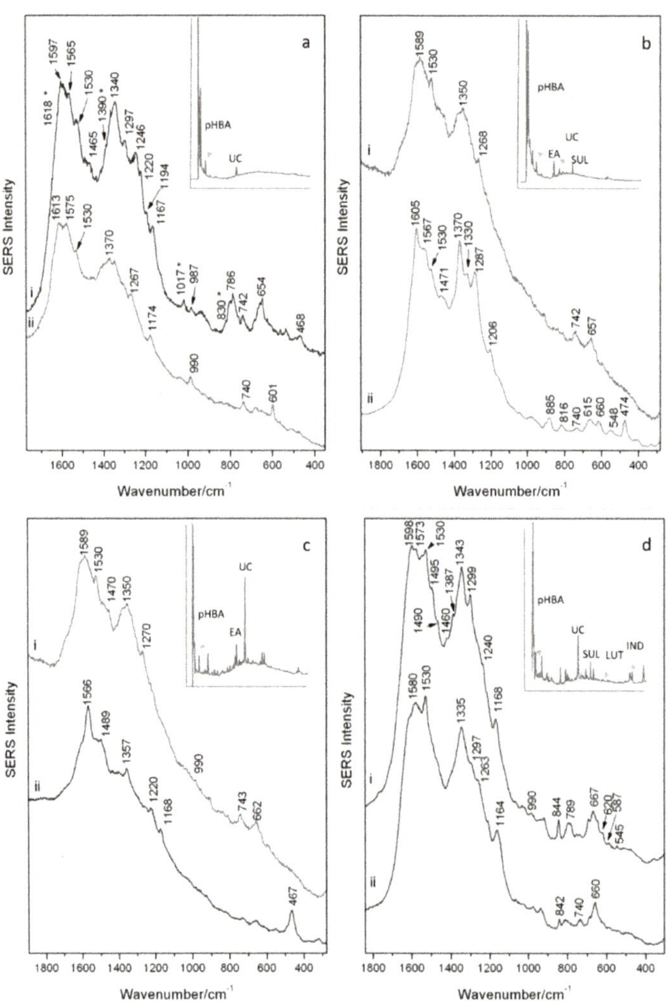

Figure 4. SERS spectra obtained (i) with regular colloidal interaction and (ii) following HF pre-treatment respectively for samples: (**a**) wool S#5, (**b**) silk S#11; (**c**) linen S#27 and (**d**) silk S#9. The corresponding HPLC-DAD chromatograms of the extracts of the samples @275 nm are inserted with abbreviations: pHBA = 4-hydroxybenzoic acid; EA = ellagic acid; UC = urolithin C; LUT = luteolin; and SUL = sulfuretin and are available enlarged in supporting information S1.

In the chromatogram (insert Figure 4c) of the extract of the linen sample S#27 shown at 275 nm, the peak for urolithin C is the most intense, while several other dye materials are also present. These include minor components typical of brazilwood and logwood dyes (sappanol; hematein; brazilein; hematoxylin; and 4-hydroxybenzoic acid) [14,31]. This sample also contains ellagic acid, suggesting the co-presence of tannins, again either used to modify the color or as a mordant during dyeing [7].

For the linen sample S#27, acquisitions performed with the colloidal SERS directly on the sample and after the hydrofluoric acid pre-treatment gave rise to the spectra shown in Figure 4c(i,ii), respectively. By direct colloidal SERS, the main features—albeit quite noisy—of the brazilwood marker urolithin C can be seen, with a band at 1530 cm^{-1} accompanied by broad bands at 1589, 660 and 990 cm^{-1} (very weak). The band at 1350 cm^{-1} possibly indicates the co-presence of brazilin/brazilein or even haematin from logwood, a further neoflavonoid which would not have been taken into consideration had it not been for the HPLC results. Following HF pre-treatment, the bands at 1566, 1357 and 467 cm^{-1} are slightly shifted from the characteristic bands reported here for brazilin/brazilein (shown in Figure 2), and the peak at 1377 cm^{-1} attributed to brazilein cannot be observed. It is plausible to hypothesize the presence of logwood also given the similarity of the SERS spectral features [25]. A few further weak bands can be seen in Figure 4c(i,ii), namely those in the 1470–1490 cm^{-1} range, and a band at 1270 cm^{-1} seen only in Figure 4c(i). These bands cannot be specifically assigned, but most likely belong to the additional minor components, 4-hydroxybenzoic acid and ellagic acid, as identified by HPLC. In this sample the urolithin C component is readily available to interact with the colloid without any pre-treatment steps and despite the confirmed presence of other coloring components. Once these other components were subsequently liberated from the mordant/textile complex following acid hydrolysis, they can be preferentially observed and distinguished. This underlines the necessity of the two-step process for a fuller sample characterization.

In the chromatogram of the extract of the deeper orange silk sample S#9 (insert Figure 4d) shown at 275 nm, urolithin C is the most intense peak. In addition, sappanol and 4-hydroxybenzoic acid are identified, as are sulfuretin, a sulfuretin-glucoside and fisetin, all ascribable to the use of young fustic [36]. Luteolin was also observed, suggesting the co-presence of a second yellow flavonoid dye. The presence of indigotin suggests that indigo or woad was also used to obtain the desired hue.

The SERS spectra obtained from S#9 can be observed in Figure 4d(i,ii) for direct colloidal SERS and after HF pre-treatment, respectively. Interestingly, in both spectra the main bands of urolithin C (1530, ~1335 and ~660 cm^{-1}) can be observed. In this case, the acid pre-treatment does not enhance the quality of the spectra. Given the delicate nature of the substrate and acid hydrolysis conditions, it is worth nothing that no bands relating to the protein content of the silk matrix are observed by SERS [37]. Furthermore, no bands attributable to brazilein are apparent either. Beyond the marker component, a number of other weaker bands can be observed relating to the flavonoid content in the sample highlighted by HPLC analysis. Specifically, in Figure 4d(i), the bands at 1573 and 1240 cm^{-1} relate to luteolin, the band at 844 cm^{-1} is suggestive of components with a flavonoid structure, while those at 587 and 620 cm^{-1} are features which correlate with the suggested fustic content [26]. HPLC analysis also indicates the presence of indigo, which could be observed instead by conventional Raman pre-SERS (spectrum not shown). The identification of the marker component is apparent in this sample in the presence of flavonoid and indigoid dyes and is evidence of the use of brazilwood even when the brazilwood colorants have not survived.

4. Conclusions

In this work, the spectral features of the brazilwood marker component urolithin C have been characterized by conventional and surface-enhanced Raman spectroscopy, and compared to those of the main brazilwood chromophores in the pure reference material and in a brazilwood lake pigment sample. The results obtained suggested that the SERS approach was suitable for use on historical textile samples and it was subsequently successfully applied to minute samples from four historical linen, silk and wool textiles dyed with brazilwood and showing varying degrees of fading. The SERS findings were correlated with HPLC-DAD and HPLC-ESI-Q-ToF-MS analyses which offered a full description of the major and minor components in each sample. The combined use of the direct colloidal approach followed by a HF pre-treatment was demonstrated to provide a fuller characterization protocol than using either approach alone. Using direct colloidal SERS methods, it was possible to observe the presence of the marker component, urolithin C, in each textile sample even when no brazilwood colorant survived and in the co-presence of other dyestuffs and textile degradation products such as 4-hydroxybenzoic acid. The additional acidic hydrolysis pre-treatment has been shown to promote the release of any remaining colorants from the mordant/textile fiber which allowed characterization of minor tannin (ellagic acid) and flavonoid dyestuff components (luteolin, fisetin) also present in the samples, although this treatment could make it harder to detect the brazilwood marker component. The results presented here confirmed the original use of brazilwood in each textile even when a further neoflavonoid-based dye, logwood, is hypothesized. This SERS methodology has the potential to be used as an in situ method for the examination of textiles where brazilwood colorant is potentially present at the fiber surface. Other artefacts where a brazilwood-based colorant is suspected in exposed surface layers could be examined using a similar approach.

Supplementary Materials: The following are available online at https://www.mdpi.com/article/10.3390/heritage4030078/s1. Figure S1: The HPLC-DAD chromatograms featured in Figure 4.

Author Contributions: Conceptualization, B.D., A.R., C.H., D.P. and C.M.; data curation, B.D. and I.D.; formal analysis, B.D. and I.D.; methodology, A.R., M.P.C. and C.M.; supervision, M.P.C. and C.M.; writing—original draft, B.D., I.D., A.R., C.H. and D.P.; writing—review and editing, B.D., I.D., A.R., C.H., D.P., M.P.C. and C.M. All authors have read and agreed to the published version of the manuscript.

Funding: This research was funded by the research activities in the IPERION HS project, funded by the European Commission, H2020-INFRAIA-2019-1 (Grant Agreement n. 871034).

Data Availability Statement: All raw and processed data are in the process of being made available to accompany this paper, please contact the corresponding author for updates.

Conflicts of Interest: The authors declare no conflict of interest.

References

1. Cardon, D. *Natural Dyes: Sources, Tradition, Technology and Science*; Archetype Publications Ltd: London, UK, 2007.
2. Kirby, J.; Saunders, D.; Spring, M. The Object in Context: Crossing Conservation Boundaries. In Proceedings of the Contributions to the Munich Congress, Munich, Germany, 28 August–1 September 2006; Studies in Conservation. Volume 51, pp. 236–243. [CrossRef]
3. Kirby, J.; Spring, M.; Higgitt, C. The Technology of Red Lake Pigment Manufacture: Study of the Dyestuff Substrate. *Natl. Gallery Tech. Bull.* **2005**, *26*, 71–88.
4. Bomford, D.; Kirby, J.; Roy, A.; Rüger, A.; White, R. *Art in the Making: Rembrandt*; National Gallery Company Ltd: London, UK, 2006; pp. 41–44.
5. Centeno, S.A.; Hale, C.; Carò, F.; Cesaratto, A.; Shibayama, N.; Delaney, J.; Dooley, K.; van der Snickt, G.; Janssens, K.; Stein, S.A. Van Gogh's Irises and Roses: The contribution of chemical analyses and imaging to the assessment of color changes in the red lake pigments. *Herit. Sci.* **2017**, *5*, 18. [CrossRef]
6. Kirby, J.; Van Bommel, M.; Verhecken, A. *Natural Colorants for Dyeing and Lake Pigments Practical Recipes and Their Historical Sources*; Archtype Publications Ltd: London, UK, 2014.
7. de Graaff, J.H.H. *The Colourful Past: Origins, Chemistry and Identification of Natural Dyestuffs*; Abbeg-Stiftung: Riggisberg & Archetype Publications Ltd: London, UK, 2004.
8. De Oliveira, L.F.C.; Edwards, H.G.M.; Velozo, E.S.; Nesbitt, M. Vibrational spectroscopic study of brazilin and brazilein, the main constituents of brazilwood from Brazil. *Vib. Spectrosc.* **2002**, *28*, 243–249. [CrossRef]
9. Edwards, H.G.M.; De Oliveira, L.F.; Nesbitt, C.M. Fourier-transform Raman characterization of brazilwood trees and substitutes. *Analyst* **2003**, *128*, 82–87. [CrossRef]

10. Nowik, W. The Possibility of Differentiation and Identification of Red and Blue Soluble Dyewoods. Determination of Species used in Dyeing and Chemistry of their Dyestuffs. Dyes in History and Archaeology16/17. Presented at the 16th Meeting, Lyons, France, 1997 and the 17th Meeting, Greenwich, UK, 1998; Kirby, J., Ed.; Archetype Publications Ltd.: London, UK, 2001; pp. 129–144.
11. Lluveras-Tenorio, A.; Parlanti, F.; Degano, I.; Lorenzetti, G.; Demosthenous, D.; Colombini, M.P.; Rasmussen, K.L. Spectroscopic and mass spectrometric approach to define the Cyprus Orthodox icon tradition—The first known occurrence of Indian lac in Greece/Europe. *Microchem. J.* **2017**, *131*, 112–119. [CrossRef]
12. Peggie, D.A.; Kirby, J.; Poulin, J.; Genuit, W.; Romanuka, J.; Wills, D.F.; De Simone, A.; Hulme, A.N. Historical mystery solved: A multi-analytical approach to the identification of a key marker for the historical use of brazilwood (Caesalpinin spp.) in paintings and textiles. *Anal. Methods* **2018**, *10*, 617–662. [CrossRef]
13. Nirma, N.P.; Rajput, M.S.; Prasad, R.G.S.V.; Ahmad, M. Brazilin from Caesalpinia sappan heartwood and its pharmacological activities: A review. *Asian Pac. J. Trop.* **2015**, *8*, 421–430. [CrossRef] [PubMed]
14. Tamburini, D. Investigating Asian colourants in Chinese textiles from Dunhuang (7th–10th century AD) by high performance liquid chromatography tandem mass spectrometry—Towards the creation of a mass spectra database. *Dye. Pigment.* **2019**, *163*, 454–474. [CrossRef]
15. Yin, P.; Zhang, J.; Yan, L.; Yang, L.; Sun, L.; Shi, L.; Ma, C.; Liu, Y.; Urolithin, C. A gut metabolite of ellagic acid, induces apoptosis in PC12 cells through a mitochondria-mediated pathway. *RSC Adv.* **2017**, *7*, 17254–17263. [CrossRef]
16. Espín, J.C.; Larrosa, M.; García-Conesa, M.T.; Tomás-Barberán, F. *Evidence-Based Complementary and Alternative Medicine*; Hindawi Publishing Corporation: London, UK, 2013. [CrossRef]
17. Dillmann, P.; Bellot-Gurlet, L.; Nenner, I. (Eds.) *Nanoscience and Cultural Heritage*; Atlantis Press: Paris, France, 2016.
18. Leona, M.; Stenger, J.; Ferloni, E. Application of Surface-Enhanced Raman scattering techniques to the ultrasensitive identification of natural dyes in works of art. *J. Raman Spectrosc.* **2006**, *37*, 981–992. [CrossRef]
19. Cleland, E.; Marjorie, E.; Wieseman, E. *Renaissance Splendor—Catherine de' Medici's Valois Tapestries*; With contributions by de Luca, F., Griffo, A., Perrone Da Zara, C., Beyer, C., Eds.; The Cleveland Museum of Art: Yale University Press: New Haven, CT, USA, 2019; ISBN-13: 978-0300237061.
20. Lee, P.C.; Meisel, D. Adsorption and surface-enhanced Raman of dyes on silver and gold sols. *J. Phys. Chem.* **1982**, *86*, 3391–3395. [CrossRef]
21. Le Ru, E.; Etchegoin, P. *Principles of Surface-Enhanced Raman Spectroscopy and Related Plasmonic Effects*, 1st ed.; Elsevier: Amsterdam, The Netherlands, 2008.
22. Idone, A.; Gulmini, M.; Henry, A.I.; Casadio, F.; Chang, L.; Appolonia, L.; Van Duyne, R.P.; Shah, N.C. Silver colloidal pastes for dye analysis of reference and historical textile fibers using direct, extractionless, non-hydrolysis surface-enhanced Raman spectroscopy. *Analyst* **2013**, *138*, 5895. [CrossRef] [PubMed]
23. Pozzi, F.; Zaleski, S.; Casadio, F.; Van Duyne, R.P. SERS Discrimination of Closely Related Molecules: A Systematic Study of Natural Red Dyes in Binary Mixtures. *J. Phys. Chem. C* **2016**, *120*, 21017–21026. [CrossRef]
24. Alajtal, A.I.; Edwards, H.G.M.; Elbagerma, M.A.; Scowen, I.J. The effect of laser wavelength on the Raman Spectra of phenanthrene, chrysene and tetracene: Implications for extra-terrestrial detection of polyaromatic hydrocarbons. *Spectrochim. Acta A* **2010**, *76*, 1–5. [CrossRef]
25. Bruni, S.; Guglielmi, V.; Pozzi, F. Historical organic dyes: A surface-enhanced Raman scattering (SERS) spectral database on Ag Lee–Meisel colloids aggregated by $NaClO_4$. *J. Raman Spectrosc* **2011**, *42*, 1267–1281. [CrossRef]
26. Fazio, E.; Neri, F.; Valenti, A.; Ossi, P.M.; Trusso, S.; Ponterio, R.C. Raman spectroscopy of organic dyes adsorbed on pulsed laser deposited silver thin films. *Appl. Surf. Sci.* **2013**, *278*, 259–264. [CrossRef]
27. Martina, I.; Wiesinger, R.; Jembrih-Simbürger, D.; Schreiner, M. Micro-Raman characterisation of silver corrosion products: Instrumental set up and reference database. *E Preserv. Sci.* **2012**, *9*, 1–8.
28. Arivazhagan, M.; Subhasini, V.P.; Kavitha, R. Density functional theory investigations on the conformational stability, molecular structure and vibrational spectra of 6-methyl-2-chromenone. *Spectrochim. Acta* **2014**, *128*, 527–539. [CrossRef] [PubMed]
29. Wang, X.; Xu, Q.; Hu, X.; Han, F.; Zhu, C. Silver-nanoparticles/graphene hybrids for effective enrichment and sensitive SERS detection of polycyclic aromatic hydrocarbons. *Spectrochim. Acta A* **2020**, *228*, 117783. [CrossRef]
30. Murray, C.A.; Bodoff, S. Depolarization effects in Raman scattering from cyanide on silver island films. *Phys. Rev. B* **1985**, *32*, 671. [CrossRef]
31. Hulme, A.N.; McNab, H.; Peggie, D.A.; Quye, A. Negative ion electrospray mass spectrometry of neoflavonoids. *Phytochemistry* **2005**, *66*, 66–70. [CrossRef]
32. Degano, I.; Biesaga, M.; Colombini, M.P.; Trojanowicz, M. Historical and archaeological textiles: An insight on degradation products of wool yarns. *J. Chromatogr. A* **2011**, *1218*, 5837–5847. [CrossRef]
33. Gao, J.; Hu, Y.; Li, S.; Zhang, Y.; Chen, X. Adsorption of benzoic acid, phthalic acid on gold substrates studied by surface-enhanced Raman scattering spectroscopy and density functional theory calculations. *Spectrochim. Acta A* **2013**, *104*, 41–47. [CrossRef]
34. Peggie, D.A. *The Development and Application of Analytical Methods for the Identification of Dyes on Historical Textiles*; University of Edinburgh: Edinburgh, Scotland, 2006. Available online: http://hdl.handle.net/1842/15623 (accessed on 7 December 2020).
35. Lee, S.J.; Cheong, B.S.; Cho, H.G. Surface-enhanced Raman Spectroscopic Studies of Ellagic Acid in Silver Colloids. *Bull. Korean Chem. Soc.* **2015**, *36*, 1637–1644. [CrossRef]

36. Valianou, L.; Stathopoulou, K.; Karapanagiotis, I.; Magiatis, P.; Pavlidou, E.; Skaltsounis, A.L. Phytochemical analysis of young fustic (Cotinuscoggygria heartwood) and identification of isolated colourants in historical textiles. *Anal. Bioanal. Chem.* **2009**, *394*, 871–882. [CrossRef]
37. Pozzi, F.; Lombardi, J.R.; Bruni, S.; Leona, M. Sample treatment considerations in the analysis of organic colorants by surface-enhanced Raman scattering. *Anal. Chem.* **2012**, *84*, 3751–3757. [CrossRef]

heritage

MDPI

Article

Yellow Lake Pigments from Weld in Art: Investigating the Winsor & Newton 19th Century Archive

Maria Veneno [1], Paula Nabais [2,*], Vanessa Otero [2,3,*], Adelaide Clemente [4], M. Conceição Oliveira [5] and Maria João Melo [2]

1 Department of Chemistry, Faculty of Sciences and Technology, NOVA University of Lisbon, 2829-516 Caparica, Portugal; m.veneno@campus.fct.unl.pt

2 LAQV-REQUIMTE and Department of Conservation and Restoration, Faculty of Sciences and Technology, NOVA University of Lisbon, 2829-516 Caparica, Portugal; mjm@fct.unl.pt

3 VICARTE and Department of Conservation and Restoration, Faculty of Sciences and Technology, NOVA University of Lisbon, 2829-516 Caparica, Portugal

4 cE3c-Centre for Ecology, Evolution and Environmental Changes, Faculdade de Ciências, Universidade de Lisboa, 1749-016 Lisboa, Portugal; maclemente@fc.ul.pt

5 Centro de Química Estrutural, Instituto Superior Técnico, Universidade de Lisboa, 1049-001 Lisboa, Portugal; conceicao.oliveira@tecnico.ulisboa.pt

* Correspondence: p.nabais@campus.fct.unl.pt (P.N.); van_otero@fct.unl.pt (V.O.)

Citation: Veneno, M.; Nabais, P.; Otero, V.; Clemente, A.; Oliveira, M.C.; Melo, M.J. Yellow Lake Pigments from Weld in Art: Investigating the Winsor & Newton 19th Century Archive. *Heritage* **2021**, *4*, 422–436. https://doi.org/10.3390/heritage4010026

Academic Editor: Diego Tamburini

Received: 1 February 2021
Accepted: 21 February 2021
Published: 25 February 2021

Abstract: Weld (*Reseda luteola*) was one of the main sources of yellow dyes used for dyeing textiles and to prepare artists' pigments in Europe until the 19th century. For the first time, this work explores the technology of preparing weld lake pigments in the 19th century by Winsor & Newton (W&N), a renowned supplier of artists' materials. Five recipes were discovered in the W&N 19th century Archive Database and reconstructed in the laboratory. W&N was extracting weld in neutral and basic media, and preparing the insoluble lake by complexation with Al^{3+} in the form of alum ($KAl(SO_4)_2{\bullet}12H_2O$) or hydrated alumina ($Al(OH)_3$). Five yellow lake pigments were successfully obtained and characterized by High-Performance Liquid Chromatography-Diode Array Detector (HPLC-DAD) and Fourier Transform Infrared Spectroscopy (FTIR). Their chromatographic profiles display as main yellows, luteolin 7-*O*-glucoside (Lut-7-*O*-glu) or both Lut-7-*O*-glu plus luteolin 3′,7-*O*-glucoside (Lut-3′,7-*O*-glu). In two of the processes, the presence of gypsum ($CaSO_4{\bullet}2H_2O$) was unequivocally detected by FTIR, being formed as a by-product. This work offers the first identification of weld lake pigments' characteristic infrared bands. The W&N Database proved again to be a unique source of information on 19th-century artists' materials and their commercial preparation. The knowledge gain is essential to ensure effective conservation and authentication procedures.

Keywords: weld lake pigments; yellow lakes; luteolin; 19th century; Winsor & Newton; conservation

1. Introduction

Yellow dyes were used in artworks for millennia up until the advances in modern chemistry. *Reseda luteola* L., or weld, was one of the most important dyes in Europe up until the 19th century, and the primary source for organic yellows [1]. These were used in the textile industry as a source of yellow and green colors and prepared as artists' pigments to create precious masterpieces [1–3].

Although weld was possibly identified in textiles from Xinjiang [4,5], in 17th-century Arraiolos carpets, Portugal [6] and in Southern Swedish painted wall hangings from the 18th–19th centuries [7], assessing its conservation condition and the causes of degradation in artworks is still in its early stages. To understand the degradation mechanisms that are in play in such complex matrices as found in our cultural heritage, it is necessary to have reference materials prepared with as much historical accuracy as possible. These are used to assess the natural evolution of these colors and simulate by accelerated ageing experiments with a limited number of variables the aging of these systems.

Methanol:water extracts of *Reseda luteola*, which gave the highest flavonoid yield, have shown that the main chromophore is luteolin 7-*O*-glucoside (Lut-7-*O*-glu), followed by luteolin 3′,7-*O*-glucoside (Lut-3′,7-*O*-glu). Luteolin (Lut), apigenin 7-*O*-glucoside (Api-7-*O*-glu), chrysoeriol glycoside (Chry-gly), luteolin 4′-*O*-glucoside (Lut-4′-*O*-glu), are also found, with apigenin (Api), apigenin-6,8-di-*C*-glucoside (Api-6,8-*C*-glu), and a luteolin di-*O*-glucoside (Lut-di-*O*-glu) found in lower amounts [6,8–10]. When analyzing dyed textiles, weld is identified as the yellow used by the presence of "luteolin-type" flavonoids. In 17th-century Arraiolos carpets, the yellow historical samples analyzed with LC–MS contained primarily Lut-7-*O*-glu, small amounts of apigenin-6,8-di-*C*-glucoside, Lut-3,7-*O*-glu and its isomer, as well as Api-7-*O*-glu and Lut, see Figure 1 [6]. Moreover, for the dyed textiles from Xinjiang, the identification of Lut-7-*O*-glu, along with other "luteolin-type" and "apigenin-type" flavonoids, led to the proposal of the use of weld in the samples analyzed [4,5].

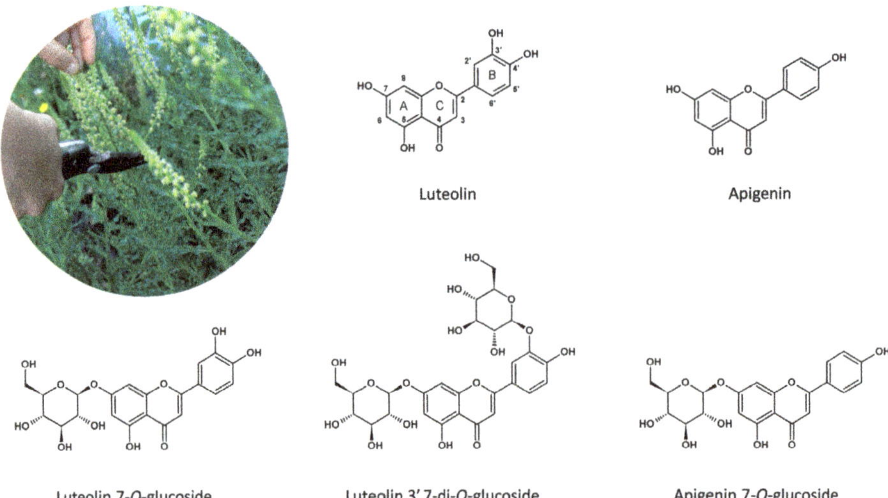

Figure 1. Collection of *Reseda luteola* in its native environment; structures for the main chromophores found in *Reseda luteola* yellows: luteolin, apigenin, luteolin 7-*O*-glucoside, luteolin 3′,7-di-*O*-glucoside, and apigenin 7-*O*-glucoside.

Generally, most dyes were applied as lake pigments, formed by the colorant's precipitation with a complexing agent, such as alum, hence becoming a non-soluble pigment, in a process analogous to the mordanting of textiles [2]. Although these dye-metal complexes' exact structure is still unknown for most lake pigments, there are some proposals for luteolin-metal complexes [11–13]. Following a DFT/TDDFT study of the complexation sites of luteolin and apigenin, Amat et al. proposed that luteolin is preferentially co-precipitated or absorbed with Al^{3+}, or other metals in a bi-dentate mode involving the 4-keto-5-hydroxy site and with a Al:Luteolin 1:1 stoichiometry [11]. The same structure was proposed by Gao et al. for luteolin-Cr(III) complexes, while Dong et al. proposed the same coordination sites for complexation with manganese (II), although with an Mg:Luteolin 1:2 stoichiometry, which means the existence of a complexation network is expected involving the hydroxyl groups [12,13]. On the other hand, Smith et al. found that the aluminium ion-flavonoids complexes (present in dyed textiles and lake pigments) prevent the natural efficient and non-degradative dissipation of excitation energy by an intermolecular proton transfer involving the 5-OH and the 4=O groups, hence are more susceptible to degradation [14].

Within an interdisciplinary team of chemists, botanists, and heritage scientists, with 20 years of experience in studying and retrieving the "lost knowledge" on natural dyes found in historical documents and artworks [15–19], this work will be the first step of a systematic study on the technology used in the past to produce weld lake pigments. For this first approach, we investigated recipes found in the Winsor & Newton (W&N) 19th century Archive Database, a unique primary documentary source covering handwritten formulation instructions and workshop notes of a leading artists' colormen that supplied prominent painters. The W&N Archive Database comprises a summary index-linked to digitalized page-images of 85 manuscript books (corresponding to 15.003 database records) and a digital collection of 47 W&N 19th-century trade and retail catalogues [20–22].

In a time of chemical development, especially of artificial dyestuffs, it is very interesting to note that W&N was producing at an industrial scale and selling natural yellow lake pigments during the 19th century [23]. In previous studies, we have proven that W&N was committed to primarily selling the most high-quality and durable products [18,24,25]. More importantly, we have demonstrated that research on the W&N Database enables pigment reconstructions with as much historical accuracy as possible. These references will be fundamental to advance analytical methodologies on the identification of weld lake pigments in artworks. In this work, for the first time, we disclose the infrared bands of weld lake pigments, complemented by their chromatographic profiles.

2. Materials and Methods

2.1. Materials

All solvents used were HPLC grade. For all chromatographic studies as well as dye extraction, Millipore ultrapure water was used. Luteolin ($C_{15}H_{10}O_6$), luteolin-7-O-glucoside ($C_{21}H_{20}O_{11}$) and luteolin-3′,7-di-Oglucoside ($C_{27}H_{30}O_{16}$) analytical standards were purchased from Extrasynthese®. Potassium aluminium sulfate ($AlK(SO_4)_2 \cdot 12H_2O$), potassium bicarbonate ($KHCO_3$), sodium borate ($Na_2B_4O_7 \cdot 10H_2O$), calcium carbonate ($CaCO_3$), and hydrated alumina ($Al(OH)_3$), were purchased from Sigma-Aldrich®, while potassium carbonate (K_2CO_3) was purchased from Merck®. Gum arabic to prepare paint references was purchased in pieces from Kremer Pigmente®.

Flowering branches of *Reseda luteola* were collected in June 2020 by A. Clemente, from wild populations near Bucelas, north of Lisbon, Portugal (38°54′21″ N–9°6′30″ W). The plant material was spread in a tray and air-dried in the dark, in a ventilated area at 20 °C.

2.2. Synthesis Methods for Weld Lake Pigments

Research on the W&N 19th century Archive Database was carried out under the sub-topic weld, which resulted in 12 database records. Among these, it was possible to identify seven records for the production of weld lake pigments, however, there are only five recipes as 3 of the records are copies, see Table 1. The remaining five database records include two notes on weld, two experiments to extract weld, and one recipe to prepare Yellow Carmine from weld, Persian berries and quercitron bark. The transcription of the production records used in this work may be consulted in Table S1, and the synthesis methods reproduced are described in Table 1. Interpretation of the materials used was based on our previous works [18,24,25], which allowed us to infer that the term 'whiting' used in the recipe *Yellow from Weld* corresponds to calcium carbonate ($CaCO_3$), 'Sub. Carb. Pot', used in the same recipe, is potassium bicarbonate ($KHCO_3$) and 'Pearlash', used in the other recipes, is potassium carbonate (K_2CO_3). It is important to note that the recipe *Yellow Lake. Cool tint.* refers to the use of alum (ammonia sort), however, it was chosen to use common alum ($KAl(SO_4)_2 \cdot 12H_2O$) in all recipes to facilitate a first comparison between them. For this reason, we also decided always to use weld flowers. The introduction of experimental variants such as using ammonium alum ($NH_4Al(SO_4)_2 \cdot 12H_2O$) will be investigated in the future. All materials were scaled-down from industrial to laboratory scale, and quantities in British measures were converted to SI units [22,23].

pH measurements were acquired throughout the syntheses. After 1 day left to precipitate, the lakes were centrifuged for 10 min at 2400 rpm, washed with distilled water, and centrifuged again for 5 min at 3000 rpm. The lakes were air-dried and ground in an agate mortar for 15 min each.

Table 1. Production name, recipe and pigment code of the synthesis methods for weld lake pigments, adapted from the original text transcribed in Table S1.

Production Name	Unique Recipe Code §	Pigment Code	Synthesis Methods
Yellow from Weld	4PP148AL01 (copy in P4P088L01)	WL1	A. To 10 mL of boiling water add 0.2 g of $CaCO_3$ and then slowly add 0.86 g of $KAl(SO_4)_2{\cdot}12H_2O$, always stirring. Leave it to rest and decant the solution; keep the precipitate. B. To 50 mL of boiling water add 2 g of weld flowers. Then add 0.014 g of $KHCO_3$ and leave it to boil during 20 min. Filter it and keep the solution. Put the solution B to boil. When boiling, add the precipitate A and leave it boiling for 1 h, always stirring. Leave it to rest for 1 day and filter the yellow lake pigment.
	4PP148AL14 (copy in P4P089L14)	WL2	A. To 10 mL of boiling water add 0.86 g of $KAl(SO_4)_2{\cdot}12H_2O$ and then slowly add 0.2 g of $CaCO_3$, always stirring. Leave it to rest and decant the solution; keep the precipitate. B. To 50 mL of boiling water add 2 g of weld flowers. Then add 0.014 g of $KHCO_3$ and leave it to boil during 20 min. Filter it and keep the solution. Put the solution B to boil. When boiling, add the precipitate A and leave it boiling for 1 h, always stirring. Leave it to rest for 1 day and filter the yellow lake pigment.
Yellow Lake. Cool tint.	P1P348AL01 (copy in X6P228L01 ¥)	WL3	To 50 mL of boiling water add 0.43 g of K_2CO_3. When dissolved add 2 g of weld flowers and leave it to boil during 20 min. Filter it and keep the solution. To the yellow solution add 0.86 g of $KAl(SO_4)_2{\cdot}12H_2O$, always stirring. Leave it to rest and filter the yellow lake pigment.
Weld Yellow	P4P100L10	WL4	To 50 mL of boiling water add 0.018 g of K_2CO_3 and then 4 g of weld flowers. Boil 10 min. Filter it and keep the solution. To the yellow solution add 0.107 g of $KAl(SO_4)_2{\cdot}12H_2O$ and then 0.07 g of $Na_2B_4O_7.10H_2O$, always stirring. Leave it to rest and filter the yellow lake pigment.
		WL5	To 50 mL of boiling water add 0.018 g of K_2CO_3 and then 2 g of weld flowers. Boil 10 min. Filter it and keep the solution. To the yellow solution, add 0.177 g of $Al(OH)_3$, always stirring. Leave it to rest and filter the yellow lake pigment.

§ The unique recipe code is the code from the W&N Database that identifies a database record. ¥ Although this record is a copy, its title is "Experiment with Weld for Yellow Lake for Water Colours" and is dated 6 October 1854.

2.3. Paint References

Paint references were prepared using gum arabic as a 20% solution; the pieces were ground and then added to pure water. The lake pigments were first ground in a glass mortar with pure water and then ground with the binder. The paints were applied on filter paper with a paintbrush and allowed to dry. Filter paper was selected because no additives are present such as brighteners; this was confirmed by checking the filter paper under an UV-lamp (280 nm). The paint references were analyzed by colorimetry.

2.4. Equipment and Characterization Methods

2.4.1. Colorimetry

For measuring color, a portable spectrophotometer colorimetry Data Color International was used. Its measuring head's optical system uses diffuse illumination from a pulsed Xenon arc lamp over the 8mm-diameter measuring area, with 0° viewing angle geometry. Color coordinates were calculated defining the D65 illuminant and the 10° observer. The calibration was performed with a white bright standard plate and a total black standard. Color, as perceived by the human eye, may be represented in a three-dimensional system. The color data are presented in the CIE-Lab system. In the Lab cartesian system,

L*, relative brightness, is represented by the z-axis. Variations in relative brightness range from white (L* = 100) to black (L* = 0). The (a*, b*) pair represents the hue of the object. The red/green y-axis plots a* ranging from negative values (green) to positive (red). The yellow/blue x-axis reports b* going from negative (blue) to positive numbers (yellow).

2.4.2. High-Performance Liquid Chromatography with a Diode Array Detector (HPLC-DAD)

For HPLC-DAD analysis, *Reseda luteola* plant was extracted by placing 1 g of the dry plant material (as collected from nature) with 100 mL of methanol:water (70:30, *v:v*) and heating in a water bath at 60 °C for one hour, as described in [26]. The dye from lake pigments was extracted by placing in an eppendorf, 10 mg of powder with a 1 mL solution of oxalic acid (0.2 M):methanol:acetone:water (0.1:3:3:4, *v:v*), as described in [27].

Prior to HPLC-DAD analysis, all extracts were centrifuged at 12,000 rpm for about 10 min. The supernatant liquid was gently removed and filtered through a 0.45 µm filter. Before analysis, the solution was diluted with methanol:water (70:30, *v:v*) if necessary.

The analysis was carried out in a Thermofinnigan Surveyor® HPLC-DAD system with a Thermofinnigan Surveyor PDA (Thermofinnigan, San Jose, CA, USA), an autosampler, and a gradient pump. The sample separations were performed in a reversed-phase column, RP-18 Nucleosil column (Macherey-Nagel) with 5 µm particle size column (250 mm × 4.6 mm), with a flow rate of 1.7 mL/min with the column at a constant temperature of 35 °C. The samples were injected via a Rheodyne injector with a 25 µL loop. The elution gradient consisted of two solvents, A: methanol and B: 0.1% (*v/v*) perchloric acid aqueous solution. A gradient elution program was used of 0–2 min isocratic 7% A, 2–8 min linear gradient to 15% A, 8–25 min linear gradient to 75% A, 25–27 min linear gradient to 80% A, 27–29 min linear gradient to 100% A, and 29–30 min isocratic 100% A (10 min re-equilibration time). The eluted peaks were monitored at 350 nm.

The peaks were integrated and the area of each peak was recorded as well as the percentage. Peak area calculation was done by defining the time intervals for each peak. The area below the peak was integrated within this interval is measured and the percentage of each area is calculated by dividing by the sum of all the peak areas. For this analysis, it was considered the area of the nine peaks visible at λ = 350 nm, between 14 and 24 min, as shown in Figure 2.

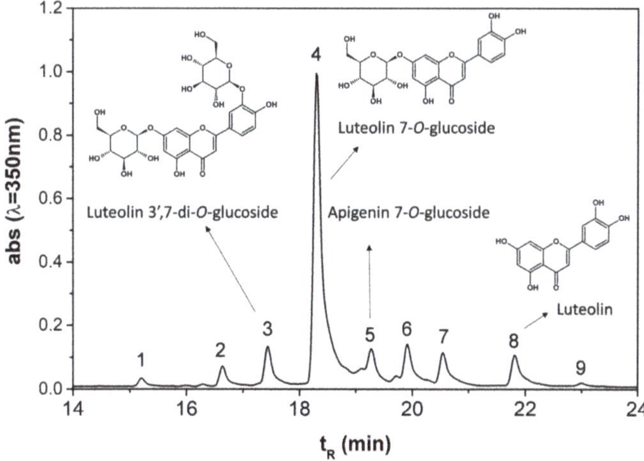

Figure 2. Chromatogram of an extract in MeOH:H_2O (70:30, *v:v*) of *Reseda luteola*, λ=350 nm: (1) apigenin-6,8-di-*C*-glucoside, (2) luteolin di-*O*-glucoside; (3) luteolin 3′,7-*O*-glucoside; (4) luteolin 7-*O*-glucoside, (5) apigenin 7-*O*-glucoside, (6) chrysoeriol glycoside, (7) luteolin 4′-*O*-glucoside, (8) luteolin, (9) apigenin.

2.4.3. Ultra-High Performance Liquid Chromatography-High Resolution Mass Spectrometry (UHPLC-DAD-HRMS)

Aliquots of 3 μL of plant material were analyzed on a UHPLC Elute system coupled on-line with a quadrupole time-of-flight Impact II mass spectrometer equipped with an ESI source (Bruker Daltoniks, Bremen, Germany). Chromatographic separation was carried out on an RF-C18 Halo column (150 mm × 2.1 mm, 2.7 μm particle size, Advanced Material Technology). The mobile phase consisted of water (A) and acetonitrile (B), containing 0.1% formic acid, at a flow rate of 600 μL/min. The elution conditions were as follows: 0–18 min, linear gradient to 50% B; 18–20 min, linear gradient to 90% B; 20–23 min, isocratic 90% B; and 23–24 min, linear gradient to 0% B (followed by 11 min re-equilibration time). The column and the autosampler were maintained at 45 °C and 8 °C, respectively. High-resolution mass spectra were acquired in the ESI negative mode. Internal calibration was achieved with an ammonium formate 10 mM solution introduced to the ion source via a 20 μL loop at the beginning of each analysis, using a six-port valve. The mass spectrometric parameters were set as follows: end-plate offset: 500 V; capillary voltage: −2.5 kV; nebulizer: 4 bars; dry gas: 8 L/min; heater temperature: 200 °C; m/z range 100–1000 Da; acquisition mode: data-dependent analysis (Auto MS/MS), acquisition rate of 3 Hz, and using a dynamic method with a fixed cycle time of 3, and an isolation window of 0.03 Da. Data acquisition and processing were performed using Data Analysis 4.2 software.

2.4.4. Fourier Transform Infrared Spectroscopy (FTIR)

Infrared analyses were carried out with a Nicolet Nexus spectrophotometer. The pigments were prepared as KBr pellets, and spectra were collected in transmission mode between 4000 cm^{-1} and 650 cm^{-1}, with a resolution of 8 cm^{-1} and 64 scans. The spectra are shown here as acquired, without corrections or any further manipulations.

3. Results and Discussion

3.1. Weld Lake Pigment Recipes in the Winsor & Newton 19th Century Archive Database

From a total of 1511 database records for yellow pigments, 42% pertains to yellow lakes. Although the majority of these records pertain to quercitron-based products, 12 of these mention weld, as referred above. Interestingly, the five recipes to prepared weld lake pigments were discovered in manuscript books belonging to the founder Henry Charles Newton and his son Arthur Henry Newton. The recipes were found under the names: "Yellow from weld", "Yellow Lake. Cool tint." and "Weld Yellow", as described in Tables 1 and 2. It is important to note that the pigments prepared from the first three recipes (*WL1*, *WL2* and *WL3*) were originally produced with 12.7 kgs of weld plants, whereas those from the last recipes ((*WL4* and *WL5*) were produced with 0.9 kg, which suggests the latter were experiments. Nonetheless, all recipes were reproduced, and the pHs of the extraction solution and after precipitation of the weld lake pigments obtained are also presented in Table 2.

As Table 1 shows, W&N was extracting weld in a neutral media for most recipes, excluding *WL3* recipe, which involved a basic media. This was accomplished by the addition of carbonate compounds (($KHCO_3$ and K_2CO_3) that "assist the extraction of the colouring matter" as stated in the recipe *Yellow from Weld (WL1* and *WL2*). The latter recipe also includes the preparation of what W&N called the "body" of the pigment formulation, which involves mixing calcium carbonate ($CaCO_3$) and alum ($KAl(SO_4)_2{\cdot}12H_2O$. *WL1* and *WL2* differ in the order of addition of these ingredients. According to W&N, the pigment resulting from *WL1* was "rather pale because the body was not thoroughly homogeneous" and the improved process *WL2* resulted in a color "deeper & looked brighter". Experimentally, we observe a yellow with a stronger red component, see Table 3. Curiously, they also refer "the quantity of yellow was less" for the *WL2*, however, we did not obtain this result as the yields are very similar, as presented in Table 2. In fact, those that experimentally presented the best yield of all five recipes were *WL1* and *WL2*, while *WL4* had the worst yield. Regarding recipe *WL3*, the yield experimentally obtained was very similar to W&N.

According to them, this recipe produced "a lively kind of yellow lake for sale &c. Greenish in hue, bright in the drop & full coloured" and when "tried in oil it produces a very beautiful yellow lake_very cool_bright & strong". Although the resulting pigment presents one of the lowest red component values ($a^* \approx 2$), it did not show a greenish hue. This may be related to incomplete precipitation of all coloring matter as in the original recipe (see Table S1) is claimed that the quantity of alum used by W&N "was found to precipitate the colour entirely, leaving only a very faint tinge of yellow in the supernatant". This was not observed in our experiment and will also be addressed in future work.

Table 2. Production name, ingredients, synthesis methods, final pHs and yields for W&N's weld lake pigments.

Dye Source	Extraction Method		Complexing Agent			Additives
Weld	Potassium bicarbonate $KHCO_3$	Potassium carbonate K_2CO_3	Alum $KAl(SO_4)_2$	Hydrated alumina $Al(OH)_3$	Calcium carbonate $CaCO_3$	Sodium Borate $Na_2B_4O_7$
	Ext	Ext	C	C	C	A

Production name	Code	Synthesis	pH extraction [1]	pH final	η (%) [2]
Yellow from Weld	WL1	Ext Filtration → C C	6.18	4.05	$\eta_{W\&N} = 25\%$ $\eta_{EXP} = 31\%$
	WL2	Ext Filtration → C C	6.27	3.36	$\eta_{W\&N} = 18\%$ $\eta_{EXP} = 32\%$
Yellow Lake. Cool tint.	WL3	Ext Filtration → C	9.46	3.63	$\eta_{W\&N} = 12.5\%$ $\eta_{EXP} = 12\%$
Weld Yellow	WL4	Ext Filtration → C A	6.02	5.07	$\eta_{W\&N}$ = n.a. $\eta_{EXP} = 6\%$
	WL5	Ext Filtration → C	6.35	6.15	$\eta_{W\&N}$ = n.a. $\eta_{EXP} = 9\%$

[1] The pH extraction is related to the pH after extracting the plant material and subsequent filtration. [2] The observed yield (η) was calculated as follows: (*final lake amount*) ÷ (*weld amount*) × 100. $\eta_{W\&N}$–observed yield calculated by the quantities given by W&N, see Table S1; η_{EXP}–observed yield calculated by the quantities obtained experimentally.

Table 3. Colorimetry, HPLC chromatograms and infrared spectra of the weld lake pigments synthesized applied over filter paper, with gum arabic media. HPLC chromatograms were obtained from the extracts of the lake pigments. For more details, please see text.

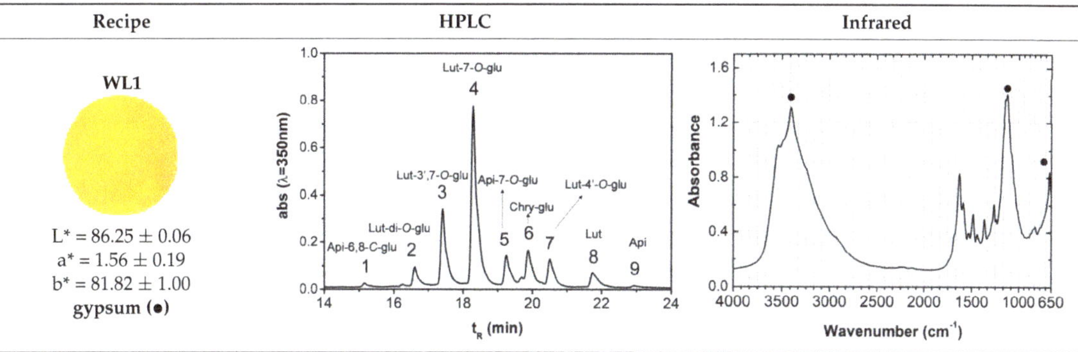

Table 3. *Cont.*

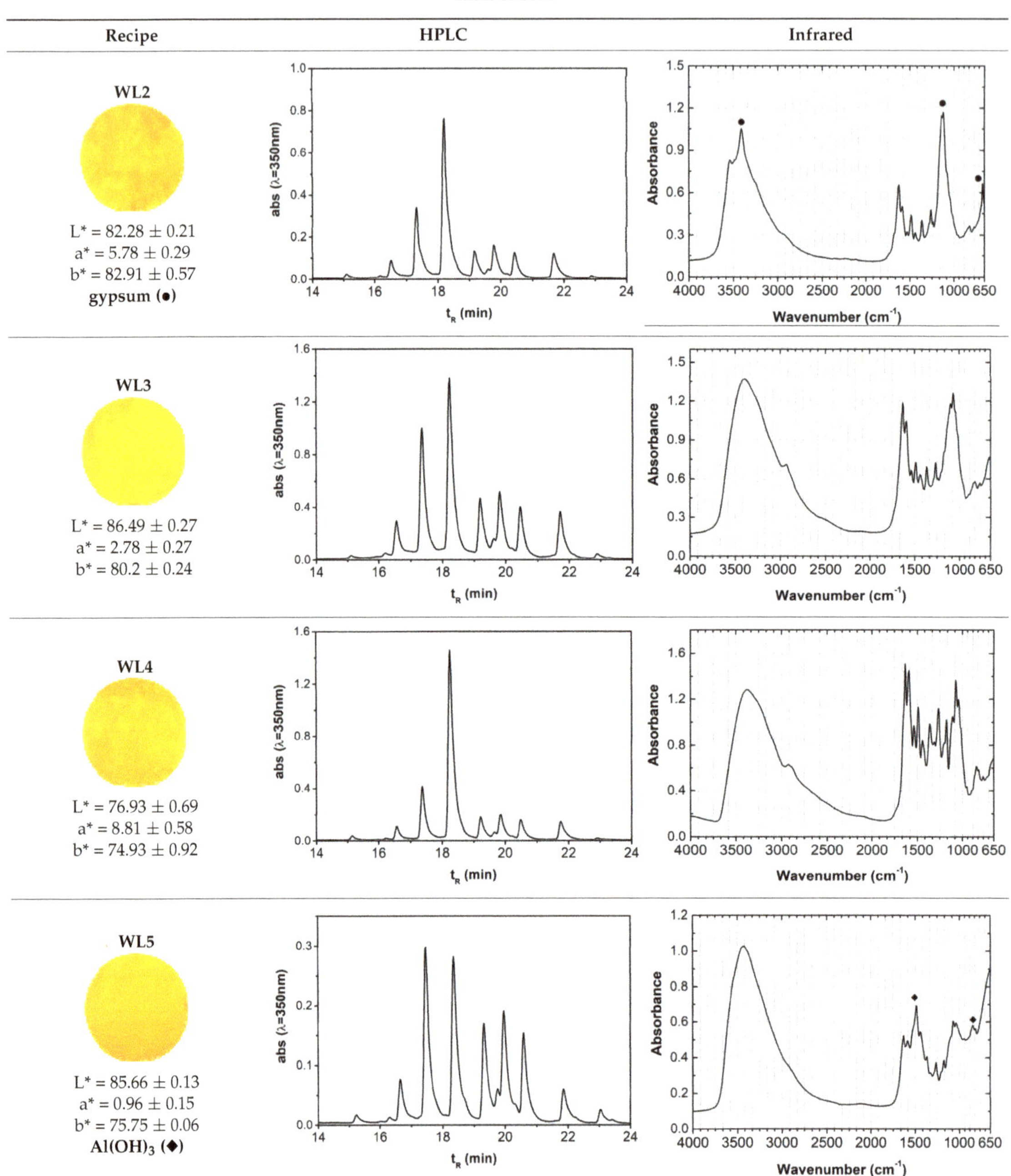

The complexing agent always used was Al^{3+} in the form of alum ($KAl(SO_4)_2 \cdot 12H_2O$) in the majority of the recipes and hydrated alumina ($Al(OH)_3$) in *WL5* recipe; however, complexation with Ca^{2+} cannot be excluded as this has been observed for W&N 19th century cochineal lake pigments [18]. The addition of borax ($Na_2B_4O_7$) was also experimented with alum in *WL4* recipe. The pH after precipitation was slightly acidic, between 3 and 6, but always resulted in bright yellow lake pigments, as shown in Table 3. The reasoning for the production methods found is discussed below.

3.2. Extraction Method

The analysis of the plant extract was done by HPLC-DAD-HRMS. It was possible to find as a major chromophore Lut-7-*O*-glu, and minor compounds, Api-6,8-*C*-glu, Lut-di-*O*-glu, Lut-3′,7-*O*-glu; Api-7-*O*-glu, Chry-gly, Lut-4′-*O*-glu, Lut and Api, see Figure 2. This is in accordance with what has been reported in the literature [6,8,9]. It was confirmed that the chromophores identified by HPLC-HRMS were the same as observed by HPLC-DAD.

For both *WL1* and *WL2* recipes, potassium bicarbonate is added to the plant material' solution, raising the pH to a neutral media (pH ~6). For the rest of the recipes, potassium carbonate was added previously to the weld, also resulting in a neutral media (pH ~6), except for the *WL3* recipe that remained at a basic media. Although the recipes present two different extraction methods, they do not influence the chromophores extracted as the chromatographic profiles are similar, see Figure 2 and Figure S1.

It is very interesting the use of carbonates ($KHCO_3$ and K_2CO_3) for the extraction of the flavonoids. The use of such extract solutions instead of water was possibly to allow the highest amount of lake pigment. Favaro et al. used fluorimetric titration to characterize the various luteolin species detected within the pH range explored (pH = 2–12) [28]: neutral form (pH < 5), mono-anion (pH ~7), di-anion (pH ~9) and tri-anion (pH ~12), and the successive deprotonations occur in the order 7-OH; 4′-OH; 3′-OH or/and 5-OH [28]. In *WL3*, *WL4* and *WL5* the extraction is carried out in a basic pH, turning neutral after the addition of the plant material. This creates the optimum conditions for the metal chelation through the OH at C^5 and the carbonyl at C^4, since the first is deprotonated only at pH $\approx$ 10.3, as mentioned above.

3.3. Characterization of the Weld Lake Pigments

A summary of the multi-analytical results of Colorimetry, High-Performance Liquid Chromatography-Diode Array Detector (HPLC-DAD) and Fourier Transform Infrared Spectroscopy (FTIR) for the weld lake pigments prepared may be observed in Table 3.

3.3.1. HPLC-DAD Analysis

Other authors have done an extensive analysis of the characterization of weld by HPLC, including quantitation of the chromophores [6,8–10,29,30]. Based on this, in this work, we only compared the chromatographic profiles of the lake pigments using HPLC-DAD, which is preferable to perform a semi-quantification.

When analyzing the HPLC chromatograms of the weld lake pigment extracts, some differences are visible, as shown in Table 3. The two variants of recipe *Yellow from Weld* (*WL1* and *WL2*) present the same chromatographic profile, indicating that the order in which alum and calcium carbonate are added does not affect the chromatographic profile, i.e., the percentage of chromophores present, as seen in Table S2. However, when compared with the extract, in Figure 2, it is possible to see that both lake pigments present a higher percentage of Lut-3′,7-*O*-glu than the plant extract (15.35–17.25% in the lake pigment when compared with 4.92% of the extract). Considering that the extracts in K_2CO_3 and $KHCO_3$ presented the same chromatographic profile as in MeOH:H_2O (see Figure S1), the difference is not due to different extraction methods, but possibly to a higher preference of complexation for the di-glucoside. Interestingly, this difference is even higher in the lake pigment from *WL3* and *WL5*, where the Lut-3′,7-*O*-gluc represents 20.91% and 23,05% of the total peak area, while the Lut-7-*O*-glu represents 30.26% and 18.63%, respectively. *WL3* is the only recipe with the addition of alum to an extraction solution of *Reseda luteola* at a basic pH of around 9. Moreover, *WL5* also has a higher percentage of Api-7-*O*-glu, representing

11.22% of the total peak area. *WL5* is the only recipe where alumina is added. Regarding the recipe *WL4*, it has the closest chromatographic profile to that of the extraction, with 13.17% of luteolin 3′,7-di-*O*-glucoside and 52.551% of luteolin 7-*O*-glucoside.

3.3.2. FTIR Analysis

Both yellow lake pigments from the recipe *Yellow from Weld* (*WL1* and *WL2*) have shown similar FTIR results as observed by HPLC-DAD. Notably, the formation of gypsum ($CaSO_4 \cdot 2H_2O$) by FTIR was detected, due to its characteristic absorption bands for νOH at 3405 cm^{-1}, $\nu_{as}(SO_4{}^{2-})$ at 1132 cm^{-1} and $\delta_{as}(SO_4{}^{2-})$ at 670 cm^{-1} [31], as observed in Table 3. Gypsum was not directly added but was rather a product of the reaction between alum ($KAl(SO_4)_2$) and calcium carbonate ($CaCO_3$). The reason why W&N chose to create gypsum through a reaction rather than adding it directly is still unclear at the moment. Further experimentation will be performed using gypsum directly in the recipe to understand the role of this reaction.

As may be seen in Table 3, FTIR analysis of all yellow lake pigments shows bands attributed to flavonoids-metal complexes, which is very clear in the infrared spectra of *WL3* and *WL4*, where only alum was added, plus borax in the latter recipe; the role of this ingredient is also still to be investigated. A more thorough analysis of the infrared data of the flavonoids-metal complexes is offered below. Besides identifying gypsum in *WL1* and *WL2* pigments, it was also detected the presence of hydrated alumina in the *WL5* lake pigment, due to its characteristic absorption bands for $\delta(H_2O)$ at 1485 cm^{-1} and the δ(Al-OH) at 852 cm^{-1} [31], as shown in Table 3.

3.4. *Infrared Markers of Weld Lake Pigments*

Although infrared spectroscopy has rarely been used to characterize flavonoids, several studies have proven its effectiveness in studying flavonoids-metal complexes [12,13]. Machado et al. [32] did DFT calculations of IR and Raman spectroscopies of hydroxyflavones, and their assignments are summarized in Table 4.

As mentioned above, flavonoids-metal complexes are observed in the infrared spectra of *WL3* and *WL4*. Figure 3 compares the two lake pigments and references of luteolin and luteolin 7-*O*-glucoside and shows the similarity between the infrared spectra, namely between *WL4* and luteolin 7-*O*-glucoside. These data are corroborated by the HPLC-DAD analysis, since this recipe has 53.31% of luteolin 7-*O*-glucoside, when compared with the 30.38% of the recipe *WL3*. The presence of a glucoside in position 7-OH shifted the vibrational frequency from 1186 to 1179 cm^{-1}. The fact that both lakes show lower frequencies, at 1178 and 1173 cm^{-1}, corroborates with the HPLC data. The reference of luteolin 3′,7-di-*O*-glucoside will provide more insight into the effect of glucosides in the infrared spectra, and further studies are underway.

Regarding the presence of an organometallic complex, the stretching vibration of C=O of luteolin at 1666 cm^{-1} is shifted to 1632 cm^{-1}. According to Dong, this shift is characteristic of the existence of a complex [12]. It is the co-ordination of carbonyl oxygen with metal ion bonded to 5-OH group of A ring and 4-CO carbonyl group of C ring [12,33,34]. This is also visible in the stretching at 1613 cm^{-1}. Moreover, the OH bending of C^5 shifts from 1509 cm^{-1} to 1484-9 cm^{-1}, probably also due to the metal coordination. Another possible indication of metal complexation in positions OH (C^5) and CO (C^4) is the decrease from 1096 cm^{-1} to 1077-68 cm^{-1} from the stretching of C^3-C^4. In fact, the extraction solution of *WL3* had pH ~9.4, the optimum conditions for the metal chelation in the OH at C^5 and the carbonyl at C^4. The analysis of luteolin, both as aglycone and with glucosides, complexed with Al^{3+} is ongoing.

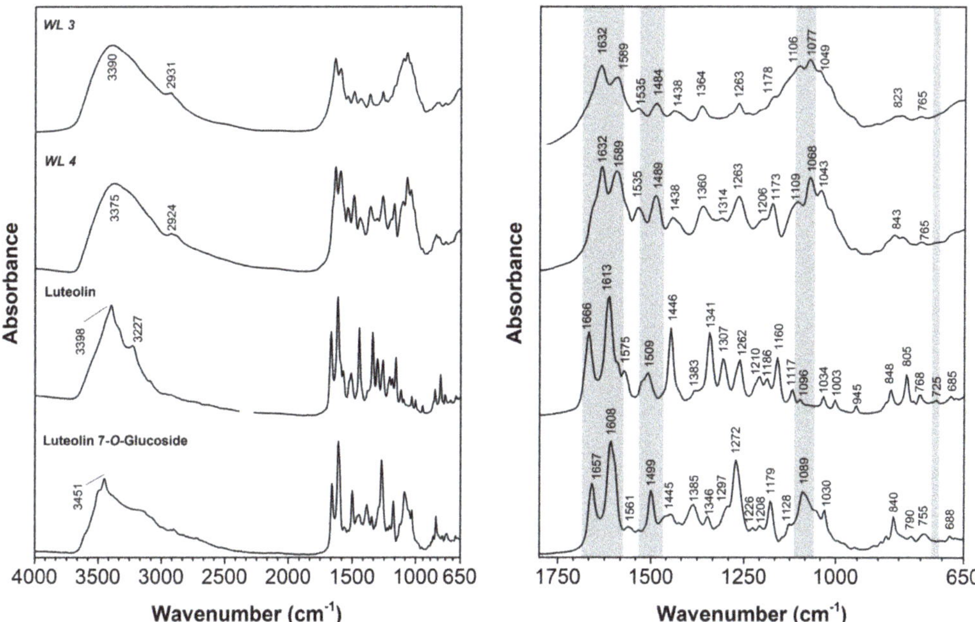

Figure 3. Infrared spectra of weld lake pigments *Yellow Lake. Cool tint.* (*WL3*) and *Weld yellow* (*WL4*), and references of Lut and Lut-7-*O*-glu.

Table 4. Infrared assignments for luteolin, luteolin 7-*O*-glucoside and weld lake pigments *WL3* & *WL4*. Highlighted in grey are the vibrations correlated with the A ring and positions C^5-OH and C^4=O, metal chelating groups.

Lut	Lut-7-*O*-glu	WL3	WL4	Literature [32] Lut	Assignments
3398	3451	3390	3375	~3400	$\nu(O^{3'}—H)$, $\nu(O^{7}—H)$, $\nu(O^{4'}—H)$
3227	-	-	-		$\nu(O^{5}—H)$
-	-	2931	2924		
1666	1657	1632	1632	1656	$\nu(C=O)$, $\nu(C^2—C^3)$, $\delta(O^3—H)$
1613	1608	1589	1589	1612	$\nu(C=O)$, $\delta(O—H)$, $\delta(O^{4'}—H)$
1575	-	-	-	1575	$\nu(C=O)$, $\delta(O—H)$, $\delta(O—H)$
-	1561	-	-	1561	$\nu(C^2=C^3)$, $\delta(O^5—H)$, $\delta(O^{4'}—H)$, $\delta(O^7—H)$
-	-	1535	1535	1518	$\delta(O^7—H)$, $\delta(O—H)^A$
1509	1499	1484	1489	1507	$\delta(O^5—H)$

Table 4. *Cont.*

Lut	Lut-7-*O*-glu	WL3	WL4	Literature [32] Lut	Assignments
1446	1445			1456	δ(O—H)^B
-	-	1438	1438	1439	δ(O—H)^B, δ(O^3—H)
1383	1385	1364	1360	1367	ν_s(C—O^1—C^2), δ(O$^{4'}$—H)
1341	1346	-	-	-	n.a.
		-	1314	1313	δ(O—H), ν(C$^{4'}$—O)
1307	-	-	-	1303	ν(C$^{4'}$—O)
-	1297	-	-	1284	ν(C$^{4'}$—O)
1262	1272	1263	1263	1263	δ(C^2—H), ν(C^2—O^1), ν(C—C), δ(O^5—H)
-	1226	-	-	-	n.a.
1210	1208	-	1206	1210	δ(O—H)^A, ν(C^2—O^1)
1186	1179	1178	1173	1194	δ(C^6—H), δ(O^7—H)
1160	-	–	-	1162	δ(O—H)^B, δ(O^7—H)
1117	1128	1106	1109	1120	δ(C^2—H), δ(O—H)^B
1096	1089	1077	1068	1094	ν_s(C—O^1—C^2), ν(C—O^1), ν(C^3—C^4), $\phi_{ip}{}^{A,B}$, δ(O—H), δ(O^7—H)
1034	1030	1049	1043	1031	ν_s(C—O^1—C^2), $\phi_{ip}{}^{A,B}$, δ(C^2—H)
1003	-	-	-	999	ϕ^A + $\phi_{ip}{}^B$
945	-	-	-	946	$\phi^{A,B}$, Δ(C^3—C^4=O), ν(O^1—C^2)
848	840	823	843	839	γ(C^2—H)
805	790	-	-		n.a.
768	755	765	765	766	ϕ^A
725	-	-	-	728	$\phi_{op}{}^A$, Γ(C^3—C^4=O), γ(O^5—H)
685	688	-	-		n.a.

Annotations: ν—stretching, ϕ—aromatic ring normal vibrations, δ—in-plane deformation, γ—out-of-plane deformation, Γ—out-of-plane deformation of skeleton atoms, Δ—in-plane deformation of skeleton atoms, ip—in-plane, op—out-of-plane, s—symmetric mode.

4. Conclusions

The W&N 19th century Archive Database has proven, once more, to be an exceptional source of information on 19th-century artists' materials and their commercial preparation, enabling the first study of five W&N manufacturing processes for yellow lake pigments from weld.

This investigation showed that W&N 19th-century methods for preparing weld lake pigments involved extracting the dye in neutral-basic media by adding carbonate compounds ($KHCO_3$ and K_2CO_3) and complexation of flavonoid compounds was always achieved by the addition of Al^{3+}. Five bright yellow lake pigments were obtained and characterized by a multi-analytical approach. Their chromatographic profiles display as main yellows, luteolin 7-*O*-glucoside (Lut-7-*O*-glu) or both Lut-7-*O*-glu plus luteolin 3′,7-*O*-glucoside (Lut-3′,7-*O*-glu). In two of the processes, the presence of gypsum ($CaSO_4{\cdot}2H_2O$) was detected by FTIR, being formed as a by-product of the reaction of alum and $CaCO_3$. This work also offers the first identification of weld lake pigments' characteristic infrared bands: the stretching vibration of C=O at 1632 cm^{-1}, the OH of C^5 bending at c. 1484-9 cm^{-1}, the stretching of C^3-C^4 at 1077-68 cm^{-1}, all clear indications of metal complexation in positions OH (C^5) and CO (C^4) of flavonoid compounds.

The five recipes result in two types of lake pigments: yellows in which a filler, such as gypsum, is present (*WL1* and *WL2*) and yellows in which the lake pigment was found in an aluminate matrix (*WL3* as well as *WL4* and *WL5*). In the first type, the paints' mechanical performance is controlled by the filler [35], and the pigments are more opaque when applied as oil paints. On the other hand, the second type allows the preparation of translucent paints that can be applied as glazes.

The yields obtained experimentally were very similar or better than those of W&N, excluding recipes *WL4* and *WL5*, which were considered experiments. In the future, we intend to investigate variants of the processes and explore the full precipitation of the coloring matter as described by W&N in *WL3*. Moreover, since infrared spectroscopy revealed a powerful technique for the characterization of flavonoids-metal chelation, further work is ongoing with the analysis of other luteolin and apigenin "type-chromophores" complexed with Al^{3+}.

The pigment reconstructions will be fundamental to advancing on degradation studies and supporting analytical methodologies useful for identifying weld lake pigments in artworks, contributing to ensuring effective conservation and authentication procedures.

Supplementary Materials: The following are available online at https://www.mdpi.com/2571-9408/4/1/26/s1.

Author Contributions: Conceptualization, P.N. and V.O.; methodology, P.N. and V.O.; synthesis, M.V.; investigation, M.V., P.N., V.O., A.C., M.C.O.; writing—original draft preparation, P.N. and V.O.; writing—review and editing, P.N., V.O., A.C., M.J.M.; funding acquisition, M.J.M. All authors have read and agreed to the published version of the manuscript.

Funding: This work was supported by the Associate Laboratory for Green Chemistry-LAQV, which is financed by national funds from FCT/MCTES (UIDB/50006/2020 and UIDP/50006/2020). FCT/MCTES also funded the project "Polyphenols in Art: chemistry and biology hand in hand with conservation of cultural heritage" (PTDC/QUI-OUT/29925/2017), in which this work was developed.

Institutional Review Board Statement: Not applicable.

Data Availability Statement: Not applicable.

Conflicts of Interest: The authors declare no conflict of interest.

References

1. Cardon, D. *Natural Dyes: Sources, Tradition, Technology and Science*; Archetype Publications: London, UK, 2007.
2. Kirby, J.; van Bommel, M.; Verhecken, A. *Natural Colorants for Dyeing and Lake Pigments: Practical Recipes and Their Historical Sources*; Archetype Publications: London, UK, 2014.
3. Clarke, M. Colours versus colorants in art history: Evaluating lost manuscript yellows. *Rev. História Arte* **2011**, *1*, 138–151.
4. Zhang, X.; Good, I.; Laursen, R. Characterization of dyestuffs in ancient textiles from Xinjiang. *J. Archaeol. Sci.* **2008**, *35*, 1095–1103. [CrossRef]
5. Liu, J.; Guo, D.; Zhou, Y.; Wu, Z.; Li, W.; Zhao, F.; Zheng, X. Identification of ancient textiles from Yingpan, Xinjiang, by multiple analytical techniques. *J. Archaeol. Sci.* **2011**, *38*, 1763–1770. [CrossRef]

6. Marques, R.; Sousa, M.M.; Oliveira, M.C.; Melo, M.J. Characterization of Weld (*Reseda luteola* L.) and spurge flax (*Daphne gnidium* L.) by high-performance liquid chromatography-diode array detection-mass spectrometry in Arraiolos historical textiles. *J. Chromatog. A* **2009**, *1216*, 1395–1402. [CrossRef]
7. Nyström, I. Spectroscopic analysis of artists' pigments and materials used in southern Swedish painted wall hangings from the eighteenth and nineteenth centuries. *Stud. Conservat.* **2015**, *60*, 353–367. [CrossRef]
8. Moiteiro, C.; Gaspar, H.; Rodrigues, A.I.; Lopes, J.F.; Carnide, V. HPLC quantification of dye flavonoids in *Reseda luteola* L. from Portugal. *J. Sep. Sci.* **2008**, *31*, 3683–3687. [CrossRef] [PubMed]
9. Cristea, D.; Bareau, I.; Vilarem, G. Identification and quantitative HPLC analysis of the main flavonoids present in Weld (*Reseda luteola* L.). *Dyes Pigment.* **2003**, *57*, 267–272. [CrossRef]
10. Ferreira, E. New Approaches Towards the Identification of Yellow Dyes in Ancient Textiles. Ph.D. Thesis, University of Edinburgh, Edinburgh, Scotland, 2002.
11. Amat, A.; Clementi, C.; Miliani, C.; Romani, A. Complexation of apigenin and luteolin in weld lake: A DFT/TDDFT investigation. *Phys. Chem.* **2010**, *12*, 6672–6684. [CrossRef] [PubMed]
12. Dong, H.; Yang, X.; He, J.; Cai, S.; Xiao, K.; Zhu, L. Enhanced antioxidant activity, antibacterial activity and hypoglycemic effect of luteolin by complexation with manganese (II) and its inhibition kinetics on xanthine oxidase. *RSC Adv.* **2017**, *7*, 53385–53395. [CrossRef]
13. Gao, L.G.; Wang, H.; Song, X.L.; Cao, W. Research on the chelation between luteolin and Cr (III) ion through infrared spectroscopy, UV-vis spectrum and theoretical calculations. *J. Mol. Struct.* **2013**, *1034*, 386–391. [CrossRef]
14. Smith, G.J.; Thomsen, S.J.; Markham, K.R.; Andary, C.; Cardon, D. The photostabilities of naturally occurring 5-hydroxyflavones, flavonols, their glycosides and their aluminium complexes. *J. Photoch. Photobio. A* **2000**, *136*, 87–91. [CrossRef]
15. Castro, R.; Miranda, A.; Melo, M.J. Interpreting lac dye in medieval written sources: New knowledge from the reconstruction of recipes relating to illuminations in Portuguese manuscripts. In *Sources in Art Technology: Back to Basics*; Clarke, M., Townsend, J., Stijnman, A., Eds.; Archetype Publications: London, UK, 2016; pp. 88–99.
16. Vitorino, T.; Melo, M.J.; Carlyle, L.; Otero, V. New insights into brazilwood manufacture through the use of historically accurate reconstructions. *Stud. Conserv.* **2015**, *61*, 255–273. [CrossRef]
17. Melo, M.J.; Castro, R.; Nabais, P.; Vitorino, T. The book on how to make all the colour paints for illuminating books: Unravelling a Portuguese Hebrew illuminators' manual. *Herit. Sci.* **2018**, *6*, 44. [CrossRef]
18. Vitorino, T.; Otero, V.; Carlyle, L.; Melo, M.J.; Parola, A.J.; Picollo, M. Nineteenth-century cochineal lake pigments from Winsor & Newton: Insight into their methodology through reconstructions. In Proceedings of the ICOM-CC 18th Triennial Conference Preprints, Copenhagen, Denmark, 4–8 September 2017; Bridgland, J., Ed.; Paper 0107.
19. Nabais, P.; Oliveira, J.; Pina, F.; Teixeira, N.; de Freitas, V.; Brás, N.F.; Clemente, A.; Rangel, M.; Silva, A.M.S.; Melo, M.J. A 1000-year-old mystery solved: Unlocking the molecular structure for the medieval blue from Chrozophora tinctoria, also known as folium. *Sci. Adv.* **2020**, *6*, eaaz7772. [CrossRef] [PubMed]
20. Clarke, M.; Carlyle, L. Page-image recipe databases, a new approach for accessing art technological manuscripts and rare printed sources: The Winsor & Newton archive prototype. In Proceedings of the ICOM-CC 14th Triennial Meeting, The Hague, The Netherlands, 12–16 September 2005; Bridgland, J., Ed.; James & James: London, UK, 2005; Volume 1, pp. 24–29.
21. Clarke, M.; Carlyle, L. Page-image recipe databases: A new approach to making art technological manuscripts and rare printed sources accessible. In *Art of the Past—Sources and Reconstructions*; Clarke, M., Townsend, J., Stijnman, A., Eds.; Archetype Publications: London, UK, 2005; pp. 49–52.
22. Carlyle, L.; Alves, P.; Otero, V.; Melo, M.J.; Vilarigues, M. A question of scale and terminology, extrapolating from past practices in commercial manufacture to current laboratory experience: The Winsor & Newton 19th century artists' materials archive database. In Proceedings of the ICOM-CC 16th Triennial Meeting, Lisbon, Portugal, 19–23 September 2011; Bridgland, J., Ed.; Paper 0102.
23. Carlyle, L. *The Artist's Assistant: Oil Painting Instruction Manuals and Handbooks in Britain, 1800–1900, with Reference to Selected Eighteenth-Century Sources*; Archetype Publications: London, UK, 2001.
24. Otero, V.; Pinto, J.V.; Carlyle, L.; Vilarigues, M.; Cotte, M.; Melo, M.J. Nineteenth century chrome yellow and chrome deep from Winsor & Newton^{TM}. *Stud. Conservat.* **2017**, *62*, 123–149.
25. Otero, V.; Campos, M.F.; Pinto, J.V.; Vilarigues, M.; Carlyle, L.; Melo, M.J. Barium, zinc & strontium yellows in late 19th-early 20th century oil paintings. *Herit. Sci.* **2017**, *5*, 46. [CrossRef]
26. Mouri, C.; Mozaffarian, V.; Zhang, X.; Laursen, R. Characterization of flavonols in plants used for textile dyeing and the significance of flavonol conjugates. *Dyes Pigment.* **2014**, *100*, 135–141. [CrossRef]
27. Zhang, X.; Laursen, R.A. Development of mild extraction methods for the analysis of natural dyes in textiles of historical interest using LC-diode array detector-MS. *Anal. Chem.* **2005**, *77*, 2022–2025. [CrossRef] [PubMed]
28. Favaro, G.; Clementi, C.; Romani, A.; Vickackaite, V. Acidichromism and ionochromism of luteolin and apigenin, the main components of the naturally occurring yellow weld: A spectrophotometric and fluorimetric study. *J. Fluoresc.* **2007**, *17*, 707–714. [CrossRef] [PubMed]
29. Villela, A.; van der Klift, E.J.C.; Mattheussens, E.S.G.M.; Derksen, G.C.H.; Zuilhof, H.; van Beek, T.A. Fast chromatographic separation for the quantitation of the main flavone dyes in Reseda luteola (weld). *J. Chromatogr. A* **2011**, *1218*, 8544–8550. [CrossRef] [PubMed]

30. Villela, A. Textile Dyeing with a Flavonoid Dye: Photo-Stability and Analytical Chemistry Methods. Ph.D. Thesis, Wageningen University, Wageningen, The Netherland, 2020.
31. Nakamoto, K. *Infrared and Raman Spectra of Inorganic and Coordination Compounds*; Wiley: Hoboken, NJ, USA, 2009.
32. Machado, N.F.L.; de Batista Carvalho, L.A.E.; Otero, J.C.; Marques, M.P.M. A conformational study of hydroxyflavones by vibrational spectroscopy coupled to DFT calculations. *Spectrochim. Acta A Mol. Biomol. Spectrosc.* **2013**, *109*, 116–124. [CrossRef] [PubMed]
33. Baranović, G.; Šegota, S. Infrared spectroscopy of flavones and flavonols. Reexamination of the hydroxyl and carbonyl vibrations in relation to the interactions of flavonoids with membrane lipids. *Spectrochim. Acta A Mol. Biomol. Spectrosc.* **2018**, *192*, 473–486. [CrossRef] [PubMed]
34. Heneczkowski, M.; Kopacz, M.; Nowak, D.; Kuzniar, A. Infrared spectrum analysis of some flavonoids. *Acta Pol. Pharm. Drug Res.* **2001**, *58*, 415–420.
35. Sanches, D.; Ramos, A.M.; Coelho, J.F.J.; Costa, C.S.M.F.; Vilarigues, M.; Melo, M.J. Correlating thermophysical properties with the molecular composition of 19th century chrome yellow oil paints. *Polym. Degrad. Stab.* **2017**, *138*, 201–211. [CrossRef]

heritage

MDPI

Article

Characterizing the Dyes of Pre-Columbian Andean Textiles: Comparison of Ambient Ionization Mass Spectrometry and HPLC-DAD

Jennifer Campos Ayala [1], Samantha Mahan [1], Brenan Wilson [1], Kay Antúnez de Mayolo [2], Kathryn Jakes [3], Renée Stein [4] and Ruth Ann Armitage [1,*]

1 Department of Chemistry, Eastern Michigan University, Ypsilanti, MI 48197, USA; camposajenn@gmail.com (J.C.A.); smahan2@emich.edu (S.M.); bwilson1@emich.edu (B.W.)
2 Box 77, Eagleville, CA 96110, USA; kayantunez@gmail.com
3 The Ohio State University, Columbus, OH 43201, USA; kajtxtl@gmail.com
4 Department of Art History and Michael C. Carlos Museum, Emory University, Atlanta, GA 30322, USA; rastein@emory.edu
* Correspondence: rarmitage@emich.edu; Tel.: +1-734-487-0290

Abstract: The complex and colorful textiles of ancient Peru have long been a focus of technical study, particularly to characterize the sources of the wide variety of dyes utilized by these Andean artisans. This manuscript describes the characterization of the dyes of both primary (red, blue, and yellow) and secondary (purple, orange, and green) colors sampled from textiles spanning five major civilizations: the Paracas Necropolis, the Nazca, the Wari, the Chancay, and the Lambayeque, all from Peru. All but the Paracas Necropolis samples were part of technical conservation studies of the ancient South American textiles collections of the Michael C. Carlos Museum. Analysis of the dyes was carried out utilizing direct analysis in real time time-of-flight mass spectrometry (DART-MS) and paper spray MS. To validate these ambient ionization MS methods, the samples were further investigated using high-performance liquid chromatography (HPLC) with ultraviolet-visible diode array detection (DAD). These results show that ambient ionization MS methods are simple and fast for characterization of the general classes of dyes, e.g., plant reds vs. insect reds, and indigoids in blues and greens. Due to the myriad possible sources of yellow dyes and their tendency to undergo oxidative decomposition, positively identifying those components in these yarns was difficult, though some marker compounds and flavonoid decomposition products were readily identified by ambient ionization mass spectrometry.

Keywords: textiles; dyes; Peru; ambient ionization mass spectrometry; DART-MS; paper spray MS; HPLC

Citation: Campos Ayala, J.; Mahan, S.; Wilson, B.; Antúnez de Mayolo, K.; Jakes, K.; Stein, R.; Armitage, R.A. Characterizing the Dyes of Pre-Columbian Andean Textiles: Comparison of Ambient Ionization Mass Spectrometry and HPLC-DAD. *Heritage* **2021**, *4*, 1639–1659. https://doi.org/10.3390/heritage4030091

Academic Editor: Diego Tamburini

Received: 29 June 2021
Accepted: 3 August 2021
Published: 7 August 2021

1. Introduction

The ancient weavers of Peru are known for their complex multicolored textiles, preserved in burial contexts primarily due to the dry and cold conditions of these high-altitude locations. These textiles, which span a wide range of time and geography across the Andes, have long been admired for their intricate workmanship and complex designs. Determining the sources of the dyes used to achieve the colors in these priceless objects has been the object of a great amount of study over the years, dating back to work by dye chemists Fester [1] and later Saltzman [2,3] in the mid-20th century.

The characterization of dyes in ancient textiles is challenging for numerous reasons, including difficulty in obtaining material in sufficient quantity for analysis while minimizing damage to objects. The high tinting strength of most dyestuffs means that while the color of the sample may be readily visible, the actual amount of colorant present may be below the limits of detection of many analytical approaches. Degradation may complicate the identification of dye colorants, particularly those present in the fugitive yellow dyes, due to photo-oxidation [4] or other decomposition pathways [5] that may occur. Further, decomposition—either during burial or after excavation—may complicate determining color, as what appears blue today may have once been green, but the yellow colorants have decomposed. Colorants, or dye chromophores, are the molecules that come from the dye preparation to impart color to a textile. It should be noted, though, that some mordants and additives can also alter the apparent color of the dye. In some cases, identifying a colorant also identifies the dye source, as in the case of carminic acid, which is indicative of cochineal from *Dactylopius coccus* and various subspecies. Other colorants, such as luteolin and quercetin, are found in numerous plant dye sources. Thus, characterizing the dyes in textile samples to identify the colorants may or may not identify the dye sources that were used by their makers.

Yellow dyes have long been considered the most difficult to identify because the sources of yellow dyes are myriad and have many overlapping colorants in common. A wide variety of different plants have been suggested as possible sources of yellow dyes in pre-Columbian Peruvian textiles by previous researchers [6–8]: *Bidens andicola* and other Asteraceae (including commercially available *Coreopsis* and *Dahlia*); *Baccharis* species, including *B. floribunda* and *B. genistelloides*; and *Alnus jorullensis*. Previous work has also investigated the composition of *Kageneckia lanceolata* and *Hypericum larcifolium* [9]. The Asteraceae contain the chalcone colorants butein and okanin and their glycosides, while the others, along with *Chlorophora tinctoria* (locally called *insira*), are primarily flavonoid dyes. The flavones luteolin and apigenin and the flavonols quercetin and kaempferol are the most abundant colorants, along with their corresponding methyl ethers [10]. *Bocconia pearcei* contains the benzoisoquinoline alkaloids sanguinarine and chelerythrine [11,12]. *Chuquiraga espinosa* and *Dicliptera hookeriana* are not yet well characterized, but both appear to be flavonoids, as well.

Early work on the identification of Andean dyestuffs relied predominantly on UV-visible spectroscopy, and occasionally on separations by thin layer chromatography [8,13]. More recently, the bulk of research on the identification of dyes in textiles has been carried out with chromatographic separations, primarily by high-performance liquid chromatography (HPLC), considered the "gold standard" method for such materials. Wouters and Rosario-Chirinos [6] reported an extensive study in which they used HPLC with diode array detection (DAD) to separate, quantify, and identify all components of purple, red, and blue dyes in textiles from a range of samples from pre-Columbian to Inca period civilizations. Niemeyer et al. [14] performed HPLC dye analysis on South American red and blue textiles, finding evidence of trade throughout the region due to the presence of non-local dye sources. Degano and Colombini [15] described the analysis of blue and red dyes in 11th and 13th century materials using HPLC with a variable wavelength UV-vis detector to determine what dyes were present in textiles associated with a mummy from a pre-Columbian site in Peru. Price et al. [16] described the identification of dyes in materials from a pre-Columbian weaver's toolkit. Most recently, Boucherie et al. [17] reported on HPLC analyses of Nazca textile dyes, and Sabatini et al. [7] combined HPLC-MS/MS with X-ray fluorescence to characterize both the mordants and dyes in Paracas textiles. HPLC is the most effective way to separate components of dyes. Using UV-vis spectroscopy alone to identify the components can be difficult, particularly when reference standards of the pure colorants are unavailable, as is often the case.

Mass spectrometry (MS) has the advantage of identifying components based on their molecular masses and, in some cases, fragmentation patterns, which are generally more specific than UV-vis spectra. Michel et al. [18] utilized high-resolution mass spectrometry to characterize red, blue, and purple textiles from the Atacama desert of northern Chile. They found evidence that the Pachacamác people used both shellfish purple and combinations of red and blue dye sources to color their textiles. This early work required a relatively large 1-centimeter-long sample to be destructively extracted in hot dimethyl sulfoxide for direct exposure electron impact mass spectrometry. The potential of some spectroscopic methods for non-invasive identification of dyes in Andean textiles is also of note, particularly applications of fiber-optic reflectance spectroscopy and Raman spectroscopy [19–21]. With secure contexts for the materials being characterized, a great deal can be learned about the culture and technology of the textile artists, as demonstrated by Boytner [22].

Another method that shows potential for dye analysis is direct analysis in real time mass spectrometry (DART-MS) [23] with a time-of-flight mass analyzer. Armitage et al. [24] undertook analysis of the dyes in red textile samples from the Paracas Necropolis period, which had been previously identified as *Relbunium*. A major limitation of all soft ionization mass spectrometric methods is the inability to distinguish structural isomers, such as indigotin and indirubin, or alizarin and xanthopurpurin, based solely on molecular mass. Yellow colorants are particularly difficult to differentiate with DART-MS alone, as in-source collision-induced dissociation, useful for pure substances, yields complex results for mixtures [25]. DART-MS and another "ambient" ionization method, paper spray, [26] have shown potential efficacy for characterizing a wide variety of dyes in reference collections [9], though HPLC with MS/MS detection is necessary for identifying true unknowns.

The work reported herein stems from projects carried out over several years, with a focus on the primary colors of red, blue, and yellow as well as the secondary colors of purple, green, and orange in dyed textiles from several pre-European contact civilizations of South America, ranging in time from 800 BCE to 1460 CE. The sampled objects are either from the Paracas Necropolis or have been attributed stylistically to the Nazca, Wari, and Chancay, all of which rose and fell throughout that period within the geographic area that is present-day Peru. Here, we compare the results of two approaches—ambient ionization mass spectrometry and high-performance liquid chromatography—for the identification of these dyes in yarns from ancient Peruvian textiles. We compare the capabilities and limitations of these approaches in terms of (1) the amount of sample necessary, (2) the limits of detection, (3) the selectivity, (4) the speed of analysis, and (5) the skill required of the analyst. This study serves as a step in the process of validating the use of ambient ionization mass spectrometry for characterizing dye colorants in ancient textiles.

2. Materials and Methods

2.1. Materials

From August to December 2017, the *Threads of Time: Tradition and Change in Indigenous American Textiles* exhibition was on display at the Michael C. Carlos Museum at Emory University. The objects in the exhibition had entered the museum from historical private collections, and lacking archaeological context, they were attributed on the basis of style. As part of the technical study in tandem with this exhibition, we undertook mass spectrometry analyses to answer questions about the colorants. Encouraged by the exhibition curator, Dr. Rebecca Stone, we chose to focus on the secondary colors: green, orange, and purple. The central question was whether these colors were obtained from a single dye source or if they were produced through overdyeing or mixing of dyes in the dye pot, or through dyeing naturally colored yarns. The colors of dyed objects selected for this study were determined by eye, with multiple individuals interpreting and agreeing on the colors as orange, green, or purple. Organic color is fugitive and subject to deterioration (fading, color shift, color loss). Even if the visible color of a dye had deteriorated, evidence of the original dye would be expected to remain detectable through chemical analysis for characteristic compounds. Secondary colors can be achieved by the use of mordants with a single primary dye. In these examples, we would expect only one dye source to be identified. The secondary colors of the objects were judged visually without assuming how the object might have originally appeared. If an object appeared blue, it was not assumed that the yellow had faded from a mixed green, and thus, the object was not sampled for green. In the case of orange, in particular, it is possible that a red dye could fade to appear orange. The color was evaluated throughout the object, confirming that protected areas (seams, insides, backs) did not reveal evidence of color loss or shift before a sample was removed to represent a given color.

The objects, along with the relevant samples and the historic time periods from which they originate, are listed in Table 1. Each sample consisted of anywhere from a few plucked fibers of the color of interest to a snippet of yarn from a frayed edge. Yarn snippets varied in size from a few millimeters to 2 cm or more in cases where the edge was already heavily damaged such that sample was readily separating from the object. The samples were collected at the Carlos Museum conservation laboratory, and the colors were determined by consensus of those selecting and observing the collection process. The original collection of Nazca, Wari, and Chancay samples were collected in collaboration with Dr. Cathy Selvius DeRoo, formerly of the Detroit Institute of Arts. These samples were placed into labeled polypropylene microcentrifuge tubes, or, in the case of the plucked fiber samples, placed between two glass microscope slides and taped closed for transport back to the laboratory at Eastern Michigan University (EMU) for analysis. A separate technical study was later undertaken of another Peruvian textile from the Pacatnamú site in northern Peru, attributed to the Lambayeque culture of the Central Andes. The Lambayeque samples were collected in glass vials and shipped to EMU for analysis. Images of the textiles and samples are provided in Supplementary Figures S1–S10.

For comparison through time, samples of similar primary and secondary colors from the Paracas Necropolis period were also included in this study. These were collected at the Museo Nacional de Arqueología, Antropología, e Historia del Perú (MNAAHP) in Lima, Peru, by Anne Paul in 1985 from loose fibers and yarns closely associated with mummy textiles; no fibers were removed from the textiles themselves. The colors of these samples were determined visually by Paul when the samples were collected, and transcribed and confirmed by Jakes [27], also through visual observation. These samples have been stored in a climate-controlled museum storage environment at The Ohio State University, Columbus, OH, USA, since they were collected. The samples come from a total of four different mummy bundles in order to sample a wide range of colors and are listed in Table 2. Images of these samples are provided in the Supplementary Figures S11–S14.

Reference samples from the Peruvian dyes collection prepared by Kay and Erik Antúnez de Mayolo in the 1970s in collaboration with Max Saltzman were included as comparative materials [28] in the latter stages of the project with an emphasis on the yellow dyes. The materials in this collection were prepared on spun sheep's wool, cotton, and Peruvian alpaca yarns, all of unknown origin, both unmordanted and with alum mordant. This collection was also the basis for the widely cited report on Peruvian dye plants [29]. Previous studies of portions of this collection have been reported elsewhere [9]; additional samples not included in the reference collection attributed to Saltzman stored at UCLA, Los Angeles, CA, USA, have been included here. The samples analyzed here were only the alum-mordanted sheep's wool yarns. Additional reference materials were prepared from commercial sources as necessary, particularly for secondary colors; details can be found elsewhere [30].

The solvents used in the methods described here were acetonitrile (ACN) of HPLC grade (Sigma-Aldrich, St. Louis, MO, USA); deionized water (H_2O) from a Barnstead Diamond water polishing system; methanol, N,N-dimethylformamide (DMF), and dimethyl sulfoxide (DMSO), all of HPLC grade (VWR Chemicals, Radnor, PA, USA); formic acid (FA) of high purity grade (VWR); and hydrochloric acid (HCl) and Na_2EDTA both from laboratory stocks. Trifluoroacetic acid (TFA) of ReagentPlus purity (99%) was obtained from Sigma-Aldrich. Dye colorant reference compounds were obtained commercially from a variety of sources. The xanthopurpurin standard was synthesized by T. Friebe and S. Augustin, EMU Chemistry Department.

Table 1. Textile samples from the Michael C. Carlos Museum; images in Supplementary Figures S1–S10.

Museum Accession Number	Object Description	Civilization and Time Period	Primary Colors	Secondary Colors	Other Colors
2002.1.100	Snake band	Nazca (Early Intermediate Period), ca. 200 CE	Pink (light red)	Purple, green	none
2002.1.3	Hummingbird border	Nazca, ca. 100–200 CE	none	Purple, green	Black
2002.40.4 A–C	People beans	Nazca (Early Intermediate Period), ca. 200 CE	none	Green, orange	none
2002.1.148	Tie-dye textile fragment	Wari (Middle Horizon), ca. 500–800 CE	Yellow (2), red, red-purple	Purple (2), green (2)	none
2002.1.1	Tie-dye textile fragment	Wari (Middle Horizon), ca. 500—800 CE	Blue/purple, red, white, yellow (2)	Green, yellow-green	none
2002.1.16	Wari tunic fragment	Wari (Middle Horizon), ca. 500–800 CE	Pink-red	Orange, green	Gold, salmon, brown
2002.1.83	Fragment	Wari (Middle Horizon), ca. 500–800 CE	Red	Green	none
2003.40.5	Wari band, frayed	No additional information available.	Red, yellow	Purple, greenish-yellow	Turquoise
2002.1.126 A–U	Yarn balls from weaver's basket	Chancay (Late Intermediate Period), ca. 1000–1450 CE	Blue (2), white	Green (2)	Brown, yellow-brown
2004.64.1 A–B	Tasseled tunic	Lambayeque (Late Intermediate Period), ca. 900–1300 CE	Pink, red, dark blue, light blue	Green, orange	Golden brown, light green, brown (warp), cream (warp)

Table 2. Paracas Necropolis samples from Museo Nacional de Arqueología, Antropología, e Historia del Perú (MNAAHP), Lima, Peru, collected in 1985 by A. Paul. Identification numbers refer to a specific mummy bundle, specimen, and subspecimen (bundle-specimen subspecimen) as recorded by Paul. Images of samples are found in Supplementary Figures S11–S14.

Identification Number	Object Description	Primary Colors	Secondary Colors	Other Colors
Cave #5 12-5236 0093	Cloth (oldest)	Yellow	none	none
310-58C 02912	Decorative border	none	Orange	none
382-10 05904	Mantle, loose threads	none	Green	none
382-45 02763	Skirt, loose threads and fibers	Red	none	none
382-48 01017	Skirt fringe	Blue	Black (purple)	none
382-49 (23808) 03174	Skirt ground cloth	none	Orange	none
382-54 02846	Turban, loose threads	Yellow, blue	Green	none
382-68 02929	Poncho, braid	none	Purple, green	none
382-72 02519	Skirt embroidery threads and fringe	Red	Black	none
421-39 03083	Mantle, loose fibers	Red	none	none
421-132 02096	Disintegrated fibers	none	none	Brown

2.2. Methods: Ambient Ionization Mass Spectrometry

2.2.1. DART-MS

A short section (approximately 1 mm) from the end of each yarn sample was cut off using a clean razor blade on a clean glass microscope slide. The resulting loose fibers were divided roughly in half for analysis, with each clump of fibers collected in the tip of cleaned tweezers. The fibers were then introduced into the gap between the DART ionization source (IonSense, Saugus, MA, USA) and Orifice 1 of the AccuTOF mass spectrometer (JEOL USA, Peabody, MA, USA). The helium DART gas was heated to 350–450 °C as needed to obtain good signal-to-noise ratios, and spectra were collected in both positive and negative ion modes using the default grid voltages in the DART controller software. The "peaks voltage" on the AccuTOF was selected for maximum intensity in the range of interest (150–1000 Da), while the ring lens and Orifice 2 voltages were set to ±5 V, depending on whether the spectra were being collected in positive (+) or negative (−) ion modes. Orifice 1 was set to ±30 V (again corresponding to the ionization mode) to minimize fragmentation. One half of the sample was analyzed directly without any preparation, and the second half was treated with approximately 1 μL of formic acid (88%) prior to introduction into the DART ion source gap. Previous studies have shown that acid treatment yields a stronger signal for most dye colorants during DART-MS analysis [25,31].

2.2.2. Paper Spray MS

Paper spray mass spectrometry (PS-MS) was carried out using a home-built ion source consisting of an alligator clip and filter paper (Whatman No. 4, from laboratory stocks) electrode that was powered by the AccuTOF electrospray port. For the PS-MS, +3500 V was applied to the paper electrode, with the Orifice 1 and ring lens voltages set to +80 V and +10 V, respectively. The tip of the paper electrode was positioned in front of Orifice 1 on the AccuTOF. A 3 μL aliquot of the extract was placed on the paper electrode for analysis. PS-MS was chosen for analysis as the electrospray ionization source available in the EMU Chemistry Department required large volumes (20–50 μL) of sample.

Calibration of each data file collected on the AccuTOF from both DART and PS-MS was carried out with PEG-600 in methanol, and files were processed using TSS Pro 3.0 software (Shrader Analytical and Consulting Laboratories, Detroit, MI, USA). Further data analysis was undertaken with Mass Mountaineer software (various versions, RBC Software, provided by R. B. Cody).

2.3. *Methods: HPLC-DAD*

2.3.1. Sample Preparation (HPLC and PS-MS)

Indigoids in the blue and purple samples were initially extracted following an adaptation of the procedure [15] used by Degano and Colombini. Roughly 1–2 mm of yarn was treated with 50 μL of dimethylsufoxide (DMSO) at 60 °C for 60 min in a sonicator bath, resulting in Solution I. Originally, the anthraquinone colorants were extracted from red yarns using a 1:1 (*v/v*) mixture of dimethylformamide (DMF) with 0.1% Na_2EDTA in water, as reported elsewhere [32]. This mixture purportedly preserves the glycosides and results in a more accurate representation of all the compounds present in a dyed yarn. A 50 μL portion of the EDTA/DMF solution was added to a similar size fragment of yarn in a microcentrifuge tube, which was then sonicated at 60 °C for 60 min to yield Solution II.

A third protocol [33] was later determined to be more reliable, wherein yarn fragments were extracted with 50–100 μL of 30:1 (*v/v*) methanol:HCl solution, hereafter referred to as Solution III. Fibers were then sonicated for 60 min at room temperature to assist extraction. While the EDTA/DMF extractant did likely preserve the glycosides, the chromatograms for the textile yarns were of low quality, possibly due to a problem with the detector controller board that was identified after a portion of the analyses were completed. As the project spanned several years and numerous operators, the methanol/HCl extract was selected as it appeared to be more reliable for the identification of the characteristic colorants in these samples by HPLC and PS-MS.

2.3.2. HPLC Methods

The HPLC system consisted of a Shimadzu (Kyoto, Japan) LC-20AT pump paired with an SPD-M20A photodiode array detector. The flow rate was 1 mL/min and a 20 μL injection volume was used for each of the methods employed. In the method used initially for indigoids, separations were conducted at controlled temperature (25 °C using a heating tape plugged into a variable AC source, monitored with a thermocouple) on an ACE Equivalence 5 C18 150 × 4.6 mm analytical column (VWR, Radnor, PA, USA) attached to a C18 guard column. The analytical column used for separation of Solution III preparations was a Discovery C18 column from Supelco (Bellefonte, PA, USA, 25 cm × 4.6 mm and 5 µm particle size) without temperature control. For the indigoid method, deionized water (solvent A) and acetonitrile (ACN, HPLC grade from Sigma-Aldrich, St. Louis, MO, USA, as solvent B), each at 0.1% (*v/v*) TFA served as mobile phases. The program for separation of the indigoids was 15% B for 5 min, followed by a linear gradient to 50% B in 25 min, then to 70% B in 10 min, followed by 90% B in 10 min held for 10 min (total run time 60 min); re-equilibration time was 15 min. For the Solution III preparations in methanol/HCl (after Manhita et al. [32]), the mobile phase consisted of acetonitrile (A) and 2.5% (*v/v*) aqueous acetonitrile with 0.5% (*v/v*) formic acid (B) in a gradient of 0–100% A from 0–10 min, 100% A from 10–15 min and a 5-minute rinse of 100% solvent B between runs. UV-vis spectra were acquired in the range of 200–650 nm with a resolution of 4 nm. Data were processed using Shimadzu LabSolutions software.

2.4. Methods: Microscopy and Fourier-Transform Infrared Spectroscopy

Each of the ancient samples was visually inspected and photographed under magnification (10×–200×) using a Dino-Lite AM4815ZT digital microscope. The samples from the Lambayeque textile were further examined using scanning electron microscopy with energy dispersive spectroscopy thanks to the short-term loan of a JEOL (JEOL USA, Peabody, MA, USA) JCM-7000 NeoScope benchtop instrument, used in our curriculum-based undergraduate research experience (CURE) in September 2019. SEM images were collected under high-vacuum conditions with a landing voltage of 10.0 kV at a working distance of approximately 12 mm. Animal hair—presumably camelid wool—was identified on the basis of visible scales and the presence of S and N, while cellulose—presumably cotton—was identified based on visible morphological features including convolutions similar to other cotton species and a marked lack of nodal structures, along with a lack of significant S and N signals.

When SEM-EDS was not available and fibers were not readily identifiable visually by optical microscopy, fibers were characterized using attenuated total reflectance Fourier-transform infrared spectroscopy. Measurements were obtained using a Pike Technologies (Madison, WI, USA) MIRacle ATR attachment with a ZnSe crystal in a Shimadzu IRTracer spectrometer (Shimadzu Corp, Kyoto, Japan). Spectra were collected by averaging 32 scans at a resolution of 4 cm^{-1} over the range of 4000–600 cm^{-1}. The presence of the amide I and II bands around 1600 and 1500 cm^{-1}, respectively, were considered characteristic of animal hair protein. Absorbance bands at 1020 cm^{-1} combined with the presence of an –OH stretching band around 3300 cm^{-1} were indicative of cellulose.

3. Results

The purpose of this study was twofold. First, we sought to identify the colorants present in the yarn samples from the ancient Peruvian textiles to add to the existing body of knowledge about the dyes used over the time periods described above, ideally to determine the relationships between the primary and secondary colors. Second, we applied ambient ionization mass spectrometry methods for these identifications, specifically to compare this approach to HPLC with diode array detection, which is the most commonly reported means of identifying dyes in archaeological and historical textiles. The two approaches differ significantly in how much material is needed for analysis, how many colorants can be detected and identified, and the length of time needed to complete the analysis. The results from the analyses of the ancient Peruvian textiles are summarized in Table 3 for the samples from the collections of the Michael C. Carlos Museum and in Table 4 for the Paracas samples from the MNAAHP. Results of the analyses of the Peruvian dyes reference collection are presented in Table 5. Complete results, including those for the fiber identifications, are found in Supplementary Tables S1 and S2.

Table 3. Results for MCCM textile samples. Details in Supplementary Table S1.

Museum Accession Number	Primary Colors	Secondary Colors	Other Colors
2002.1.100 Snake band, Nazca	Light red, *Relbunium*	Purple, *Relbunium* + indigo Green, indigo + flavonoid	none
2002.1.3 Hummingbirds, Nazca	none	Purple, *Relbunium* + indigo Green, indigo + flavonoid	Black, indigo
2002.40.4 A–C People beans, Nazca	none	Green, indigo + oxidized flavonol Orange, *Relbunium* + oxidized flavonol	none
2002.1.148 Tie-dye textile fragment, Wari	Yellow (2), both unknown/oxidized flavonol Red, *Relbunium* Red-purple, indigo + *Relbunium*	Purple (2), both indigo + *Relbunium* Green (2), both indigo + oxidized flavonol	none
2002.1.1 Tie-dye textile fragment, Wari	Blue/purple, indigo + *Relbunium* Red, *Relbunium* White, undyed Yellow (2), both oxidized flavonol	Green, indigo + unknown Yellow-green, indigo + oxidized flavonol	none
2002.1.16 Wari tunic fragment	Pink-red, cochineal	Orange, *Relbunium* Green, indigo + oxidized flavonol	Gold, oxidized flavonol Salmon, cochineal Brown, tannin or *Juglans*
2002.1.83 Fragment, Wari	Red, cochineal	Green, indigo + oxidized flavonol	none
2003.40.5 Wari band, frayed	Red, cochineal Yellow-brown, undyed or oxidized flavonol	Purple, indigo + cochineal greenish-yellow, indigo + oxidized flavonol	Turquoise, indigo + oxidized flavonol
2002.1.126 A–U Yarn balls from Chancay weaver's basket	Blue (2), indigo + oxidized flavonol White, undyed	Green (2), indigo + oxidized flavonol	Red-brown, *Relbunium* + oxidized flavonol Yellow-brown, undyed
2004.64.1 A–B Lambayeque tasseled tunic	Pink and red, both cochineal Dark blue, indigo + tannin? Light blue, indigo	Light green, indigo + tannin or oxidized flavonol Orange, cochineal + flavonol	Golden brown and brown (warp), oxidized flavonol w/tannin + unknown (laccaic acid?) Cream (warp), undyed

Table 4. Results of the analysis of the dyes in the Paracas Necropolis samples from Museo Nacional de Arqueología, Antropología, e Historia del Perú (MNAAHP), Lima, Peru. Details in Supplementary Table S2.

Identification Number	Object Description	Primary Colors	Secondary Colors	Other Colors
Cave #5 12-5236 0093	Cloth (oldest)	Yellow, decomposed flavonoid	none	none
310-58C 02912	Decorative border	none	Orange, *Relbunium* + decomposed flavonols	none
382-10 05904	Mantle, loose threads	none	Green, indigo + flavonoid yellow	none
382-45 02763	Skirt, loose threads and fibers	Red-orange, *Relbunium*	none	none
382-48 01017	Skirt fringe	Blue, indigo	Black (purple), indigo + *Relbunium*	none
382-49 (23808) 03174	Skirt ground cloth	none	Orange, *Relbunium* + flavonoid yellow	none
382-54 02846	Turban, loose and embroidery threads	Blue, indigo	Green, indigo	Gold, *Bidens*
382-68 02929	Poncho braid and ground cloth	none	Purple, indigo + *Relbunium* Green, indigo + flavonoid (unique)	none
382-72 02519	Skirt embroidery threads and fringe	Red, *Relbunium*	Black, indigo + *Relbunium*	none
421-39 03083	Mantle, loose fibers	Red, *Relbunium*	none	none
421-132 02096	Disintegrated fibers	none	none	Brown, *Bidens* + *Baccharis?*

Table 5. Results of analysis of reference samples from the Antúnez de Mayolo Peruvian dye plant collection.

Genus and Species	Dye Sample ID	DART Negative Ion Results	PS-MS Results	HPLC-DAD Results
Galium antuneziae	G 21/248	purpurin, xanthopurpurin/alizarin, munjistin, rubiadin, pseudopurpurin	rubiadin, xanthopurpurin/alizarin, lucidin	xanthopurpurin/alizarin, purpurin
Relbunium ciliatum	Rc	xanthopurpurin/alizarin, purpurin, rubiadin, lucidin, munjistin, pseudopurpurin	purpurin, xanthopurpurin/alizarin, rubiadin, lucidin	xanthopurpurin/alizarin, rubiadin, purpurin
R. hypocarpium	Rh 16/301,313	purpurin, xanthopurpurin/alizarin, rubiadin, lucidin, munjistin, pseudopurpurin	purpurin, munjistin, lucidin	purpurin, xanthopurpurin/alizarin, rubiadin
Relbunium sp.	R 10/253	purpurin, xanthopurpurin/alizarin, rubiadin, munjistin, lucidin, pseudopurpurin	purpurin, lucidin	purpurin, rubiadin, xanthopurpurin/alizarin
Dactylopius coccus	D 26/356	nothing detected	carminic acid, dcII, flavokermesic acid, kermesic acid	carminic acid
Alnus jorullensis	Aj 40/65	gallic acid, methyl gallate, luteolin	nothing detected	nothing identifiable
Baccharis floribunda	Bf 9/255	quercetin, quercetin methyl ether, apigenin methyl ether, luteolin dimethyl ether, apigenin	quercetin methyl ether, di-O-caffeoylquinic acid, quercetin, apigenin methyl ether, luteolin	quercetin, isorhamnetin, rhamnetin
Baccharis genistelloides	Bg 29/320	apigenin, luteolin, luteolin methyl ether, quercetin methyl ether	apigenin, luteolin methyl ether, quercetin/hydroxyluteolin, luteolin	apigenin, quercetin, kaempferol
Bidens andicola	Ba 2/236	luteolin/fisetin, okanin, apigenin, butein	luteolin/fisetin, okanin, apigenin, butein	okanin, butein, luteolin
Bocconia pearcei	Bp 5/245	no benzoisoquinoline alkaloids (sanguinarine and chelerythrine in DART positive ion mode)	sanguinarine, chelerythrine, norsanguinarine, protopine	sanguinarine, chelerythrine, unknown pks at 12.3 (λ_{max} 282) and 12.6 min (λ_{max} 284)
Chlorophora tinctoria	insira 33/350	kaempferol, morin	morin, kaempferol	morin, kaempferol
Chuquiraga espinosa	Ce 24/332	kaempferol, quercetin, isorhamnetin	kaempferol, quercetin, isorhamnetin	kaempferol, quercetin
Dicliptera hookeriana	D 26/356	quercetin	nothing detected	quercetin
Hypericum larcifolium	Hl 25/329	quercetin, quercetin methyl ether	quercetin, quercetin methyl ether	quercetin, possible quercetin methyl ether
Kageneckia lanceolata	Kl 12a/263	methyl apigenin, (313), (299), ursolic acid	quercetin, apigenin methyl ether, luteolin methyl ether, luteolin	apigenin, quercetin and/or luteolin

3.1. Primary Colors

3.1.1. Blue

Indigo, derived from *Indigofera* species, including *I. suffructosa* and *I. truxillensis*, *Cybistax antisyphilitica*, or other such species native to South America, are the most likely sources of the blue colorants indigotin and indirubin in all of the investigated samples, which is consistent with previous studies of pre-Columbian Andean textiles [6–8,14,15,17,18]. DART-MS cannot differentiate between these two isomers, and the result is a single peak at *m/z* 262.07 for the M^- ion in negative ion mode. DART-MS is extremely sensitive to indigoids, and this peak (or the MH^+ in positive ion mode at *m/z* 263.08 [34]) often dominates the spectra of the samples in which it is present. Due to solubility issues, the signal is quite variable in paper spray mass spectra, making this approach less than ideal for identification of indigoids. Only HPLC-DAD was able to separate the isomers and clearly identify each of them based on their different UV-visible spectra. While the ratio of indigotin to indirubin has been suggested to be indicative of the dyeing process or even the altitude at which the dyes were prepared due to the lower concentration of atmospheric oxygen content at high altitudes [14], the peak areas for the isomers are variable depending on the wavelength used for integration. Further, we observed that the dark blue Lambayeque sample extracted into dimethyl sulfoxide (one of the few solvents in which indigoids are readily soluble) changed color from blue to reddish-purple during a delay between preparation and HPLC analysis, indicating some shift from indigotin to indirubin in solution. Indeed, HPLC analysis of this sample showed only indirubin, while a re-extracted sample showed both isomers in a ratio indicating more of the indigotin and less indirubin, as expected for the dark blue solution. Based on these observations, the ratio of the isomers should not be interpreted as particularly meaningful. Both the DMSO and methanol/HCl extractants were appropriate for the HPLC analyses of the indigoids. Example chromatograms, UV-vis spectra, and mass spectra for a blue sample are shown in Supplementary Figure S15.

3.1.2. Reds

The source of red dyes in ancient South American textiles has long been of interest [2,24,27,35,36] and, as such, has been widely studied. Plant reds consistent with those derived from the roots of *Relbunium* or *Galium* species were used throughout the Paracas and Nazca periods, as indicated by the presence of purpurin and other related anthraquinones. Carminic acid from cochineal insects (*Dactylopius coccus*) appears later, specifically in the Wari and Lambayeque textile fragments. Samples from the discontinuous warp and weft textiles attributed to the Wari period showed no evidence of cochineal, only anthraquinone compounds characteristic of the plant reds. This may be due to the difference in preparation of these textiles, where tie- and over-dyeing were used to create the unique patterns of "white" undyed fabric with primary and secondary color designs. The other red yarns from the Wari tapestry weave textiles were prepared with cochineal, as indicated by the presence of carminic acid in those samples. While the Chancay would have had access to cochineal, the reddish-brown yarn from the weaver's basket was clearly dyed with a plant red consistent with a *Relbunium* or *Galium*. Example chromatograms, UV-vis and mass spectra for a *Relbunium* red sample are shown in Supplementary Figure S16. Chromatograms and mass spectra for red yarns from the Antúnez de Mayolo Peruvian dye plant collection reference samples are part of another manuscript in preparation with Degano et al. at the University of Pisa. Detailed results are shown in Supplementary Tables S1 and S2.

Of the anthraquinone aglycones present in the plant reds, purpurin was the major component by HPLC. A peak at *m/z* 239.04 in the negative ion DART mass spectra may be indicative of either xanthopurpurin or alizarin, both dihydroxyanthraquinones differing only in the position of the –OH groups: xanthopurpurin is 1,3-dihydroxyanthraqui-none, while alizarin is 1,2-dihydroxyanthraquinone. Both give nearly identical UV-vis spectra, and under the conditions used for the majority of the HPLC analyses, both also give identical retention times as determined from multiple analyses of standards; the synthesized xanthopurpurin used as a standard may have contained trace impurities. Under collision-induced dissociation conditions (with the AccuTOF Orifice 1 set to 90 V), reference samples of alizarin and xanthopurpurin yielded different mass spectra consistent with what has been reported [24,37]. However, under these conditions, dyed yarn samples yield complex spectra that do not clearly differentiate between these compounds. Dutra Moresi [38] carried out significant work to characterize the anthraquinones found in various species of *Relbunium*, finding that they contain no alizarin, as determined by HPLC. Negative ion DART mass spectra of reference dyeings prepared from three *Relbunium* species and *Galium antunezìae* [39,40] from Peru showed no consistent differences in the distribution of the characteristic anthraquinone compounds, as indicated in Table 5. Differences that have been attributed to different species may actually be due to differences in dye preparation conditions such as temperature, age of the roots, pH, and mordants. Further complicating the situation is the fluidity of the classification of these plants at the genus level [39,41]. Thus, we have not attempted to differentiate the plant reds and consider all to be "*Relbunium*" for simplicity, though further studies are warranted.

Carminic acid and other glycosides are not well identified with DART-MS [9,25], likely because they cannot be desorbed from the fibers solely by heat. For red samples that do not show the characteristic peaks for the plant red anthraquinones (purpurin, rubiadin, etc.) by DART-MS, we followed up with paper spray MS on the extracted solution used for HPLC analysis. Carminic acid readily ionizes by this method, as do the other glycosides characteristic of cochineal. HPLC of the same extracts showed primarily carminic acid; the other glycosides were either not detected or not identified, as no standards were available for comparison. Example chromatograms, UV-vis spectra, and mass spectra for a cochineal red sample are shown in Supplementary Figure S17.

3.1.3. Yellows

The two yellow samples from the Paracas mummy fragments showed all of the hydroxybenzoic acid compounds that Zhang et al. [5] describe as characteristic of degradation products of the flavonoids quercetin, isorhamnetin, and kaempferol. While signals at *m/z* values consistent with the presence of a number of different flavonoids were observed in the Cave #5 mummy sample (Table 4), none were consistent across the various mass spectrometry methods, and none were unambiguously identifiable by HPLC. While two of the peaks in the HPLC had retention times consistent with the benzoisoquinoline alkaloids of *Bocconia*, their signal intensities were too low to yield a good spectral match. As the positive ion DART-MS did not show any signal for sanguinarine or chelerythrine, the most likely source of the yellow color is a degraded flavonoid. The gold sample from the loose threads associated with the turban in bundle 382-54 (02846) appears to be most similar to a degraded yellow obtained from *Bidens* or another Asteraceae, as indicated by signals at *m/z* values consistent with okanin ($[M\text{-}H]^-$ at *m/z* 287.06) and butein ($[M\text{-}H]^-$ at *m/z* 271.06) by negative ion DART-MS. For the reference samples, only those of Asteraceae (e.g., *Bidens, Coreopsis, Dahlia*) showed significant signals at these *m/z* values. This was confirmed by the HPLC results, which showed significant overlap in the pattern of compounds absorbing at 375 nm in the *Bidens andicola* sample from the Peruvian dye reference materials, as shown in Figure 1. Reference standards of butein and okanin were used to confirm the identities of the peaks at 7.7 min and 8.6 min, respectively.

Zhang et al. [5] described the difficulties of identifying chalcone colorants and how they may have been missed in earlier analyses where strong acid extraction protocols were used for HPLC analyses. Comparison of methods where the sample preparation differs is difficult at best. The DART mass spectra for the yellow dyes in the Wari textile samples from the Carlos collections showed primarily the degradation products of flavonoid yellows and little or nothing else. The HPLC results, in some cases, showed peaks at retention times different from that of luteolin, yet with markedly similar spectra, as was described by Wouters and Rosario-Chirinos in their extensive study of ancient Peruvian textile dyes [6]. The yellows from the Wari discontinuous warp and weft textiles were not consistent within the textiles, either due to variations in how they were prepared, degradation, or limitations of the methods to detect such low concentrations of chromophores. It is likely that pale undyed fibers were used to achieve the pale or off-white color observed. This seems to be the case for the cream, white, and beige samples from the one Lambayeque textile investigated, as well, which showed only traces of the hydroxybenzoic acids, some contamination from carminic acid, and no compounds specific to yellow colorants.

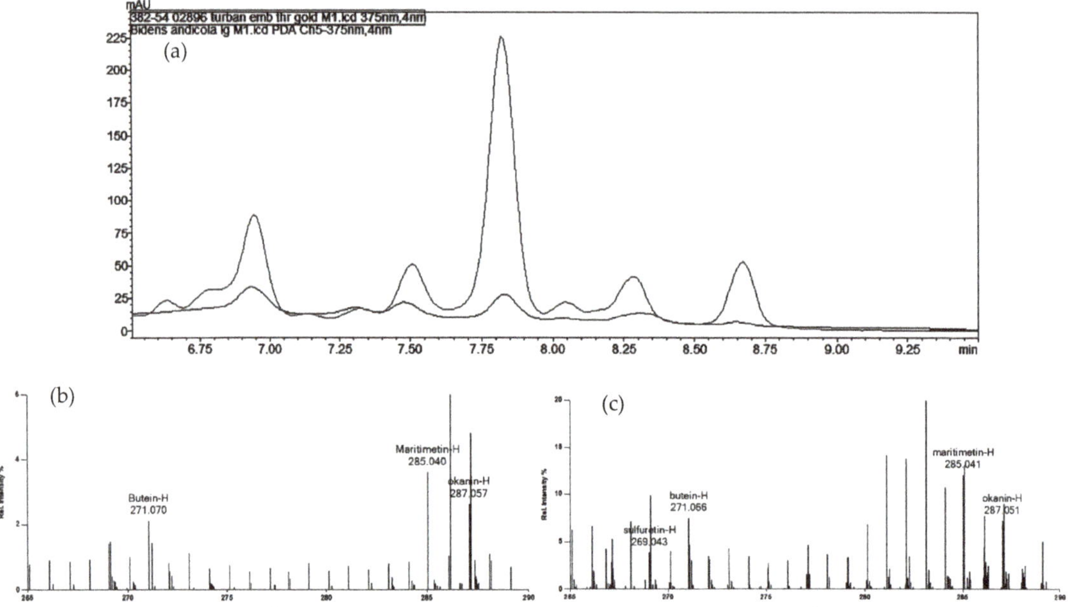

Figure 1. (**a**) Chromatograms (at 375 nm) for the gold embroidery thread from Paracas turban 382-54 02896 (black) and the reference sample of *Bidens andicola* (gray). The peak at 7.8 min was identified with a standard as okanin, and the peak at 8.6 min is consistent with butein (see Supplementary Figure S18 for spectra). (**b**,**c**) DART mass spectra for a few fibers of yarn from the same sample (**b**) and the reference sample of *Bidens andicola* (**c**), both treated with formic acid prior to introduction into the ion source. The spectra clearly show $[M\text{-}H]^-$ ions consistent with those of butein, okanin, and another aglycone, maritimetin. Sulfuretin is also present in the reference dye sample.

3.2. Secondary Colors

Because secondary colors were of the most interest when the initial sampling of the textiles was undertaken, the majority of the results fall into this category. Example chromatograms, UV-vis spectra, and mass spectra for the secondary colors purple, green, and orange are shown in Supplementary Figures S19–S21.

3.2.1. Purple

All the purple samples contained indigoids, likely obtained by applying the indigo vat process to yarns (or textiles) previously dyed red; this is definitely the case for the tie-dyed discontinuous warp and weft Wari textiles, as indicated by one yarn sample that was red on one end and purple on the other (Figure 2a). The purple sample from the frayed Wari band (2003.40.5) was the only one containing both indigo and cochineal, as shown by the presence of carminic acid that was detected by both PS-MS and by HPLC. All of the other purples, from the Paracas mummies to the Nazca cross-knit looped border fragments, contained plant reds consistent with *Relbunium*. Interestingly, this also included the Wari discontinuous warp and weft textiles with tie-dye patterning. While cochineal was available to the Wari and widely used in other textiles, perhaps the unique construction and preparation of these textiles precluded its use for some reason. No evidence of brominated indigoids obtained from shellfish were observed in any of the chromatographic or mass spectrometric results, save for a single analysis early on that indicated a trace of dibromoindigo by positive ion DART-MS, which could not be reproduced. Wallert and Boytner [8] also noted the surprising absence of dibromoindigo in purple samples from the south coast of Peru, near the source of the shellfish from which the dye is obtained. Saltzman [2], however, believed that shellfish purple was utilized in the preparation of Andean textiles; the use of these purples was confirmed by studies carried out by Michel et al. [18], showing brominated indigoids attributed to the use of *Concholepus concholepus.*

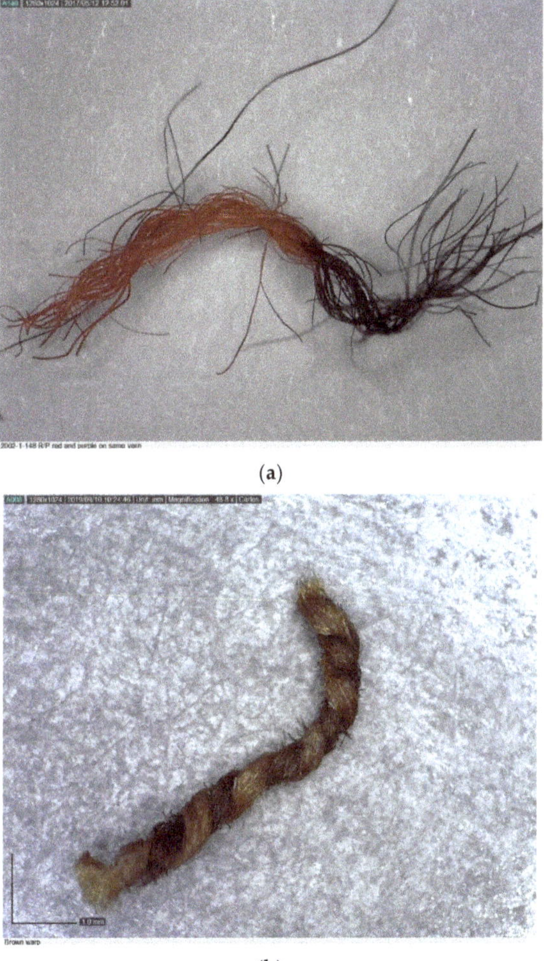

(**a**)

(**b**)

Figure 2. Examples of some of the yarn samples collected for analysis. (**a**) Red and purple yarn from MCCM 2002.1.148, a discontinuous warp and weft textile, showing evidence of overdyeing to achieve the purple color. (**b**) Brown warp yarn from MCCM 2004.64.1 from the Lambayeque textile, showing plying of two shades of brown cotton yarn.

3.2.2. Green

As expected, the green samples appear to be indigo overdyed onto yellow—either dyed or naturally yellowish—yarns. Indigotin (and usually indirubin, depending on the sample) was observed in the chromatographic profiles of all the green yarns, and was detected as the M^- ion (formed by direct Penning ionization) as well as the $[M\text{-}H]^-$ ion at *m/z* 262.08 and 261.07, respectively, in negative ion mode by DART-MS. Indigo was detected in all the green samples, as well, both by DART-MS and by HPLC. Separation of the indigoid isomers by HPLC showed that indirubin was a minor component, and in several cases, was not detected. Thus, the peak at *m/z* 262.08 in the negative ion DART spectra can be attributed primarily to indigotin, the major constituent. The yellow components in the green yarns were, as with the yellow yarns, difficult to interpret. The Paracas sample of mantle threads from 382-10 05904 indicated a luteolin-based dye by both direct mass spectrometry and by HPLC. Unfortunately, it did not directly correlate with any of the yellow reference dyes. The Paracas poncho ground cloth yarns from 382-68 02929 indicated a different source of yellow lacking evidence of luteolin, with a unique chromatogram. The DART mass spectra (negative ion mode) also did not indicate the presence of any luteolin, but peaks at *m/z* 343.07 and 373.09 correlate to the $[M\text{-}H]^-$ ions of eupatorin or of cirsilineol and gardenin D, respectively. The corresponding MH^+ ions were observed in the positive ion PS mass spectra of this sample, as well. With no reference standards for these compounds, their presence could account for the unique chromatographic profile observed. According to Cardon [10], these compounds are found in some species of Baccharis, though the lack of luteolin, also expected in dyes prepared from Baccharis, complicates the identification.

The Nazca samples were all extremely small, as they were obtained from intact objects. Only negative ion DART-MS and HPLC analyses were performed on the green samples, showing little in addition to indigo. Luteolin and its methyl ether were observed in two of the samples, the snake band and the hummingbird border, by DART-MS. Luteolin-like components, differing significantly in retention time from luteolin and luteolin methyl ether standards, were observed by HPLC. The snake band sample also showed traces (less than 1% relative abundance) of the same ions observed in the green Paracas poncho ground cloth. A tentative identification of Baccharis seems probable for this Nazca sample. The yellow component in the green yarn from the anthropomorphic human "bean people" border was significantly different from the other two Nazca samples and may be completely oxidized.

Similarly, the green samples from the Wari and Lambayeque textiles showed no significant evidence of yellow dyes. Only the yellow-green sample from the frayed Wari band (2003.40.5) showed evidence of quercetin and kaempferol by HPLC; neither was detected by negative ion DART-MS. Both of these compounds have been proposed to oxidize to form the hydroxybenzoic acids described by Zhang et al. [5], all of which were found in the Wari and Lambayeque green samples based on the presence of molecular ions of the same mass in the negative ion DART mass spectra. The most likely explanation is that the yellow component of the greens in these textiles has decomposed over time. As the objects sampled at the Carlos Museum textiles came from collectors, it is reasonable to presume that the objects were displayed and may have experienced at least some light exposure, leading to photo-oxidation. It is also possible that the green coloration was obtained by applying indigo to naturally yellowish or buff-colored yarns, which may themselves contain these hydroxybenzoic acids. Further studies on naturally colored wools are needed.

3.2.3. Orange

The orange yarn from the Lambayeque textile and the salmon fibers from the Wari tunic contained carminic acid. All of the other orange samples had components characteristic of plant reds from *Relbunium* in both the negative ion DART mass spectra and by the presence of purpurin in the chromatograms. Of the Paracas samples investigated, only the orange ground cloth from the skirt in 382-49 (23808) 03174 showed sufficient evidence to suggest that *Bidens* may be the source of the yellow, based on the negative ion DART mass spectra showing a small peak at *m/z* 287.06, consistent with the $[M\text{-}H]^-$ ion of okanin. Confirmation with HPLC was not possible due to the low signal intensity. The yellow component in the only orange sample from the Nazca textiles is likely to be oxidized, as was the case with the green.

The salmon and orange fibers from the Wari tapestry tunic differed in their source of red dye, with the former being carminic acid and the latter consistent with *Relbunium* plant red. Neither showed any evidence of yellow dyes. Mordants or the dyeing process (e.g., temperature, pH, etc.) may account for the different shades of color observed without the addition of a yellow dye. The HPLC of the Lambayeque orange yarn showed primarily carminic acid but little evidence for yellow dyes. The negative ion DART-MS showed a peak at *m/z* 301.05, consistent with a number of possible flavonoids, along with the decomposition products expected for flavonols. PS-MS failed to detect carminic acid in this sample.

3.2.4. Browns

Brown yarns may be naturally colored fibers in shades of brown, or they may have been dyed with tannins, which are difficult to identify and can be sourced from a variety of different plants. Indeed, the brown yarn from the Wari tapestry weave tunic appears to be undyed, as none of the methods showed any evidence of dye chromophores. The brown fragments from the Paracas textile 421-132 02096, on the other hand, have a composition similar to that of the gold sample from 382-54 02846 described above and consistent with the reference samples of *Bidens andicola.* The presence of both butein and okanin, identified by their $[M-H]^-$ ions in negative ion DART, could not be confirmed by HPLC due to the low signal intensity; monitoring the chromatogram at 375 nm, a trace of butein may be present at 7.7 min, as shown in Supplementary Figure S22.

Both the brown warp yarn (consisting of two plies, one dyed brown and one possibly natural, Figure 2b) and the golden brown yarn from the Lambayeque textile had a peak in the HPLC at 7.4 min with a UV-vis spectrum consistent with that of one of the laccaic acids found in Kerria lacca, which is native to Asia and not found in South America (Figure 3). None of the laccaic acids were detected in the paper spray mass spectra of the extracts from these yarns, though peaks at *m/z* 331.05 and 353.04, consistent with the MH^+ and $[M+Na]^+$ ions for kermesic acid, were observed in the PS mass spectra of the EDTA/DMF extracts of both of these samples, though no carminic acid was present. Degano and Colombini [15] observed laccaic acids in red yarn from the belt of a child mummy from the transitional period between the Wari and Chancay cultures, overlapping the time period of the Lambayeque, as well. These findings further emphasize the importance of developing a database of known reference dyes available to the textile artists of ancient Peru.

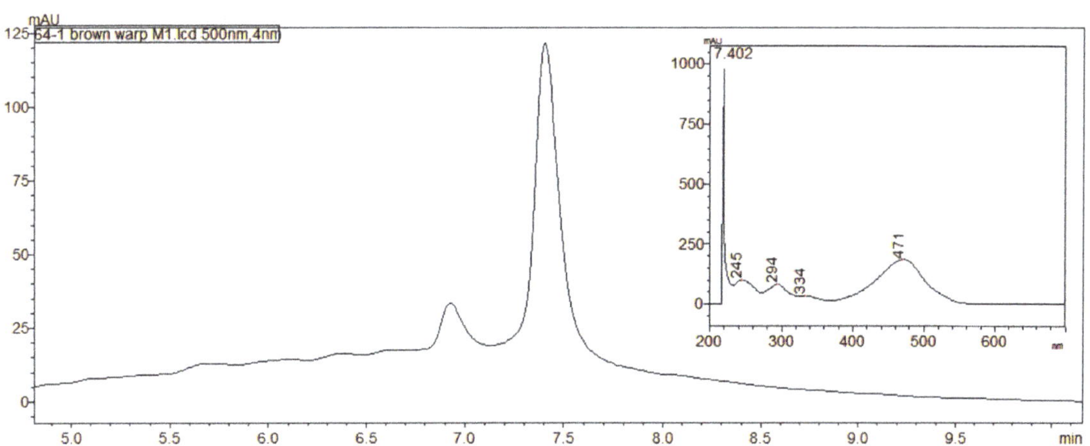

Figure 3. Unusual chromatogram (at 500 nm) and UV-vis spectrum (inset) for peak at 7.402 min in the brown warp yarn (MCCM 2002.64.1) shown in Figure 2b.

4. Discussion

These results demonstrate the successful identification of many of the dyes present in samples from pre-Columbian Peruvian textiles spanning more than two millennia. This interpretation is complicated by deterioration, the use of undyed naturally colored fibers, similarity among the wide variety of plant sources, dyeing processes, and deterioration over time. The results presented above show the difficulty of positively identifying yellow colorants due to their tendency to decompose, the many possible sources from which they can be obtained, and the overlap in compositions between many of the plant sources. These challenges affect the secondary colors, as well. If yellow colorants are not identified in an orange or green sample, it is impossible to determine if yellow colorants were present to begin with, if some other method of achieving those colors was utilized, or if a naturally yellow-colored fiber was used. Mordants, dye bath temperatures, and fermentation conditions can each affect the final color obtained from various dye plants, perhaps enabling some secondary colors to be obtained without combining dyes.

This work also compares the results obtained by ambient ionization mass spectrometry with results from high-performance liquid chromatography. We consider how much sample is needed for analysis, how sensitive and selective the methods are for the dye components, and performance characteristics such as the time required for analysis and the operator skill necessary to carry out both the analyses and data interpretation. Each of the methods has capabilities and limitations that make using them together the best choice for cases where LC-MS/MS is not available, as is the case here at EMU and many other laboratories. Ideally, dyes would be identified completely non-invasively; fiber optic reflectance spectroscopy (FORS) and surface-enhanced Raman spectroscopy have been shown to be useful for some dyes, particularly the reds [20,21,42]. However, these techniques are not yet as widely available and used as are HPLC and MS. The lower limits of detection for the colorant compounds are critical for their identification by any analytical technique and are dependent upon the amounts of sample used for analyses and, more critically, how that sample is prepared for analysis.

To roughly compare the limits of detection of the ambient ionization mass spectrometry methods and HPLC-DAD, solutions of four of the colorant compounds characteristic of the dyes identified were prepared in methanol and diluted to concentrations of 1 and 10 parts per million (ppm). These compounds included luteolin, okanin, purpurin, and carminic acid. Each solution was run by DART-MS (or paper spray MS in the case of carminic acid) and by HPLC, with the chromatogram monitored at an appropriate wavelength for each compound: 350 nm for luteolin, 380 nm for okanin, 480 nm for purpurin, and 500 nm for carminic acid. Of the 1 ppm solutions, only luteolin was reliably identified with HPLC using both retention time and UV-vis spectrum; all the rest were easily detected in the 10 ppm solutions at the appropriate wavelengths.

By DART-MS, both luteolin and purpurin were readily detected in the 1 ppm solutions applied on the closed end of a capillary melting point tube, while okanin gave a large even-electron molecular ion signal at 10 ppm, placing the LOD for okanin in solution by DART-MS somewhere between 1 and 10 ppm. Interestingly, carminic acid was not detected in positive ion mode with PS-MS at either concentration, likely due to the lack of Na^+ ions to aid in adduct formation. This may explain why this compound was not identified in the orange-colored fibers from the Lambayeque textile (2004.1.64). The use of negative ion mode would likely remedy this problem. It is important to consider how little of the chromophore can be differentiated from background by the method of choice for small and limited samples. Further, the sample preparation method—except in the case of DART-MS, where none is necessary—strongly affects both the quantity and quality of the data obtained, with different combinations of organic solvents, chelating agents such as EDTA, and acids (formic, oxalic, HCl, etc.) yielding different results even for the same samples. Acid strength can cause changes in the overall profile of the dye as glycosidic linkages are broken to yield aglycone components [32,43], and some dye components including the chalcones are decomposed by strong acids. Solubility, particularly for indigoids, further affects how much dye is in the solution being analyzed. How the extraction solvent(s) disrupt the binding of mordant metal ions also influences the results.

DART-MS, in particular, is affected by the presence of certain mordants and how the mordant and dye bind to either cellulose or proteinaceous fibers, at least for some dyes such as logwood [39] with transition metal mordants such as Fe, Sn, and Cu. A recent study of mordants in Paracas textiles with X-ray fluorescence spectroscopy [7] showed that Fe and Cu may indeed have been used as mordants in Paracas textiles, which would likely influence the signal intensity observed in the DART-MS results. The binding of dye colorants with cotton fibers appears to be quite strong, as DART-MS spectra show reliably stronger signals from dyes applied to wool and other animal hair yarn compared to those dyed on cotton [25,31]. Nearly all of the fibers investigated herein were camelid wool; two of the Paracas samples, three of the Chancay yarn balls, and three samples, including both warp threads from the Lambayeque textile, were cotton, as identified visually with optical and/or scanning electron microscopy or with attenuated total reflectance Fourier-transform infrared spectroscopy (ATR-FTIR). Two of the Paracas samples yielded FTIR spectra consistent with the presence of both cellulose and protein, which is consistent with previous reports of some of the yarns containing a mixture of fibers [27,44].

The primary advantage of ambient ionization mass spectrometry methods is the ability to characterize molecules with little or no sample preparation and very short analysis times on the order of a few seconds. However, collecting data is not the same as interpreting data, and making sense of the results takes expertise and, ideally, another method such as HPLC-DAD for confirmation. Knowing what does and does not ionize by DART-MS (e.g., glycosides such as carminic acid) and under what conditions (e.g., benzoisoquinoline alkaloids such as sanguinarine forming ions in positive mode, but not in negative) requires analysis of known materials such as the database of Peruvian plant dyes, as in this case. Comparing ambient ionization mass spectrometry results directly, even with multivariate analysis, may provide more insight into dye sources that would be difficult or even impossible otherwise. The application of these methods to art and archaeological materials is in its infancy, and much work remains.

5. Conclusions

Here, we have shown the characterization of the dyes present in both the primary and secondary colors found in ancient Peruvian textiles from multiple cultural periods over a span of nearly 1800 years, including the Paracas Necropolis, the Nazca, the Wari, the Chancay, and the Lambayeque. The dyes were identified by both direct analysis in real time time-of-flight mass spectrometry (DART-MS) and paper spray MS, and these results were compared to ones obtained from extraction and separation with high-performance liquid chromatography (HPLC) with ultraviolet-visible diode array detection (DAD). The ambient ionization MS methods were simple and fast: DART-MS required no sample preparation at all, and paper spray results were obtained in a few seconds of analysis time once the samples were extracted into an appropriate solvent system (30:1 methanol:HCl). In general, the ambient ionization MS results compared well with the more traditional HPLC-DAD analyses, which provided the advantage of separation and identification of isomeric species (e.g., indigotin and indirubin). Analysis with DART-MS yielded the general classes of dyes, either through rapid identification of marker compounds (e.g., indigotin/indirubin, purpurin, luteolin, etc.) or the marked absence of such (as was the case with carminic acid). Many of the possible yellow dyes have overlapping compositions, making their identification difficult regardless of analytical approach. Oxidative decomposition, either due to age or light exposure, further complicates the conclusive identification of yellow dyes in the Peruvian textiles. Chalcone biomarkers characteristic of *Bidens* were observed in some of the samples by use of DART-MS; these findings were confirmed with HPLC-DAD. The speed and simplicity of ambient ionization mass spectrometry holds significant promise for the identification of textile dyes using only small samples, though much work remains to understand the breadth and depth of possible sources of dyes available to the artisans of the ancient Andes.

Supplementary Materials: The Supplementary Tables and Figures are available online at https://www.mdpi.com/article/10.3390/heritage4030091/s1. Figure S1, (a) Tapestry fragment with fish and snake designs (MCCM accession number 2002.1.100), made of camelid fiber; attributed to the Nazca, Early Intermediate Period. (b–d) Samples from (a). Figure S2, (a) Mantle fragment with crossloopknit stitched embroidered border of hummingbird motifs (MCCM accession number 2002.1.3), made of camelid fiber; attributed to the Nazca, Early Intermediate Period. (b–d) Samples from (a). Figure S3, (a) Fragments of a mantle border (MCCM accession number 2002.40.4 A-C), made of camelid fiber; attributed to the Nazca, Early Intermediate Period. (b–c) Samples from (a). Figure S4, (a) Tie-dye textile fragment (MCCM accession number 2002.1.148), made of camelid and cotton fibers; Wari related, Middle Horizon. (b–h) Samples from (a). Figure S5, (a) Tie-dye textile fragment (MCCM accession number 2002.1.1), made of camelid fiber; Wari related, Middle Horizon. (b–h) Samples from (a). Figure S6, (a) Single interlocked tapestry administrator's tunic (MCCM accession number 2002.1.16), made from cotton and camelid fibers, Wari Middle Horizon. (b–g) Samples from (a). Figure S7, (a) Brocade textile fragment with winged staff-bearer figures in headdresses (MCCM accession number 2002.1.83), made from cotton and camelid fibers, Wari Middle Horizon. (b–c) Samples from (a). Figure S8, (a) Woven band (MCCM accession number 2003.40.5), no additional information available, attributed to Wari. (b–f) Samples from (a). Figure S9, (a) Weaver's work basket yarns (MCCM accession number 2002.1.126 A-U), cotton and camelid fibers, Late Intermediate Period Chancay. (b–h) Samples from (a). Figure S10, (a) Red tasseled tunic fragments (MCCM accession number 2004.64.1), cotton and camelid fibers, Late Intermediate Period Pacatnamú. (b–k) Samples from (a), brown warp shown in Figure 2b. Figure S11, Paracas Necropolis yarn fragment samples collected by Anne Paul in 1985 from the Museo Nacional de Arqueología, Anthropología, e Historiadel Perú (MNAAHP). (a) Orange yarn from mummy bundle textile designated 310-58c 02912; (b) yellow yarn from mummy bundle textile designated Cave #5 12-5236 0093, cloth (oldest). Figure S12, ParacasNecropolis yarn fragment samples collected by Anne Paul in 1985 from the Museo Nacional de Arqueología, Anthropología, e Historiadel Perú (MNAAHP), mummy bundle 382. (a) Specimen 10, subspecimen 05904, green yarn from mantle; (b) specimen 45, subspecimen 02763, orange loose threads from skirt; Specimen 48, subspecimen 01017, (c) skirt fringe blue and (d) black/purple (continued next page). Figure S12, continued. Specimen 49, subspecimen (23808) 03174, (e) orange from skirt ground cloth; Specimen 54 subspecimen 02846, (f) gold embroidery threads; (g) blue fibers from turban ground cloth; (h) green threads from turban. Figure S13, continued. Specimen 68, subspecimen02929, (i) poncho purple braid and (j) poncho border ground cloth green fibers; Specimen 72, subspecimen 02519, (k) black skirt fringe and (l) red skirt embroidery thread. Figure S14, Paracas Necropolis yarn fragment samples collected by Anne Paul in 1985 from the Museo Nacional de Arqueología, Anthropología, e Historiadel Perú (MNAAHP), mummy bundle 421. (a) Specimen 39, subspecimen 03083, red loose fibers from mantle; (b) specimen 132, subspecimen 03096, brown mantle fibers from box. Figure S14, Paracas Necropolis yarn fragment samples collected by Anne Paul in 1985 from the Museo Nacional de Arqueología, Anthropología, e Historiadel Perú (MNAAHP), mummy bundle 421. (a) Specimen 39, subspecimen 03083, red loose fibers from mantle; (b) specimen 132, subspecimen 03096, brown mantle fibers from box. Figure S16, Example chromatogram at 500 nm (a) for a red sample from 2002.1.1, yarn from tie-dye Wari textile, indicative of a plant red consistent with Relbunium. UV-vis spectrum for the major peak identified as purpurin (b) detected at 10.4 min. DART mass spectra in negative ion mode (c) and positive ion mode (d) of the same sample. Figure S17, Example chromatogram at 495 nm (a) for a red sample from 2004.61.1, yarn from Lambayeque tasseled tunic, indicative of insect red consistent with cochineal. UV-vis spectrum for the major peak compared with that of carminic acid standard (b) detected at 6.6 min. Paper spray mass spectrum (c) in positive ion mode of the same sample extracted in methanol:HCl. Figure S18, UV-vis spectra for peaks observed in the chromatogram shown in Figure 1 in the main text. Spectrum of peak at 7.8 min in the extract from the gold embroidery thread from Paracasturban 382-54 02896 (a), compared to the spectrum of the okaninstandard (b). Spectrum of peak at 8.6 min in the extract from the gold embroidery thread from Paracasturban 382-54 02896 (c), compared to the spectrum of the buteinstandard (d). Figure S19, Example chromatogram at 606 nm (a) for a purple sample from 382-68 02929, yarn from poncho braid, indicative of both Relbunium and indigo. UV-vis spectra for the major peaks (b) show that purpurin (λmax~480 nm) and indigotin (λmax at 606 nm) co-elute under the chromatographic conditions used; only a trace of indirubin (c) was observed, in part because it does not absorb strongly at 606 nm. DART mass spectrum in negative ion mode (d) of the same sample. Figure S20, Example chromatograms at 350 nm (a) and 606 nm (b) for a green sample from 382-68 02929, yarn from poncho border ground cloth, indicative of indigo and an as-yet undetermined but unique yellow. The UV-vis spectrum for the peak at 10.6 min (b, inset) is consistent with that of indigotin. The other peaks observed at 350 nm gave similar spectra (c–f), which did not correlate with any of the standards or reference dyes. Figure S21, Example chromatograms at 350 nm (a) and 450 nm (b) for an orange sample from 310-56c 02912, yarn from a decorative border from a Paracas Necropolis period textile. The UV-vis spectrum for the peak at 10.5 min (b, inset) is consistent with that of purpurin from Relbunium. No major peaks were observed at 350 nm and none could be identified as any of the known yellow components. No signal consistent with any of the yellow colorants was observed above 1% relative abundance by DART-MS; the negative ion spectrum (c) shows only purpurin and caffeic acid. Figure S22, Example chromatogram at 375 nm (a) for the deteriorated brown fibers from 421-132 03096 (brown). The chromatogram from the gold yarn consistent with Bidens or Coreopsis is shown in gray for comparison. A trace of butein may be present around 8.6 min, but the poor signal-to-noise ratio limits the identification. Several signals consistent with yellow colorants from Bidens or Coreopsis were observed both by negative ion DART-MS (b) and paper spray mass spectrometry (c); no glycosides were detected, likely due to the sample extraction in methanol:HCl. Table S1, Detailed results from analysis of yarns from ancient Peruvian textiles from the Michael C. Carlos Museum collections. Table S2, Detailed results from analysis of yarns from Paracas Necropolis samples collected in 1985 by Anne Paul at the MNAAHP in Lima, Peru.

Author Contributions: Funding acquisition, R.S. and R.A.A.; Investigation, J.C.A., S.M., B.W. and R.A.A.; Methodology, R.A.A.; Project administration, R.S. and R.A.A.; Resources, K.A.d.M. and K.J.; Validation, R.A.A.; Writing—original draft, J.C.A., R.S. and R.A.A.; Writing—review & editing, J.C.A., S.M., B.W., K.A.d.M., K.J., R.S. and R.A.A. All authors have read and agreed to the published version of the manuscript.

Funding: This research was funded by the Andrew W. Mellon Foundation, Grant ID# G-1806-05955. The APC was funded in part by the EMU College of Arts and Sciences Dean's Faculty Professional Development Award program.

Data Availability Statement: Data not found in the paper and Supplementary Materials can be requested from the corresponding author.

Acknowledgments: The authors acknowledge a number of undergraduate students who contributed to this work, including Clara Gonzales from Georgia State University; Bennet Dunstan and Christin Neiman from the EMU Chemistry Department; and Chelsea Van Buskirk, Ethan Burke, and Jenna Zoerman from the Fall 2019 Analytical Instrumentation curriculum-based undergraduate research experience at EMU. Support for graduate student authors was provided by the EMU College of Arts and Sciences Dean's Faculty Professional Development Award Program, the James H. Brickley Faculty Professional Development and Innovation Award, EMU Faculty Research Fellowships, the EMU Chemistry Department and Sellers Fund, and an EMU Graduate School Research Support Award to J.C.A. The AccuTOF DART mass spectrometer was purchased through NSF MRI-R^2 award #0959621, and the HPLC-DAD and FTIR instruments used for this project were purchased with funding from the Kresge Foundation. SEM-EDS analyses were made possible by JEOL USA through the loan of a JSM-7000 Neoscope in August–September 2019.

Conflicts of Interest: The authors declare no conflict of interest.

References

1. Fester, G.A. Some Dyes of an Ancient South American Civilization. *Dyestuffs* **1954**, *40*, 238–244.
2. Saltzman, M. The Identification of Dyes in Archaeological and Ethnographic Textiles. In *Archaeological Chemistry*; Carter, G., Ed.; American Chemical Society: Washington, DC, USA, 1978; Volume 171, pp. 172–185.
3. Saltzman, M.; Keay, A.M.; Christensen, J. The Identification of Colorants in Ancient Textiles. *Dyestuffs* **1963**, *44*, 241–251.
4. Ferreira, E.S.B.; Quye, A.; McNab, H.; Hulme, A.N. Photo-Oxidation Products of Quercetin and Morin as Markers for the Characterisation of Natural Yellow Dyes in Ancient Textiles. *Dye. Hist. Archaeol.* **2002**, *18*, 63–72.
5. Zhang, X.; Boytner, R.; Cabrera, J.L.; Laursen, R. Identification of Yellow Dye Types in Pre-Columbian Andean Textiles. *Anal. Chem.* **2007**, *79*, 1575–1582. [CrossRef]

6. Wouters, J.; Rosario-Chirinos, N. Dye Analysis of Pre-Columbian Peruvian Textiles with High-Performance Liquid Chromatography and Diode-Array Detection. *J. Am. Inst. Conserv.* **1992**, *31*, 237–255. [CrossRef]
7. Sabatini, F.; Bacigalupo, M.; Degano, I.; Javér, A.; Hacke, M. Revealing the organic dye and mordant composition of Paracas textiles by a combined analytical approach. *Herit. Sci.* **2020**, *8*, 1–17. [CrossRef]
8. Wallert, A.; Boytner, R. Dyes from the Tumilaca and Chiribaya Cultures, South Coast of Peru. *J. Archaeol. Sci.* **1996**, *23*, 853–861. [CrossRef]
9. Armitage, R.A.; Fraser, D.; Degano, I.; Colombini, M.P. The analysis of the Saltzman Collection of Peruvian dyes by high performance liquid chromatography and ambient ionisation mass spectrometry. *Herit. Sci.* **2019**, *7*, 1–23. [CrossRef]
10. Cardon, D. *Natural Dyes: Sources, Tradition, Technology and Science*; Archetype: London, UK, 2007.
11. Duke, J.A. *Duke's Handbook of Medicinal Plants of Latin America*; Taylor & Francis: Abingdon, UK, 2008.
12. Little, E.L.; Woodbury, R.O.; Wadsworth, F.H. *Common Trees of Puerto Rico and the Virgin Islands*; United States Department of Agriculture: Washington, DC, USA, 1974.
13. Saltzman, M. Analysis of Dyes in Museum Textiles or, You can't Tell a Dye by Its Color. In *Textile Conservation Symposium in Honor of Pat Reves*; McLean, C., Connel, P., Eds.; Conservation Center, Los Angeles County Museum of Art: Los Angeles, CA, USA, 1986; pp. 32–39.
14. Niemeyer, H.M.; Agüero, C. Dyes used in pre-Hispanic textiles from the Middle and Late Intermediate periods of San Pedro de Atacama (northern Chile): New insights into patterns of exchange and mobility. *J. Archaeol. Sci.* **2015**, *57*, 14–23. [CrossRef]
15. Degano, I.; Colombini, M.P. Multi-analytical techniques for the study of pre-Columbian mummies and related funerary materials. *J. Archaeol. Sci.* **2009**, *36*, 1783–1790. [CrossRef]
16. Price, K.E.; Higgitt, C.; Devièse, T.; McEwan, C.; Sillar, B. Tools for Eternity: Pre-Columbian Workbaskets as Textile Production Toolkits and Grave Offerings. *Br. Mus. Tech. Res. Bull.* **2015**, *9*, 65–86.
17. Boucherie, N.; Nowik, W.; Cardon, D. La Producción Tintórea Nasca: Nuevos Datos Analíticos Obtenidos Sobre Textiles Recientemente Descubiertos En Excavaciones. *Nuevo Mundo Mundos Nuevos* **2016**. [CrossRef]
18. Michel, R.H.; Lazar, J.; McGovern, P.E. Indigoid Dyes in Peruvian and Coptic Textiles of the University Museum of Archaeology and Anthropology. *Archaeomaterials* **1992**, *6*, 69–83.
19. Bernardino, N.; De Faria, D.; Negrón, A. Applications of Raman spectroscopy in archaeometry: An investigation of pre-Columbian Peruvian textiles. *J. Archaeol. Sci. Rep.* **2015**, *4*, 23–31. [CrossRef]
20. Burr, E.A. Dye Analysis of Archaeological Peruvian Textiles Using Surface Enhanced Raman Spectroscopy (Sers). Ph.D. Thesis, University of California, Los Angeles, CA, USA, 2016.
21. Sepúlveda, M.; Urzúa, C.L.; Cárcamo-Vega, J.; Casanova-Gónzalez, E.; Gutiérrez, S.; Maynez-Rojas, M.; Ballester, B.; Ruvalcaba-Sil, J.L. Colors and dyes of archaeological textiles from Tarapacá in the Atacama Desert (South Central Andes). *Herit. Sci.* **2021**, *9*, 59. [CrossRef]
22. Boytner, R. Class, Control, and Power: The Anthropology of Textile Dyes at Pacatnamú. In *Andean Textile Traditions: Papers from the 2001 Mayer Center Symposium at the Denver Art Museum*; Young-Sánchez, M., Simpson, F.W., Eds.; Denver Art Museum: Denver, CO, USA, 2006; p. 50.
23. Cody, R.B.; Laramee, J.A.; Durst, H.D. Versatile New Ion Source for the Analysis of Materials in Open Air under Ambient Conditions. *Anal. Chem.* **2005**, *77*, 2297–2302. [CrossRef]
24. Armitage, R.A.; Jakes, K.A.; Day, C.J. Direct Analysis in Real Time-Mass Spectroscopy for Identification of Red Dye Colorants in Paracas Necropolis Textiles. *Sci. Technol. Archaeol. Res.* **2015**, *1*, 60–69. [CrossRef]
25. Day, C.J.; DeRoo, C.S.; Armitage, R.A. Developing Direct Analysis in Real Time Time-of-Flight Mass Spectrometric Methods for Identification of Organic Dyes in Historic Wool Textiles. In *Archaeological Chemistry*; Armitage, R.A., Burton, J.H., Eds.; ACS: Washington, DC, USA, 2013; Volume 8, pp. 69–85. [CrossRef]
26. Liu, J.; Wang, H.; Manicke, N.E.; Lin, J.-M.; Cooks, R.G.; Ouyang, Z. Development, Characterization, and Application of Paper Spray Ionization. *Anal. Chem.* **2010**, *82*, 2463–2471. [CrossRef] [PubMed]
27. Jakes, K.A. Physical and Chemical Analysis of Paracas Fibers. In *Paracas Art and Architecture*; Paul, A., Ed.; University of Iowa Press: Iowa City, IA, USA, 1991; pp. 222–239.
28. Antúnez de Mayolo, K.K. *Report on the Collection of Peruvian Dye Plants*; 1997; p. 43. (unpublished report to the Smithsonian Museum).
29. Antúnez de Mayolo, K.K. Peruvian Natural Dye Plants. *Econ. Bot.* **1989**, *43*, 181–191. [CrossRef]
30. Campos Ayala, J. Purple Dyes from the Carlos Museum Pre-Columbian Textiles Collection: Direct Mass Spectrometry and HPLC Analyses. Master's Thesis, Eastern Michigan University, Ypsilanti, MI, USA, 2019.
31. Geiger, J.; Armitage, R.A.; DeRoo, C.S. Identification of Organic Dyes by Direct Analysis in Real Time-Time of Flight Mass Spectrometry. In *Collaborative Endeavors in the Chemical Analysis of Art and Cultural Heritage Materials*; Lang, P.L., Armitage, R.A., Eds.; American Chemical Society: Washington, DC, USA, 2012; pp. 123–129. [CrossRef]
32. Manhita, A.; Ferreira, T.; Candeias, A.; Dias, C.B. Extracting natural dyes from wool—an evaluation of extraction methods. *Anal. Bioanal. Chem.* **2011**, *400*, 1501–1514. [CrossRef]
33. Degano, I.; Magrini, D.; Zanaboni, M.; Colombini, M.P. The Saltzman Collection: A Reference Database for South American Dyed Textiles. In Proceedings of the ICOM-CC 18th Triennial Conference, Copenhagen, Denmark, 4–8 September 2017.

34. DeRoo, C.S.; Armitage, R.A. Direct Identification of Dyes in Textiles by Direct Analysis in Real Time-Time of Flight Mass Spectrometry. *Anal. Chem.* **2011**, *83*, 6924–6928. [CrossRef]
35. Martoglio, P.A.; Jakes, K.A.; Katon, J.E. The Use of Infrared Microspectroscopy in the Analysis of Etowah Textiles: Evidence of Dye Use and Pseudomorph Formation. In *Proceedings of the 50th Annual Meeting of the Electron Microscopy Society of America;* Bailey, W., Bentley, J., Small, J.A., Eds.; Microbeam Analysis Society: San Francisco, CA, USA, 1992; pp. 1534–1535.
36. Jakes, K.A.; Katon, J.E.; Martoglio, P.A. Identification of Dyes and Characterization of Fibers by Infrared and Visible Microspectroscopy: Application to Paracas Textiles. In *Archaeometry'90;* Wagner, G.A., Pernicka, E., Eds.; Birkhäuser: Basel, Switzerland, 1991.
37. Szostek, B.; Orska-Gawrys, J.; Surowiec, I.; Trojanowicz, M. Investigation of natural dyes occurring in historical Coptic textiles by high-performance liquid chromatography with UV–Vis and mass spectrometric detection. *J. Chromatogr. A* **2003**, *1012*, 179–192. [CrossRef]
38. Dutra Moresi, C.M.; Wouters, J. Hplc Analysis of Extracts, Dyeings and Lakes, Prepared with 21 Species of Relbunium. *Dye. Hist. Archaeol.* **1997**, *15*, 85–97.
39. Dempster, L.T. Three New Species of Galium from the Northern Andes. *Madroño* **1988**, *35*, 1–5.
40. Dempster, L.T. The Genus Galium (Rubiaceae) in South America. II. *Allertonia* **1982**, *3*, 211–258.
41. De Toni, K.L.G.; Mariath, J.E.A. Developmental Anatomy and Morphology of the Flowers and Fruits of Species from Galium and Relbunium (Rubieae, Rubiaceae). *Ann. Mo. Bot. Gard.* **2011**, *98*, 206–225. [CrossRef]
42. Prikhodko, S.V.; Rambaldi, D.C.; King, A.; Burr, E.; Muros, V.; Kakoulli, I. New advancements in SERS dye detection using interfaced SEM and Raman spectromicroscopy (μRS). *J. Raman Spectrosc.* **2015**, *46*, 632–635. [CrossRef]
43. Zhang, X.; Laursen, R.A. Development of Mild Extraction Methods for the Analysis of Natural Dyes in Textiles of Historical Interest Using LC-Diode Array Detector-MS. *Anal. Chem.* **2005**, *77*, 2022–2025. [CrossRef]
44. Martoglio, P.A.; Bouffard, S.P.; Sommer, A.J.; Katon, J.E.; Jakes, K.A. Unlocking the secrets of the past: The analysis of archaeological textiles and dyes. *Anal. Chem.* **1990**, *62*, 1123A–1128A. [CrossRef]

heritage

MDPI

Article

Textile Dyes from Gokstad Viking Ship's Grave

Jeannette Jacqueline Łucejko [1], Marianne Vedeler [2] and Ilaria Degano [1,*]

[1] Department of Chemistry and Industrial Chemistry, University of Pisa, I-56124 Pisa, Italy; jeannette.lucejko@unipi.it

[2] Museum of Cultural History, Department of Archaeology, University of Oslo, 0130 Oslo, Norway; marianne.vedeler@khm.uio.no

* Correspondence: ilaria.degano@unipi.it

Abstract: The grave from Gokstad in Norway, dating to ca 900 AD, is one of the best-preserved Viking Age ship graves in the world. The grave mound contained a variety of goods along with human remains, buried in a Viking ship. Several textiles, including embroideries and shreds of what might have been the ship's tent, were also found. The colors of the textile fragments are now severely faded, but the high quality of the embroidery made of gold and silk threads is still apparent. The style of the embroidery is exceptional, having no equivalents in other Scandinavian graves. The analyses by HPLC coupled with both diode array and mass spectrometric detectors revealed that the striped "tent" cloth as well as the silk thread used for the embroidery were originally dyed with anthraquinones of plant origin (alizarin, purpurin, pseudopurpurin, and anthragallol), markers of madder-type dyestuffs.

Keywords: Viking Age; dyestuff; textiles; HPLC-DAD-MS

Citation: Łucejko, J.J.; Vedeler, M.; Degano, I. Textile Dyes from Gokstad Viking Ship's Grave. *Heritage* **2021**, *4*, 2278–2286. https://doi.org/10.3390/heritage4030129

Academic Editor: Lucia Burgio

Received: 21 July 2021
Accepted: 6 September 2021
Published: 8 September 2021

1. Introduction

The production of natural dyes for textiles was an important economic factor in the early Middle Ages, and a number of written sources mention the production of madder for sale [1].

The period between ca AD 750–1050 is called the Viking Age in Scandinavia. This was a time when people from Norway, Sweden, and Denmark first began travelling to other parts of the world on a large scale and established themselves as a political factor in Europe. Profound social and political change took place in that period, when old religious ideas gave way to new ones and when the Scandinavian countries gradually became unified kingdoms [2]. The impact of long-distance trade and the establishment of urban trading settlements played an important role, bringing the Scandinavians into contact with foreign cultures all over the old world. New trends in arts and crafts fueled the expansion of long-distant trades, including textiles and dyestuffs [3–5].

Given the expansion of the Scandinavian cultural and commercial routes and possible degradation processes that have affected the textiles, it is not easy to predict which could be the natural dyes originally used in the samples retrieved from archaeological sites. The best-preserved textiles from the Viking Age come mainly from bogs or mound contexts. The textiles found in the bogs constitute one of the largest and best-preserved collections of textiles in existence [6]. The chemical study of these fabrics has permitted the identification of a narrow range of natural dyes used in these times. Plants containing luteolin, indigotin, and alizarin were determined in textiles from Søgårds Mose II and Skærsø [6]. These components of natural colorants were also frequently reported in the studies performed by Walton P. Rogers on a wide number of textile samples from the Viking Age, along with insect dyes and lichen purples [7–9]. The earliest use of indigotin in Scandinavia dates back to Rebild, Denmark, dated between the 4th and the 3rd century BC, while the earliest evidence of the use of alizarin-containing madder dye in Scandinavia is dated to the 1st century BC, and was assessed in the textiles collection from Skærsø [10]. The Scandinavian finds indicate that madder became a common dye source in Scandinavia during the Migration period (AD 400–520/540) [11].

Another source of well-preserved textiles are mounds. Some people in the uppermost social strata of society were at this time buried in ships, which were then placed in a large burial mound. The Oseberg ship, on the board of which the bodies of two women with very rich grave goods were found, is one of these. A rich array of textiles, both woolen and silken, designed for a range of uses, were unearthed within this discovery. A very important ship grave, dating to around 900 AD, was also found in Gokstad, Norway [12]. Various textiles from this grave were in good preservation conditions and included an outstanding embroidery in silk and gold (see Figure 1). The same technique has been used in embroideries found in other Scandinavian graves from this period, but the patterns found in these graves are very different [13]. The embroidery from Gokstad is formed in a flower pattern by using a combination of a gold-thread with a silk-core and threads made of pure silk. Traces of a red color are still visible on the thread used to stitch down the gold lamella. When studying the flower pattern in a stereo loupe, it looked like a red and a golden color have been used in combination to form a polychrome pattern [14]. Remnants of a ship's tent were also found in this grave [15–17]. The remnants consist of some 150 fragments of coarse 2/2 twill textiles made of wool, intervened with fragments of rope. Some of the textile fragments bear traces of red color, while others look undyed to the naked eye [14].

Figure 1. Photographs of the embroidery in silk and gold found in the ship grave in in Gokstad, Norway. Picture by Ellen C. Holte, Museum of Cultural History, UiO.

The most used method for the analysis of dyes in archaeological artefacts is liquid chromatography (HPLC) with different detectors, which, after a suitable sample treatment, allows the separation of the colored molecules and the identification of the original dyes used. The most common setup is HPLC with spectrophotometric detectors, in particular, diode-array (DAD), because it allows the reliable identification of several dyestuffs by comparison with the profiles of reference materials [18]. In the past 15 years, HPLC coupled with mass spectrometry (MS) has been successfully used for dyes analyses, providing a higher sensitivity than HPLC-DAD, which is particularly useful in the analysis of archaeological textiles, where the sample is usually small in size and contains low concentrations of time-altered colorants (see, in general, [19–21]). HPLC-MS also allows, thanks to the acquisition of tandem mass spectra, us to identify unknown species, thus providing information on dyes not available as reference materials [22–24]. Different reversed phase analytical columns have been employed for dyes analysis with mass spectrometric detection, differing in the length of the hydrophobic chains (C8 or C18), particle size (also capillary columns were employed), and end capping, or even presenting polar embedding, such as RP-amide columns [23,25]. Recently, two-dimensional liquid chromatography was also employed successfully [26].

One of the most important steps in dye analysis is the extraction/separation of the analytes from the sample. The primary objective of the sample treatment is to find the conditions in which the maximum extraction yield and minimum alteration of the molecular profile are achieved, as reviewed in [27,28]. Most extraction methods are thus based on complexation with organic acids such as formic, oxalic, or acetic, alone or combined with EDTA, where mild extraction takes place [29,30]. Various organic solvents have also been applied for the extraction of less hydrophilic compounds, e.g., dimethyl sulfoxide (DMSO), dimethylformamide (DMF), and pyridine [31]. Finally, whenever the mild extraction does not provide any result, harsher hydrolysis treatments may be applied, which were quite diffused in the past, but contraindicated for labile colorants and/or materials unstable in acidic conditions [20,31,32].

This study presents the results obtained for a set of textile samples collected from the Gokstad Viking ship's grave. The identification of the coloring materials was achieved through the application of three increasingly harsh extraction procedures, based on dimethyl sulfoxide, EDTA with DMF extraction of the dyestuff from the yarns, followed by a hydrolysis in acidic condition of the residue; the analysis of the extract was performed with high performance liquid chromatography (HPLC) with diode array detection and subsequently confirmed with HPLC coupled with a high resolution mass spectrometric detector (ESI-Q-ToF). The study led to the identification of the molecular markers of a relevant dyestuff used during the Viking Age.

2. Materials and Methods

2.1. Samples

Samples of colored textiles were subjected to specific procedures to extract the characteristic markers of the materials used during dyeing. The extraction procedures applied are specific for the target molecular markers for each sample and are described in detail in Section 2.2. Table 1 shows the samples information, the extraction procedures chosen, and the instrumental methods applied.

Table 1. Description of the samples.

Sample	Color	N° in Gokstad Collection	Description	Procedure
S1	Red/brown	C10389/37C	Ship's tent	EDTA/DMF and HCl/MeOH + AcOEt extraction
S2	Yellow (white?)	C10389/37C	Ship's tent	DMSO and HCl/MeOH + AcOEt extraction
S3	Red (red silk?)	C10459/b	Red thread silk embroidery	EDTA/DMF and HCl/MeOH + AcOEt on the residue (both fractions injected together)
S4	Red (golden silk?)	C10459/b	Golden thread silk embroidery	EDTA/DMF and HCl/MeOH + AcOEt on the residue (both fractions injected together)

2.2. Analytical Procedures

Three analytical procedures were applied to extract the colored compounds present in the textiles studied. The conditions of each procedure applied are reported below.

2.2.1. EDTA/DMF Procedure

A mild extraction by dimethylformamide (DMF) and 0.1% Na_2EDTA 1:1 (v/v) was first applied. Then, 200 µL of mixture solution were added to the sample; extraction was performed at 60 °C for 60 min in ultrasonic bath. The supernatant was filtered with PTFE filter (0.45 µm) and injected in the chromatographic system (20 µL). The procedure is particularly useful to detect the presence of labile components and in case of fragile materials and degraded textiles [33].

2.2.2. DMSO Extraction

A specific extraction by dimethyl sulfoxide (DMSO) to maximize the recovery of indigoids, and in particular, of photo-labile brominated indigoids, was applied to the yellow faded sample S2, since no hints on the original color could be drawn. The procedure consisted of an extraction assisted by ultrasounds at 60 °C for 5 min followed by filtration with PTFE syringe filter (0.45 μm). The sample was kept in the dark for 5 min and then injected into the chromatographic system.

2.2.3. HCl/MeOH Procedure

The extraction solution was a 30:1 (*v*/*v*) mixture of hydrochloric acid (HCl) in methanol (MeOH). A total of 300 μL of MeOH/HCl solution were added to the sample and the extraction was performed in an ultrasonic bath for 60 min at 60 °C, followed by liquid-liquid extraction with ethyl acetate (3 × 200 μL). The extract was dried under a gentle stream of nitrogen and dissolved in 200 μL of acetonitrile (ACN) before injection into the chromatographic systems.

2.3. *Instrumentation*

2.3.1. HPLC-DAD

For the analysis of the extracts, a High Performance Liquid Chromatography system equipped with a quaternary pump with degasser PU-2089 was used, with an AS-950 autosampler and coupled to a diode array spectrophotometric detector MD-2010 (all modules were Jasco International Co., Tokyo, Japan). Data were acquired and processed using the ChromNav software. The working conditions were as follows: acquisition of the spectra in the range 200–650 nm with a resolution of 1 nm; chromatographic separation conducted at room temperature (25 °C) on an analytical column TC-C18 (2) (4.6 × 150 mm, particle size 5 μm, Agilent) with a guard column TC-C18 (2) (4.6 × 12.5 mm, particle size 5 μm, Agilent), flow rate 1 mL/min. The injection volume was 20 μL. The two eluent solutions were: A: trifluoroacetic acid (TFA 0.1% *v*/*v*) in bi-distilled water; B: TFA (0.1% *v*/*v*) in acetonitrile (ACN, HPLC grade). The elution program started at 15% B, held for 5 min, then a linear gradient to 50% B was applied in 25 min; re-equilibration time took 15 min.

2.3.2. HPLC-ESI-Q-ToF System

HPLC-ESI-Q-ToF MS analyses were carried out using a 1200 Infinity HPLC, coupled with a Jet Stream ESI interface with a Quadrupole-Time of Flight tandem mass spectrometer 6530 Infinity Q-ToF detector (all modules were Agilent Technologies, Santa Clara, CA, USA). The ESI operating conditions were: drying gas (N_2, purity >98%): 350 °C and 10 L/min; capillary voltage 4.5 KV; nebulizer gas 35 psig; sheath gas (N_2, purity >98%): 375 °C and 11 L/min. The nozzle, skimmer, and octapole were set at 1000 V, 65 V, and 750 V, respectively. High resolution MS and MS/MS spectra were acquired in negative mode in the range 100–1000 *m*/*z* with a scan rate of 1.04 spectra/sec, employing the AutoMS/MS acquisition mode (1 precursor per cycle, CID voltage for tandem mass spectra 30 V, collision gas N_2, purity 99.999%, FWHM (Full Width Half Maximum) of quadrupole mass bandpass used during MS/MS precursor isolation 4 *m*/*z*). MassHunter® Workstation Software (B.04.00) was used to carry out mass spectrometer control, data acquisition, and data analysis. The mass spectrometer was calibrated daily using Agilent tuning mix HP0321. An Agilent Zorbax Extend-C18 column (2.1 × 30 mm, particle size 1.8 μm) with precolumn Extend-C18 column (2.1 × 12.5 mm, particle size 1.8 μm) was used for the chromatographic separation. The injection volume was 20 μL and the flow rate was 0.2 mL/min.

The two eluent solutions were: A: trifluoroacetic acid (FA, 0.1% *v*/*v*) in bi-distilled water; B: FA (0.1% *v*/*v*) in acetonitrile (ACN, HPLC grade). The elution program started at

15% B, held for 1 min, then a linear gradient to 50% B was applied in 5 min, then to 70% B in 2 min and to 90%B in 7 min; 100% was reached after further 5 min; re-equilibration time took 5 min.

3. Results and Discussion

HPLC/DAD and HPLC-ESI-Q-ToF MS analyses were performed in order to determine the dyestuff used in the Viking Age textiles from the Gokstad ship's grave.

The chromatograms obtained from the extracts after DMSO extraction did not show any relevant peaks. The extracts obtained after EDTA/DMF procedure gave satisfactory results for sample S1, providing the highest peaks in HPLC-ESI-Q-ToF MS analysis. Only the harsh extraction based on hydrolysis followed by extraction in ethylacetate provided results for sample S2, S3, and S4, and was also successfully applied to S1.

The three samples S1, S3, and S4 showed (Figures 2, 4 and 5) a very similar composition, consisting of the presence of anthraquinones typical of madder-type dyestuffs.

In detail, Figure 2 shows the Extracted Ion Chromatograms for the HPLC-ESI-Q-ToF MS chromatograms of the EDTA/DMF extract of sample S1 from the ship's tent, where anthragallol, alizarin, munjistin, purpurin, and pseudopurpurin were detected. The detected anthraquinones are typical of a madder-dye source. The overall molecular profile was not typical of a specific dye source, although the highest content in alizarin respect to the other components may point to *Rubia tinctorum* as the original raw material. The good results obtained by the mild extraction might suggest that a relatively high amount of dye survived in the sample, or that a different recipe was used to obtain the desired color, with respect to the other samples. For comparison, the inset presents the results obtained by HPLC-DAD (shown at 254 nm) relative to sample S1 after hydrolysis followed by extraction in ethylacetate, which provided the highest peaks in the UV-Vis range.

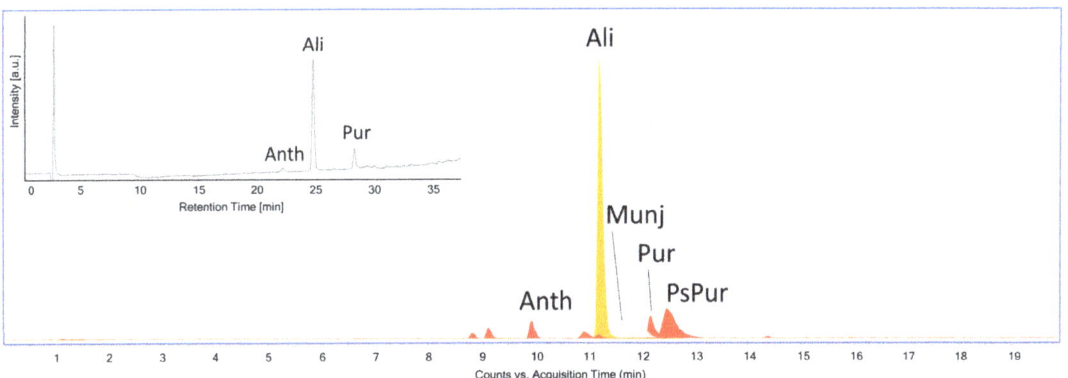

Figure 2. EIC HPLC-ESI-Q-ToF MS chromatograms of the EDTA/DMF extract of sample S1, acquired in negative mode; where Anth is anthragallol (EIC—m/z 255.029), Ali is alizarin (EIC—m/z 239.031), Munj is munjistin (EIC—m/z 283.002), Pur is purpurin (EIC—m/z 255.029), and PsPur is pseudopurpurin (EIC—m/z 299.019). In the inset, the HPLC-DAD chromatogram at 254 nm relative to sample S1 extract after HCl/MeOH + AcOEt is presented for comparison.

The second sample collected from the ship's tent (S2, Figure 3), characterized by a yellow/white color at the time of analysis, showed the presence of traces of alizarin and purpurin (only detected by HPLC-ESI-Q-ToF MS, thanks to its superior sensitivity with respect to HPLC-DAD), pointing again to the use of a madder-based dye.

Samples S3 (Figure 4) and S4 (Figure 5), collected from the embroidery, contained the anthraquinones anthragallol, alizarin, purpurin, and pseudopurpurin; rubiadin was only present in S3.

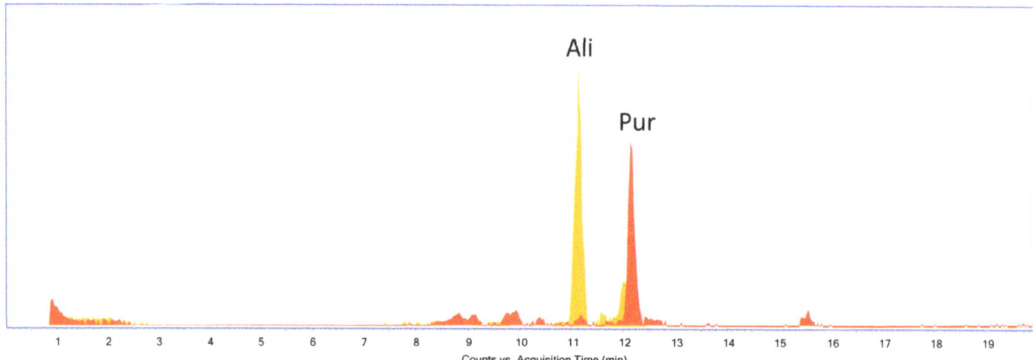

Figure 3. EIC HPLC-ESI-Q-ToF MS chromatograms of the extract of sample S2 extract after HCl/MeOH + AcOEt, acquired in negative mode; where Ali is alizarin (EIC—m/z 239.031), Pur is purpurin (EIC—m/z 255.029).

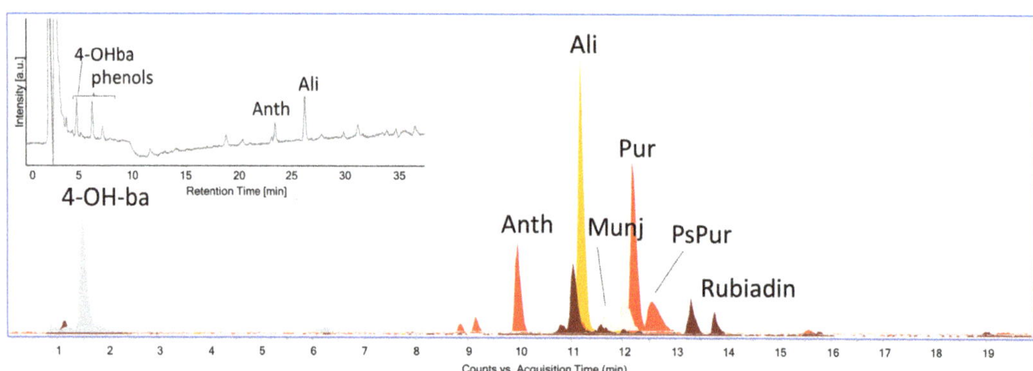

Figure 4. EIC HPLC-ESI-Q-ToF MS chromatograms of the extract of sample S3 extract after HCl/MeOH + AcOEt, acquired in negative mode; where 4-OH-ba is 4-hydroxybenzoic acid (EIC—m/z 137.024), Anth is anthragallol (EIC—m/z 255.029), Ali is alizarin (EIC—m/z 239.031), Munj is munjistin (EIC—m/z 283.002), Pur is purpurin (EIC—m/z 255.029), PsPur is pseudopurpurin (EIC—m/z 299.019), Rubiadin (EIC—m/z 253.051). In the inset, the corresponding HPLC-DAD chromatogram at 254 nm is presented for comparison.

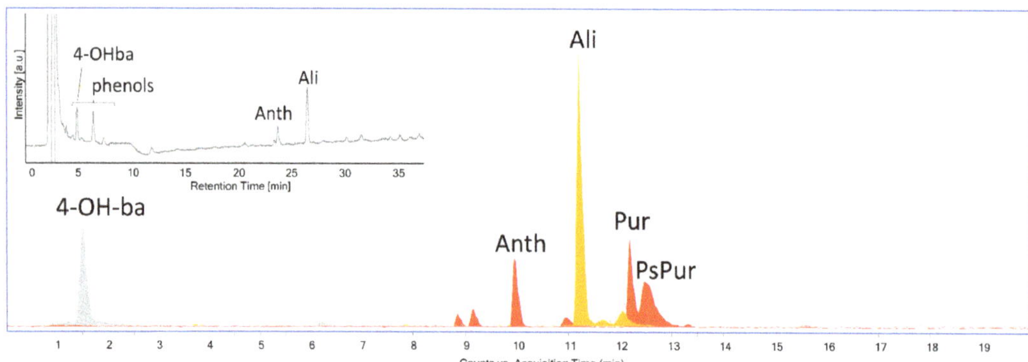

Figure 5. EIC HPLC-ESI-Q-ToF MS chromatograms of the extract of sample S4 extract after HCl/MeOH + AcOEt, acquired in negative mode; where 4-OH-ba is 4-hydroxybenzoic acid (EIC—m/z 137.024), Anth is anthragallol (EIC—m/z 255.029), Ali is alizarin (EIC—m/z 239.031), Pur is purpurin (EIC—m/z 255.029), PsPur is pseudopurpurin (EIC—m/z 299.019). In the inset, the corresponding HPLC-DAD chromatogram at 254 nm is presented for comparison.

In addition, the chromatograms feature 4-hydroxybenzoic acid, most probably due to degradation processes involving the fabric (silk) [34]. Peaks attributed to further unknown phenols based on their early retention times and maxima of absorption around 275 nm in the DAD spectra were also detected as highlighted in the insets in Figures 4 and 5, but were not confirmed by HPLC-ESI-Q-ToF MS analysis.

Table 2 summarizes the results obtained by both analytical systems (HPLC-DAD, HPLC-ESI-Q-ToF); it includes detected markers and identified dyes (associated with the presence of these compounds).

Table 2. Molecular markers detected along with identified materials in the analyzed extracts.

Sample	Molecular Markers Detected by HPLC-DAD	Molecular Markers Detected by HPLC-ESI-Q-ToF MS	Identified Materials
S1	alizarin, purpurin, anthragallol	alizarin, purpurin, pseudopurpurin, munjistin, anthragallol	madder-type dyestuff
S2	no peak identified	alizarin, purpurin	madder-type dyestuff
S3	4-hydroxybenzoic acid, phenols, anthragallol, alizarin	4-hydroxybenzoic acid, anthragallol, alizarin, purpurin, pseudopurpurin, rubiadin	madder-type dyestuff; degradation products of protein-based fibers
S4	4-hydroxybenzoic acid, phenols, anthragallol, alizarin	4-hydroxybenzoic acid, anthragallol, alizarin, purpurin, pseudopurpurin	madder-type dyestuff; degradation products of protein-based fibers

4. Conclusions

In this paper, an investigation of textile samples collected from Gokstad Viking ship's grave by high performance liquid chromatography (HPLC) with diode array (DAD) and high resolution mass spectrometric detector (ESI-Q-ToF MS) was performed. The use of ESI-Q-ToF MS as a detector was paramount in order to allow us to detect not only major but also minor compounds, and to confirm the presence of analytes in sample 2. The presence of molecular markers attributable to madder-type dyestuff, most probably from *Rubia tinctorum*, was determined in all four analyzed samples. A harsh sample treatment, followed by purification by liquid/liquid extraction with ethylacetate, was needed to retrieve significant amount of analytes from most samples (only sample S1 gave satisfactory results after treatment in mild conditions). This can be attributed to the extensive degradation of the dyestuffs or the use of specific dyeing recipes which prevented the extraction with chelating agents, requiring acidic conditions.

Madder-type dyestuffs are commonly found in graves from the Viking Age and medieval Scandinavia: a number of written sources mentioned the use and export of madder and surviving trace amounts of dyes interpreted as madder have been found in a number of Scandinavian graves dating to the Viking Age [1,5,35].

With regard to the embroidery, the two samples (S3 and S4) were collected from the threads forming a polychrome pattern [14], which looked slightly redder (S3) and more yellow (S4) respective to each other. Although both extracts had a very similar composition, consisting of alizarin, purpurin, pseudopurpurin, and anthragallol, the relative alizarin and anthragallol content with respect to all other anthraquinones is much higher in the golden sample (S4) than in the redder one (S3). Thus, the HPLC data confirmed the visual observation, being alizarin and anthragallol more yellow than the other components, featuring maxima of absorbance in the visible range around 430 and 410 nm, respectively, whereas purpurin and pseudopurpurin both have maxima of absorbance higher than 480 nm. Thus, the different amounts of the components of the dyeing material suggest that different recipes were indeed employed to obtain different shades of silk thread. Finally, protein-based fiber degradation products were identified in the silk samples, confirming the animal origin of the textiles [36].

In one sample collected from the tent (S1), madder-type markers were also identified, while in S2, characterized by a yellow/white color at the time of analysis, only traces of anthraquinones were detected related to the dyestuff. It is possible that, given the purpose of this textile material, only a small amount of dyestuff was applied, or it severely faded during the use of the fabric. On the other hand, given the different results obtained for samples S1 and S2, the tent might have been colored in two different alternating shades.

In conclusion, the application of different extraction methods along with sensitive and selective analytical techniques allowed us to detect madder-type dyestuffs used to dye both a red-and-white/undyed striped woolen cloth used as tent or sail, and for coloring a delicate embroidery of silk and gold.

These new analyses provide a small but important new piece of information about the Viking Ship grave from Gokstad, allowing us to picture the original multicolored funerary goods.

Author Contributions: Conceptualization, M.V. and J.J.Ł.; methodology, I.D. and J.J.Ł.; investigation, I.D. and J.J.Ł.; writing—original draft preparation, M.V., I.D. and J.J.Ł.; writing—review and editing, M.V., I.D. and J.J.Ł.; supervision, J.J.Ł. All authors have read and agreed to the published version of the manuscript.

Funding: Funded by the Gokstad Revitalised project, University of Oslo.

Data Availability Statement: All data presented in this paper can be made available by the authors upon request.

Acknowledgments: We thank the Museum of Cultural History, University of Oslo, for making samples of the Gokstad textiles available for our research.

Conflicts of Interest: The authors declare no conflict of interest.

References

1. Cardon, D. Natural Dyes. In *Sources, Tradition, Technology and Science*; Archetype: London, UK, 2007.
2. Skre, D. Viking-Age Economic Transformations. The West-Scandinavian Case. In *Viking-Age Transformations: Trade, Craft and Resources in Western Scandinavia*; Glørstad, Z.T., Loftsgarden, K., Eds.; Routledge: London, UK, 2016.
3. Sindbæk, S. Urbanism and Exchange in the North Atlantic/Baltic, 600–1000 ce. In *The Routledge Handbook of Archaeology and Globalization*; Hodos, T., Ed.; Routledge: New York, NY, USA, 2017; pp. 553–565.
4. Strand, E.A. Tools and Textiles—Production and Organisation in Birka and Hedeby. In Proceedings of the XVI Viking Congress, Reykjavík, Iceland, 14–23 August 2009.
5. Vedeler, M. *Silk for the Vikings*; Oxbow Books: Oxford, UK, 2014; Volume 15.
6. Vanden Berghe, I.; Gleba, M.; Mannering, U. Towards the Identification of Dyestuffs in Early Iron Age Scandinavian Peat Bog Textiles. *J. Archaeol. Sci.* **2009**, *36*, 1910–1921. [CrossRef]
7. Walton Rogers, P. Dyes and Wools in Iron Age Textiles from Norway and Denmark. *J. Dan. Archaeol.* **1988**, *7*, 144–158. [CrossRef]
8. Walton Rogers, P. Dyes of the Viking Age: A Summary of Recent Work. In *Dyes in History and Archaeology*; Walton, P., Ed.; Pangur Press: York, UK, 1988; Volume 7, pp. 14–21.
9. Walton Rogers, P. *Cordage and Raw Fibre from 16–22 Coppergate*; Addyman, P.V., Ed.; The Archaeology of York: York, UK, 1989.
10. Vanden Berghe, I.; Gleba, M.; Mannering, U. Dyes: To Be or Not to Be. An Investigation of Early Iron Age Dyes in Danish Peat Bog Textiles. In Proceedings of the North European Symposium for Archaeological Textiles X, Copenhagen, Denmark, 14–17 May 2008.
11. Hofenk de Graaf, J. *The Colourful Past: Origins, Chemistry and Identification of Natural Dyestuffs*; Archetype Publications Ltd.: London, UK, 2004.
12. Nicolaysen, N. *Langskibet Fra Gokstad Ved Sandefjord*; Cammermeyer, A., Ed.; Kristiania: Oslo, Norway, 1882.
13. Nockert, M. Silkebroderi. In *Tekstilene, Osebergfunnet Bind Iv*; Christensen, E.A., Nockert, M., Eds.; Kulturhistorisk: Oslo, Norway, 2006; pp. 325–337.
14. Vedeler, M. Golden Textiles from Gokstad. *Archaeol. Text. Rev.* **2021**, in press.
15. Hougen, B. Gulltråd Fra Gokstadfunnet. Et Fragment. In *Honos Ella Kivikoski*; Sarvas, P., Kivikoski, E., Siiriäinen, A., Eds.; Finska Fornminnesföreningen; Suomen Muinaismuistoyhdistys: Helsinki, Finland, 1973; pp. 75–81.
16. Ingstad, A.S. Archaeological Textiles: Textiles from Oseberg, Gokstad and Kaupang. In *Archaeological Textiles: 2nd NESAT Symposium: Report*; Københavns Universitet Arkaeologisk Institut: Kobenhavn, Denmark, 1988.
17. Vedeler, M. Reconstructing the Tunic from Lendbreen in Norway. *Archaeol. Text. Rev.* **2017**, *59*, 24–33.
18. Vasileiadou, A.; Karapanagiotis, I.; Zotou, A. Development and Validation of a Liquid Chromatographic Method with Diode Array Detection for the Determination of Anthraquinones, Flavonoids and Other Natural Dyes in Aged Silk. *J. Chromatogr. A* **2021**, *1651*, 462312. [CrossRef] [PubMed]
19. Degano, I. Liquid Chromatography: Current Applications in Heritage Science and Recent Developments. *Phys. Sci. Rev.* **2019**, *4*, 5. [CrossRef]

20. Degano, I.; La Nasa, J. Trends in High Performance Liquid Chromatography for Cultural Heritage. *Top. Curr. Chem.* **2016**, *374*, 20. [CrossRef]
21. Pauk, V.; Barták, P.; Lemr, K. Characterization of Natural Organic Colorants in Historical and Art Objects by High-Performance Liquid Chromatography. *J. Sep. Sci.* **2014**, *37*, 3393–3410. [CrossRef]
22. Tamburini, D. Investigating Asian Colourants in Chinese Textiles from Dunhuang (7th–10th Century Ad) by High Performance Liquid Chromatography Tandem Mass Spectrometry—Towards the Creation of a Mass Spectra Database. *Dyes Pigm.* **2019**, *163*, 454–474. [CrossRef]
23. Lech, K.; Fornal, E. A Mass Spectrometry-Based Approach for Characterization of Red, Blue, and Purple Natural Dyes. *Molecules* **2020**, *25*, 3223. [CrossRef]
24. Otłowska, O.; Ślebioda, M.; Kot-Wasik, A.; Karczewski, J.; Śliwka-Kaszyńska, M. Chromatographic and Spectroscopic Identification and Recognition of Natural Dyes, Uncommon Dyestuff Components, and Mordants: Case Study of a 16th Century Carpet with Chintamani Motifs. *Molecules* **2018**, *23*, 339. [CrossRef]
25. Degano, I.; Mattonai, M.; Sabatini, F.; Colombini, M.P. A Mass Spectrometric Study on Tannin Degradation within Dyed Woolen Yarns. *Molecules* **2019**, *24*, 2318. [CrossRef]
26. Pirok, J.B.W.; den Uijl, M.J.; Moro, G.; Berbers, S.V.J.; Croes, C.J.M.; van Bommel, M.R.; Schoenmakers, P.J. Characterization of Dye Extracts from Historical Cultural-Heritage Objects Using State-of-the-Art Comprehensive Two-Dimensional Liquid Chromatography and Mass Spectrometry with Active Modulation and Optimized Shifting Gradients. *Anal. Chem.* **2019**, *91*, 3062–3069. [CrossRef]
27. Valianou, L.; Karapanagiotis, I.; Chryssoulakis, Y. Comparison of Extraction Methods for the Analysis of Natural Dyes in Historical Textiles by High-Performance Liquid Chromatography. *Anal. Bioanal. Chem.* **2009**, *395*, 2175–2189. [CrossRef]
28. Wouters, J.; Grzywacz, C.M.; Claro, A. A Comparative Investigation of Hydrolysis Methods to Analyze Natural Organic Dyes by Hplc-Pda—Nine Methods, Twelve Biological Sources, Ten Dye Classes, Dyed Yarns, Pigments and Paints. *Stud. Conserv.* **2011**, *56*, 231–249. [CrossRef]
29. Sanyova, J. Mild Extraction of Dyes by Hydrofluoric Acid in Routine Analysis of Historical Paint Micro-Samples. *Microchim. Acta* **2008**, *162*, 361–370. [CrossRef]
30. Zhang, X.; Laursen, R.A. Development of Mild Extraction Methods for the Analysis of Natural Dyes in Textiles of Historical Interest Using Lc-Diode Array Detector-Ms. *Anal. Chem.* **2005**, *77*, 2022–2025. [CrossRef] [PubMed]
31. Koren, Z.C. Archaeo-Chemical Analysis of Royal Purple on a Darius I Stone Jar. *Microchim. Acta* **2008**, *162*, 381–392. [CrossRef]
32. Degano, I.; Lucejko, J.J.; Colombini, M.P. The Unprecedented Identification of Safflower Dyestuff in a 16th Century Tapestry through the Application of a New Reliable Diagnostic Procedure. *J. Cult. Herit.* **2011**, *12*, 295–299. [CrossRef]
33. Manhita, A.; Ferreira, T.; Candeias, A.; Dias, C.B. Extracting Natural Dyes from Wool—an Evaluation of Extraction Methods. *Anal. Bioanal. Chem.* **2011**, *400*, 1501. [CrossRef] [PubMed]
34. Dilillo, M.; Restivo, A.; Degano, I.; Ribechini, E.; Colombini, M.P. Gc/Ms Investigations of the Total Lipid Fraction of Wool: A New Approach for Modelling the Ageing Processes Induced by Iron-Gallic Dyestuffs on Historical and Archaeological Textiles. *Microchem. J.* **2015**, *118*, 131–140. [CrossRef]
35. Serjeant, R.B. Material for a History of Islamic Textiles up to the Mongol Conquest. *Ars Islamica* **1942**, *9*, 54–92.
36. Skals, I. *Identifikation Af Textilfibre Fra Gokstad-Fundet*; Nationalmuseet i København: Copenhagen, Denmark, 2015.

MDPI
St. Alban-Anlage 66
4052 Basel
Switzerland
Tel. +41 61 683 77 34
Fax +41 61 302 89 18
www.mdpi.com

Heritage Editorial Office
E-mail: heritage@mdpi.com
www.mdpi.com/journal/heritage

www.ingramcontent.com/pod-product-compliance
Lightning Source LLC
LaVergne TN
LVHW070159170726
843515LV00007B/1953

* 9 7 8 3 0 3 6 5 3 1 3 4 2 *